Gynecology:

A Textbook for Students

SPRINGER
STUDY
EDITION

medicine

Gynecology:

A Textbook for Students

Fritz K. Beller
Karl Knörr
Christian Lauritzen
Ralph M. Wynn

Springer-Verlag New York • Heidelberg • Berlin
1974

Fritz K. Beller
Department of Obstetrics and
  Gynecology
Wilhelm's University
Münster, West Germany
Formerly: Department of Obstetrics
  and Gynecology
New York University School of Medicine
New York

Christian Lauritzen
Department of Obstetrics and
  Gynecology
University of Ulm
West Germany

Karl Knörr
Department of Obstetrics and
  Gynecology
University of Ulm
West Germany

Ralph M. Wynn
Department of Obstetrics and
  Gynecology
The Abraham Lincoln School of Medicine
University of Illinois at the Medical
  Center
Chicago, Illinois

CREDITS

Figure numbers 54, 55, 61, 103, 104, 126, 129 and 175, based on G. Kern, Gynaekologie–Ein kurzgefasstes Lehrbuch, 2. Auflage, Stuttgart, Thieme 1973.
Figure 155, after K. G. Ober and H. B. Boetzelen
Figure 24, modified after F. H. Netter, MD.

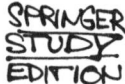 Design: Peter Klemke, Berlin

Library of Congress Cataloging in Publication Data
  Main entry under title:
  Gynecology: a textbook for students.
    Based on K. Knörr's Lehrbuch der Gynäkologie.
    1. Gynecology. I. Beller, Fritz K., ed.
    II. Knörr, K. Lehrbuch der Gynäkologie.
    [DNLM: 1. Gynecologic diseases. WP140 K72L 1974]
    RG101.G93   618.1   73-22331

ISBN-13: 978-0-387-90087-2   e-ISBN-13 978-1-4615-7128-5
DOI: 10.1007/978-1-4615-7128-5

PREFACE

This textbook has its origin in a longstanding international professional and personal relationship of the authors. The success of the German edition was a result of the interest of the German authors, Dr. K. Knörr and Dr. C. Lauritzen, in synthesizing basic and clinical information in a textbook, the size and scope of which are appropriate to the modern undergraduate medical student. A unique feature of the volume was the addition of American gynecological thinking by Dr. F. K. Beller, who spent 12 years at New York University before returning to Germany. The English edition, which is aimed at the American medical student, has enlisted the services of Dr. R. M. Wynn, whose academic career has been in the United States, although he is thoroughly familiar with principles and practices of German gynecology.

The English edition is not merely a translation of the German text. Rather, it incorporates most of the material, but with judicious deletions, additions, and substitutions it brings the teaching into the mainstream of American medical education. We acknowledge with gratitude the expert editional assistance of Dr. Henriette Knörr-Gärtner. Dr. R. Schuhmann and Mr. H. Brandt of Springer-Verlag were responsible for the histological sketches. The late Dr. H. Roemer prepared the section on psychosomatic disorders; Dr. F. W. Ahnefeld assisted in the section on shock; and Dr. H. Breinl helped to prepare the chapter on anatomy in the German edition.

The text fills the need for a book that is more than an outline and less than an encyclopedia of gynecology. Perhaps most important, it is tangible proof of the decline in insularity among Western nations. We hope that this effort will set the trend for further collaboration among multilingual medical educators.

January, 1974

Fritz K. Beller
Karl Knörr
Christian Lauritzen
Ralph M. Wynn

TABLE OF CONTENTS

CHAPTER 1

Development of the Genital Organs

The Human Karyotype

It has been known since 1956 that the diploid number of 46 chromosomes is found in all human somatic cells. The human karyotype consists of 22 homologous autosomal pairs and 2 sex chromosomes (gonosomes). The phenotypic development and somatic function in both sexes are dependent on genes that are localized on the autosomes. The sex-specific differences are determined by the sex chromosomes, represented in the female by two X chromosomes and in the male by one X and one Y chromosome.

The genetic formula for the female karyotype is 46/XX, and for the male, 46/XY.

The autosomes are identified by chromosomal analysis. They are classified by international agreement in pairs according to size, numbered consecutively from 1 to 22 and divided according to their structures into groups from A to G. The X chromosome is placed in the C group since it cannot be distinguished without autoradiographic marking, or the move recently developed banding techniques, from the larger autosomes of this group. The Y chromosome corresponds to the smaller acrocentric chromosomes of group G, although the size may vary individually; yet, it can be distinguished by morphologic characteristics from chromosomes 21 and 22. According to the Denver classification (1959), the scheme for the human karyotype is A (#1–3), B (#4–5), C (#6–12 + X), D (#13–15), E (#16–18), F (#19–20), (#6–12 + X), and G + Y (#21–22 + Y) (Fig. 1).

These schemes were recently revised on the basis of subgroups according to length and special staining of individual chromosomes, but the basic classification was not significantly altered.

The methods for diagnosing chromosomal abnormalities are usually based on preparations of cultured lymphocytes. These cells are transformed into lymphoblasts when an extract of phytohemagglutinin is added in vitro to the culture. Lymphoblasts divide rapidly in culture media. The mitoses are blocked in the metaphase before harvesting by colchicine. The chromosomes are then dispersed by hypotonic solutions; they can then be distinguished by numbers, size, and form.

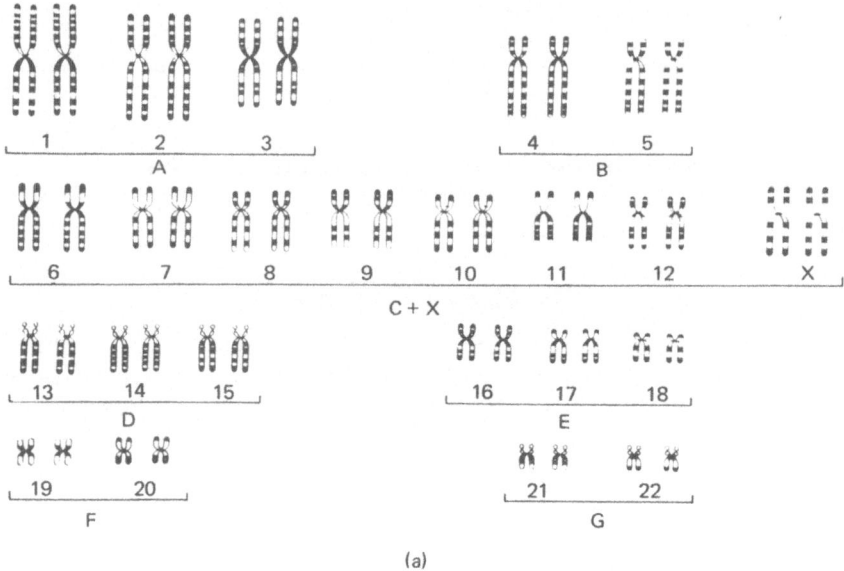

(a)

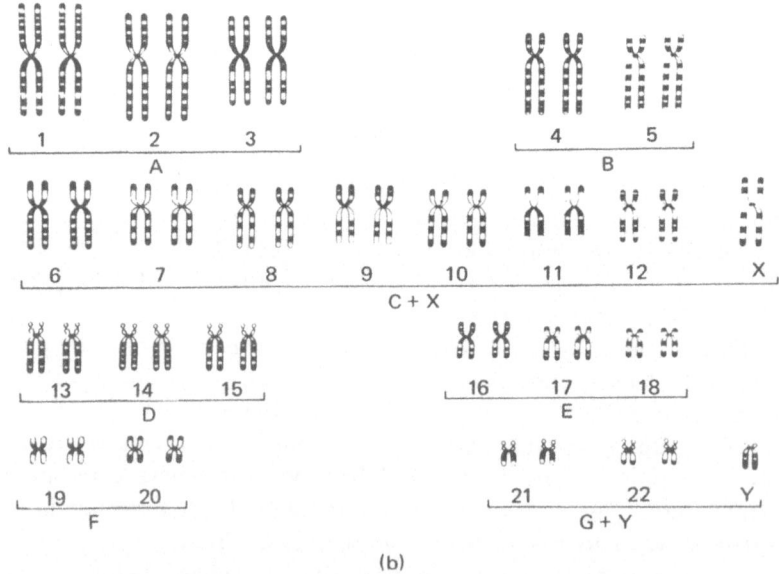

(b)

Fig. 1. (a) Normal female karotype (46, XX); (b) normal male karyotype (46, XY).

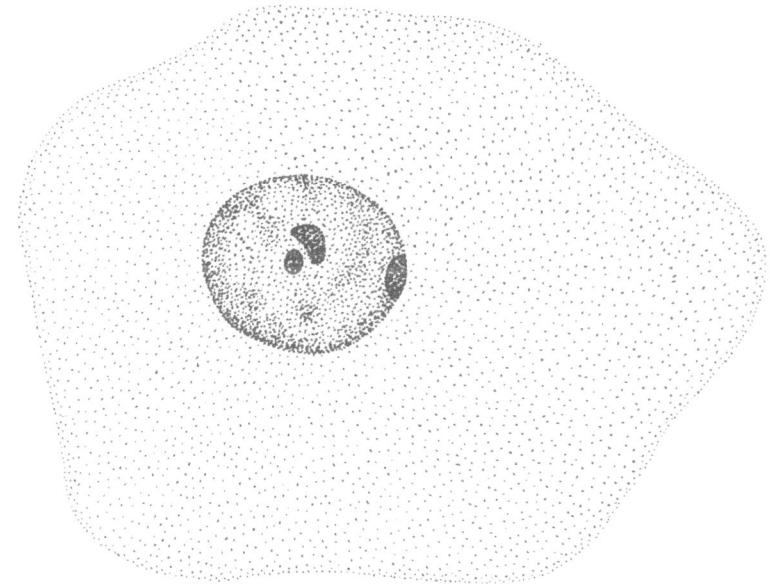

Fig. 2. Sex chromatin positive cell. Notice the Barr body on the nuclear membrane at 3 o'clock (vaginal smear).

The Sex Chromatin

One of the two X chromosomes can be demonstrated in the interphase nuclei of female cells. In a certain percentage of nuclei, chromatin masses can be seen adherent to the membrane. They are triangular or planoconvex (Fig. 2). These structures represent the inactive X chromosome (see above) and are named sex chromatin, or Barr bodies. According to the Lyon hypothesis the X-minus-1 formula indicates the relation between the number of X chromosomes of the karyotype and the number of Barr bodies. The formula implies that the number of chromatin bodies is always one less than the total number of X chromosomes. The nuclei of the normal human female cell contain 2 X chromosomes (46/XX) and, therefore, one Barr body. The Barr body is missing in normal human male cells (46/XY).

Barr body screening is done in buccal or vaginal smears, or in hair root tips. It can also be performed during the analysis of interphase cells in tissue cultures. The formation of the sex chromatin by one of the two X chromosomes is not a constant phenomenon. Barr bodies are, therefore, not present in all nuclei of female cells. They can be seen in approximately 15 to 20% of cells.

The sex chromatin is further represented in the neutrophilic leuko-

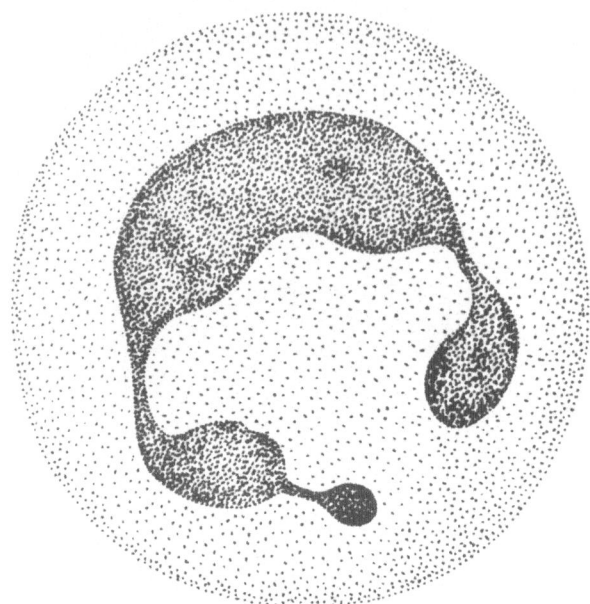

Fig. 3. Polymorphonuclear leukocyte with drumstick at 6 o'clock.

cytes by a "drumstick" adjacent to the nucleus (Fig. 3). Drumsticks can be identified in a stained blood smear in approximately 2 to 3% of female cells. Because of this small percentage, at least 500 nuclei have to be counted. These methods provide simple and at the same time reliable screening.

The X-minus-1 formula not only permits detection of numerical anomalies of sex chromosomes, but also provides a clue to the identification of mosaic constellations by showing the differences between Barr bodies and drumsticks. Abnormalities of form and size of Barr bodies and drumsticks are indicative of structural abnormalities of the X chromosome (see page 124).

The Y chromosome is identified in the buccal smear by staining with quinacrine or similar dyes that cause fluorescence. This new technique provides a screening method for the detection of numerical and structural abnormalities of the Y chromosome.

Molecular Genetic Aspects

Chromosomes are carriers of the genetic material. Deoxyribonucleic acid (DNA) contains the genetic information coded by the sequence of four bases: adenine, guanine, cytosine, and thymine (genetic code). The code words (or codons) consist of three adjacent bases, the triplets. Each cell contains the complete set of chromosomes and is, therefore,

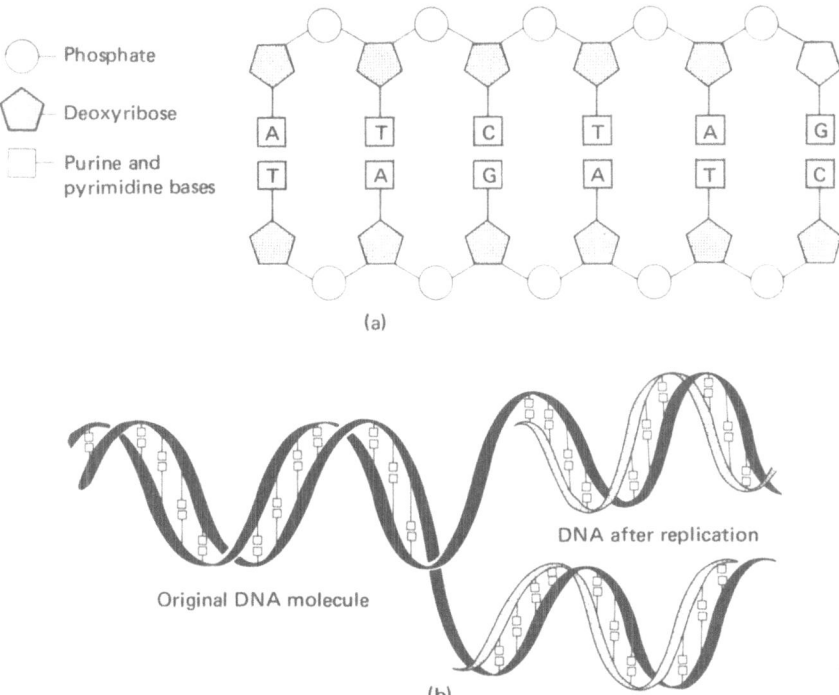

Phosphate

Deoxyribose

Purine and pyrimidine bases

(a)

DNA after replication

Original DNA molecule

(b)

Fig. 4. Structural model of DNA. (a) Schematic representation of DNA molecules. Two nucleotides are linked to the bases in the side chains in such a way that adenine (A) and thymine (T), on the one hand, and guanine (G) and cytosine (C), on the other, are next to each other. (b) Replication of the DNA molecule. *Left:* The parent molecule with the complementary helices, which are connected by base side chains. *Right:* Replication of DNA: after splitting the hydrogen bond, two new strands of DNA are synthesized. The new strands contain an arrangement of bases complementary to that of the parent DNA molecule (after Langman, 1969).

furnished with the complete amount of genetic information. The transfer of information within a cell differs from the transfer of the genetic information from one cell to the next cell generation.

The transfer of information within an individual cell proceeds in two steps. The first step, *transcription,* is the synthesis of the messenger ribonucleic acid (mRNA), which is a complete copy of one of the two DNA chains. The mRNA is used for the second step, *translation,* which regulates the synthesis of polypeptide chains. The amino acid sequence is directly influenced by the nucleotide sequence of the mRNA. The genetic unit is, therefore, equivalent to one specific DNA base sequence. According to the basic principle of genetics, one gene is responsible for one polypeptide chain—for example, a particular enzyme or protein.

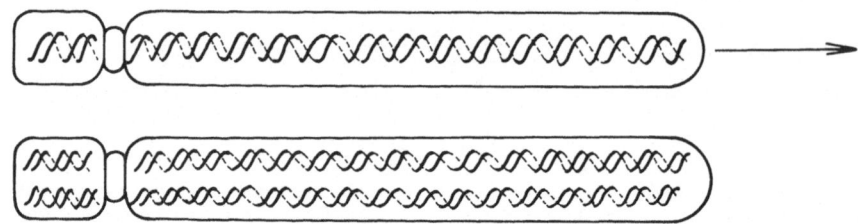

Fig. 5. Schematic representation of a chromosome before and after DNA replication (after Langman, 1969).

The complete transmission of all genetic information to the next cell generation requires reduplication, which is the de novo synthesis of DNA according to a given sequence.

The structural model of the DNA as a double helix (Figs. 4 and 5) provides the explanation of both principles: transformation of information within a given cell and passage of the total genetic information to the next cell generation. In differentiating and differentiated cells and tissues only the required parts of the genetic code are active, whereas other complexes of the gene pool are inactivated or remain out of function at different times.

It appears that only a few genes or gene complexes remain active throughout life. Additional complexes are triggered or turned off according to the needs of growth or differentiation. One example of periodicity of genetic activities during human development is the synthesis of five different hemoglobins before and after birth. The difference in gene activity is explained by five different genetic loci. The total number of genes is estimated as approximately 10,000 to 50,000, each set containing complete genetic information. A reservoir of genetic activation and inactivation is thus available for the entire life span at the right time and the right location.

Much is still unknown about the mechanisms for controlling the formation, activation, and inactivation of genes. They are, however, probably controlled by other genes.

The Development of Sexual Organs

The following steps determine the male or female sexual characteristics and the development of sexual organs: the chromosomal sex determination; the development and differentiation of the fallopian tubes, uterus, and vagina; and the development and differentiation of the external organs.

Chromosomal Sex Determinations. The sexual characteristics of the individual are determined by the X and Y chromosomes. The result of the union of maternal and paternal gametes is the combination of sex chromosomes to form male (XY) and female (XX) gonosomal complements. The germ cells of both sexes have to undergo meiosis before fertilization. This process reduces the original diploid set of 22 pairs of autosomes and two sex chromosomes to half the number (haploid). The specific diploid karyotype of 46 chromosomes is restored in the zygote after fertilization. After meiosis the haploid spermatozoa contain either a Y or an X chromosome. The haploid oocyte always contains one X chromosome. Fertilization by a spermatozoon with a Y chromosome results in a male (46 XY). Fertilization by a spermatozoon containing an X chromosome results in a zygote with 2 X chromosomes, or a female (Fig. 6). Consequently in man the sex of the offspring is determined by the paternal gamete.

The sex-determining function of the sex chromosome is more clearly understood after consideration of the evolutionary process. There is some evidence that the sex chromosomes consisted originally of a homologous pair of autosomes. These autosomes then developed divergently during evolution. It is assumed that during evolution the Y chromosome accumulated the genes that are important for the determination of the male sex, whereas the information less important for sex determination was lost for the most part. Thus, the Y chromosome became highly specialized for determination of the male sex.

The X chromosome, however, remained unchanged during evolution and contained autosomal genetic information in addition to female sex determination. It is assumed that 58 autosomal gene factors are located on the X chromosome. Some of them are known because of certain X-chromosomal anomalies as well as the characteristically sex-linked (X-linked) inborn errors of metabolism. Examples are the X-linked

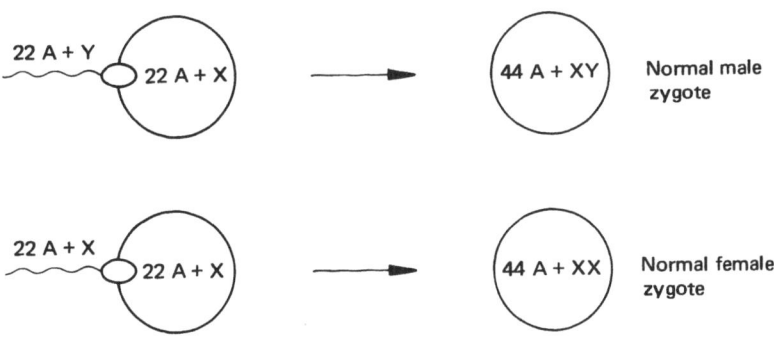

Fig. 6. Scheme of sex determination at fertilization. If the spermatozoon contains a Y chromosome, the resulting zygote will be male; if the spermatozoon contains an X chromosome, the union of egg and sperm will produce a female zygote.

form of mucopolysaccharidosis (Hunter's type), the hereditary muscular dystrophy (Duchenne type), and the Xg blood factor.

One X chromosome must be sufficient for somatic differentiation, since males have only one X chromosome. A buffer mechanism provides the balance in regard to the quantitative difference in genetic material between males with only one and females with two X chromosomes: one of the two X chromosomes in the female embryo is genetically inactivated during early development. It is still unclear whether there is a total or only a partial inactivation of one of the X chromosomes (page 3).

Female and male somatic cells are quantitatively similar, for they contain only one genetically active X chromosome. It is significant, however, that the X chromosomes can be inactivated in the embryonic cells regardless of whether they are derived from maternal or paternal chromosomes. The female can thus be regarded as containing two cell populations: one with an active paternal and one with an active maternal X. The female organism can, therefore, be considered a natural mosaic. The inactivation of one of the two X chromosomes occurs at day 12 to 20 after conception in chromosomally female embryos and, therefore, before differentiation of testes or ovaries. It is not yet clear why this significant event occurs at this stage of development, but it may indicate that both X chromosomes have to be genetically active up to that stage of development.

The Development and Differentiation of the Gonads. Although the sex of the future individual is chromosomally determined at fertilization, the primordium begins as an indifferent gonadal blastema in both sexes. The gonads appear in pairs on each side of the 4- to 5-mm embryo (fifth to sixth week after conception) as small conical prominences on the surface of the mesonephros. The celomic epithelium is regarded as the common origin of the potential follicle cells of the ovaries and the interstitial cells of the testes. *The sex chromosomal constitution of these primitive cells determines the direction of differentiation of the gonadal anlage and, therefore, the gonadal sex.* The primitive germ cells are believed to be the precursors of the oogonia and spermatogonia. They represent the uninterrupted transmission of human genetic material. They can be detected first at day 21 outside the gonadal anlage in the endoderm of the yolk sac. They move to both sides of the gonadal anlage and spread onto the surface. (Fig. 7). During this time they multiply by rapid mitotic activity. As soon as germ cells have accumulated in the primitive gonad (7 to 8 mm development), the activated blastemic cells begin to proliferate rapidly, resulting in the early differentiation into testes or ovaries.

The differentiation in male gonads is accompanied by migration of the germ cells from the periphery to the medulla of the primitive gonad.

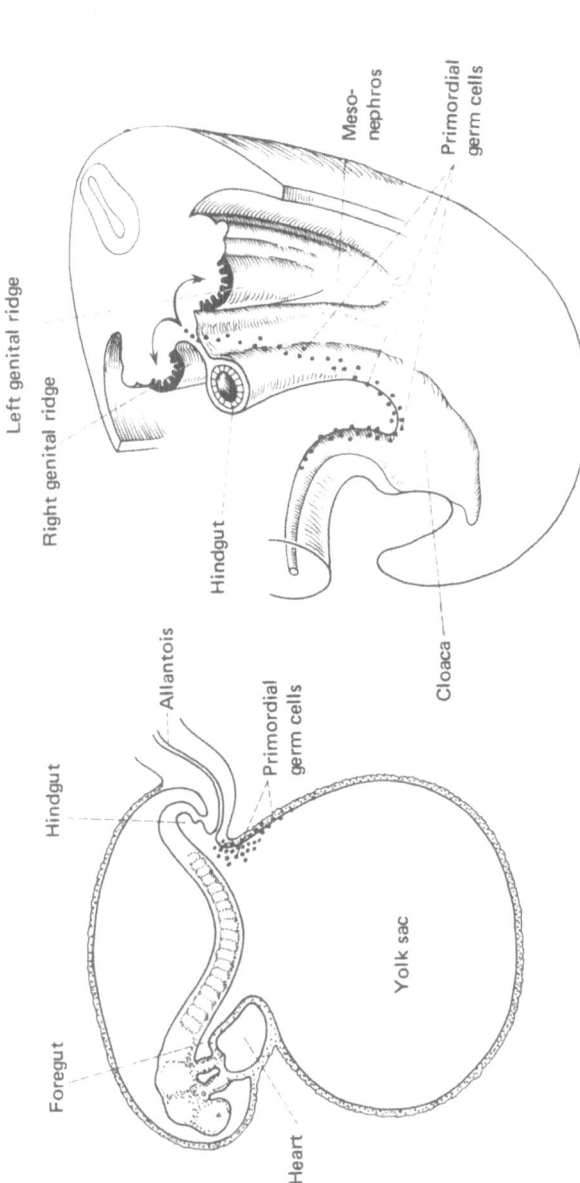

Fig. 7. Migration of the primordial germ cells. *Left:* Primordial germ cells in the wall of the yolk sac close to the attachment of the allantois. *Right:* Primordial germ cells reach the gonadal anlage.

The germ cells may now be called spermatogonia. The organization of seminiferous tubules begins in this area at the same time. Blastemic cells outside the seminiferous tubules differentiate into the androgen-producing interstitial cells. The area free of these elements in the hilus becomes the rete testis (Fig. 8). If differentiation occurs in the female direction into ovaries, the germ cells, the oogonia, accumulate below the surface of the gonad in the cortex. From there, the oogonia migrate into the deeper layers of the cortex, where cells of the blastema proliferate in the opposite direction from deep layers of the gonad, which contain precursors of the follicle cells. The follicle cells envelop the oogonia in a single layer to form the primary follicle. This process progresses from the periphery in the direction of the hilus. The central parts of the follicular cords are not reached by the germ cells. The rete ovarii develops from this germ-cell-free area of the mesenchyme. The celomic epithelium provides a single cell layer, the so-called germinal epithelium, which envelops these structures and separates them from extragonadal structures.

The oogonia begin to develop into oocytes in the follicular cords. The orderly progress of this phase of differentiation requires coordination, temporal and quantitative, of oogonia and follicle cells. The contact with follicle cells induces a functional change in the germ cell. The phase of multiplication is terminated and replaced by preparations for reproduction. Inhibition of mitotic division characterizes the differentiation from oogonia to oocytes, as does the initiation of the prophase of the first meiotic division. There is simultaneously an increase in cytoplasm. The first maturation division is arrested at the end of the first meiotic prophase, the oocytes remaining in a resting period similar to the interphase, or so-called dictyotene. The interruption of the maturation division requires the envelopment of the oocytes by a single layer of follicle cells. The transformation into the dictyotene stage seems to be controlled by the follicle cells.

The ring of follicle cells apparently provides the specific milieu necessary for arresting meiotic division in the dictyotene and for generating nutritive material in the cytoplasm of the oocyte (Fig. 9). The lack of an adequate number of follicle cells in the adjacent area can result in disorderly development. For instance, an excess of oogonia in the upper layer of the cortex prevents the optimal numerical relation between germ cells and follicle cells. This disproportion results in failure to arrest meiosis in the dictyotene phase after the end of the prophase of the first maturation division. The egg cells disintegrate in diakinesis, that is, in the metaphase of the first maturation division. This area, full of innumerable degenerated germ cells, is later replaced by connec-

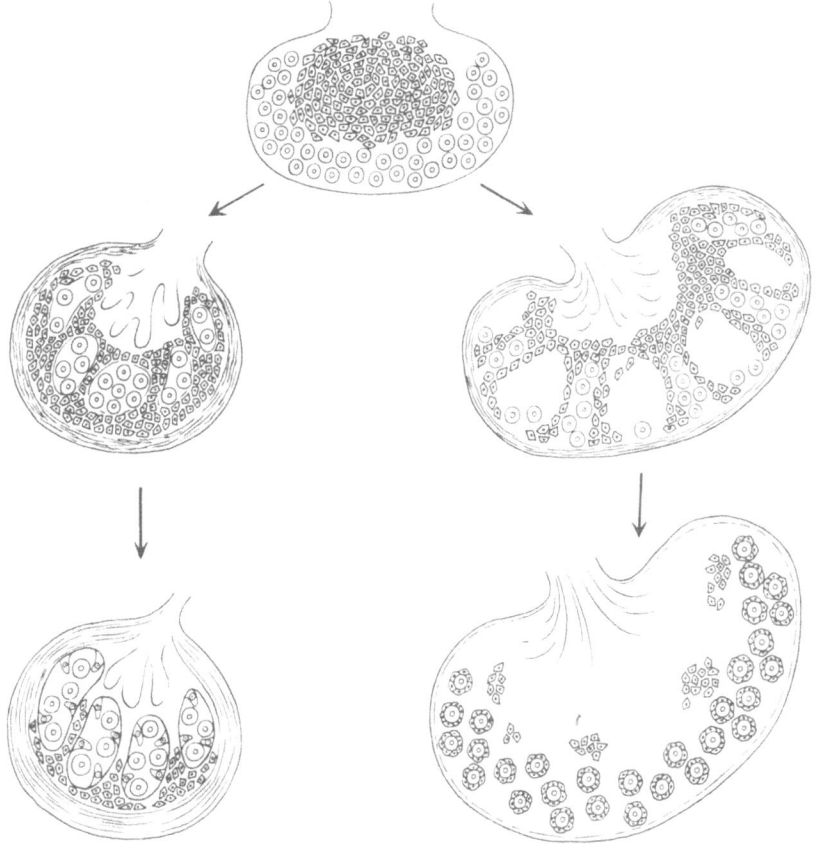

Fig. 8. Schematic representation of differentiation of gonadal anlage into testis and ovary. *Top:* The undifferentiated gonadal anlage with primordial germ cell in the periphery and somatic cells in the center. *Center:* Beginning of differentiation. *Left:* In the case of male differentiation, the primordial germ cells migrate into the mass of somatic cells, while the somatic cells are pushed somewhat toward the periphery, leaving a central zone that later becomes the rete testis in the hilus. The periphery becomes free of cells and is later transformed into a zone of connective tissue to form the tunica albuginea. *Right:* In the case of female differentiation, the somatic cells migrate toward the periphery and form pathways for the germ cells. *Bottom left:* Differentiated testis: somatic cells can be recognized as interstitial cells. *Bottom right:* Differentiated ovary. The somatic cells are recognized as follicle cells, surrounding the egg cells and forming the primary follicle. (after Ohno, 1967).

tive tissue. The tunica albuginea originates in this area and forms, together with a single layer of celomic epithelium, the covering of the

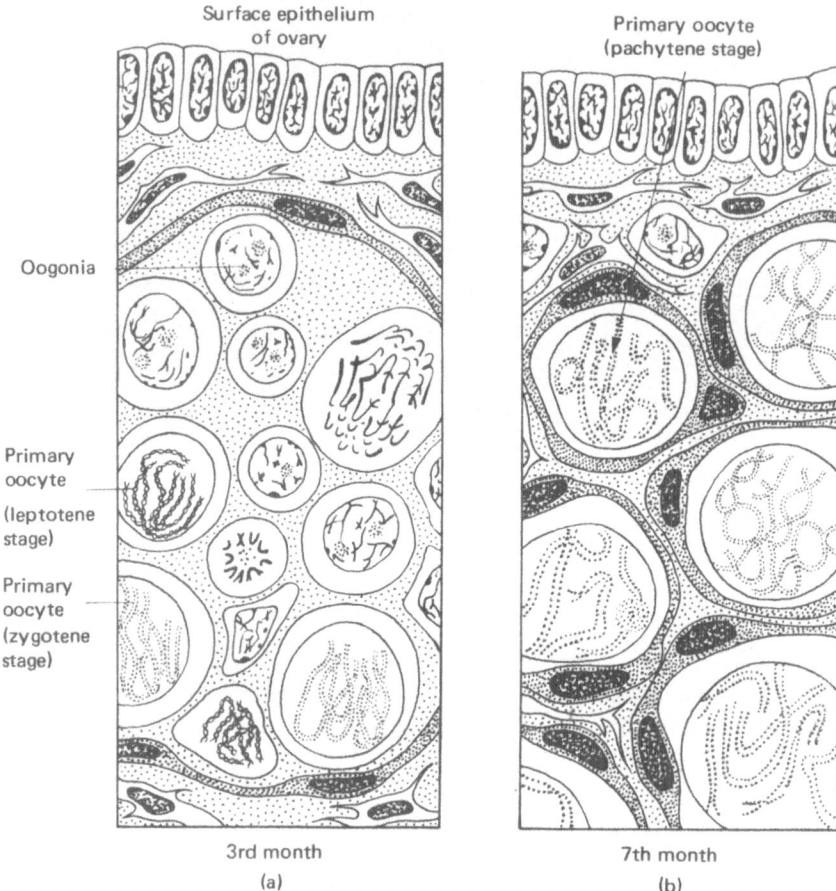

Fig. 9. Schematic representation of the histologic structure of the ovary at various stages of development. (a) Ovary in the third month of fetal development. The oogonia are grouped in the cortex of the ovary. Some are in mitosis and others have already differentiated into primary oocytes and have entered the prophase of the first meiotic division (leptotene and zygotene stages). (b) By seven months, almost all of the oogonia have been transformed into primary oocytes, which are shown in the pachytene phase of the prophase of the first meiotic division. (c) At birth, oogonia can no longer be detected. Each primary oocyte is surrounded by a single layer of follicle cells, forming the primordial follicle. The oocytes have entered the dictyotene stage, in which they remain until maturation. Only then do they enter the metaphase of the first meiotic division (modified after Ohno et al., 1962).

ovary. The so-called germinal epithelium remains as a complete covering layer and is present even into the reproductive age. It gradually degenera-

Resting primary oocyte
(dictyotene stage)

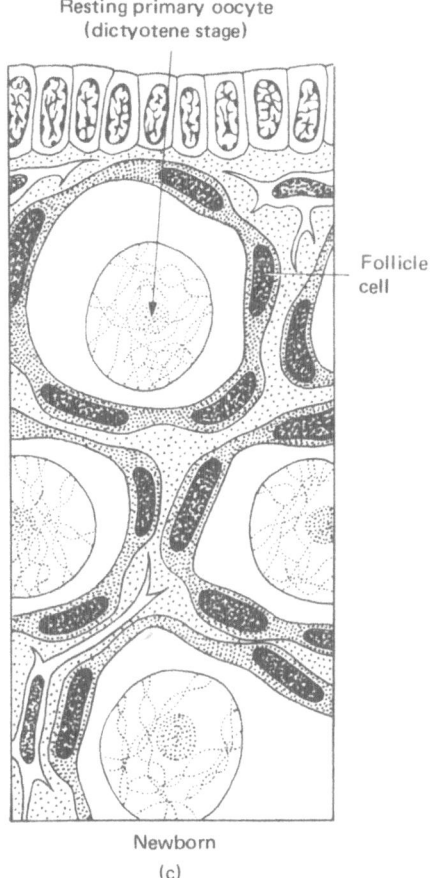

Follicle
cell

Newborn

(c)

ates as the result of consecutive ovulations. It may be the origin of certain ovarian tumors (page 331).

The sex-specific differentiation of the gonads begins approximately in the seventh week after conception. Gonadal sex is therefore determined at a very early phase in development. It is generally accepted that the differentiation of the gonads is controlled by the sex-specific genes on the sex chromosomes of the indifferent gonad.

The presence or the absence of the Y chromosome is the significant factor for this development. The presence of a Y chromosome results in testes and its absence, in ovaries. The primitive germ cells are not required for the modus of differentiation. However, the final development of testes and ovaries is dependent on a sufficient number of germ cells. Primary follicles can be seen first in embryos 2½ months after conception. Their development extends into the eighth month of fetal age. The total number of oocytes at birth in one ovary is at least 400,000 to 500,000.

The Development and Differentiation of the Internal Sex Organs. The specific differentiation of sex organs ensues, depending on the gonadal sex. It starts, therefore, after the induction of the differentiation of the gonad. Still later begins the development of the male or female external organs.

In the early indifferent state of development the embryos of both sexes have on either side two genital ducts, the primitive wolffian ducts as potential male, and laterally the müllerian ducts as potential female structures. The müllerian ducts cross the wolffian ducts at the pelvic rim and run medially.

During development, the lower parts of the müllerian ducts fuse to form a solid cord that joins the urogenital sinus. The fallopian tubes and uterine corpus develop from the upper parts of the müllerian ducts.

The fallopian tubes are arranged bilaterally at the upper margin of the broad ligament and open freely through the fimbriated ends into the abdomen. The corpus uteri develops from the upper paired parts and the cervix uteri from the fused lower part of the müllerian ducts. The uterus passes through a bicornuate stage during development. The formation of a lumen begins in the fallopian tube followed by canalization of the cervix, and finally the uterine cavity. The final stage is characterized by the unification of the cavities of both cornual ends by resorption of the septum (Fig. 10). Partial or total failure of this development results in malformations of the uterus, termed uterus septus or subseptus. The arrest of growth, differentiation, and fusion of the primordia results in paired malformed structures at different levels.

The uterovaginal canal is surrounded by a mesenchymal layer from which the myometrium forms at the fifth month of fetal development. The mucosa becomes evident at approximately the same time. The cervix uteri is separated from the uterine cavity by an internal os at the seventh fetal month and is twice as long as the corpus. Although the uterus has developed in its final form, a small depression in the fundus is present in the newborn (Fig. 10). The final form with the rounded fundus and a ratio of cervix to corpus of 1:2 is reached only at the onset of puberty.

The müllerian ducts regress in the male embryo. Epididymis, seminal vesicle, and ductus deferens form from the wolffian ducts. In the female embryo the wolffian ducts regress, but parts may persist adjacent to fallopian tubes, uterus, and vagina. They form, in the upper aspects of the broad ligament, the rudimentary structures known as the epoophoron and paroophoron. The lower parts of persistent wolffian ducts adja-

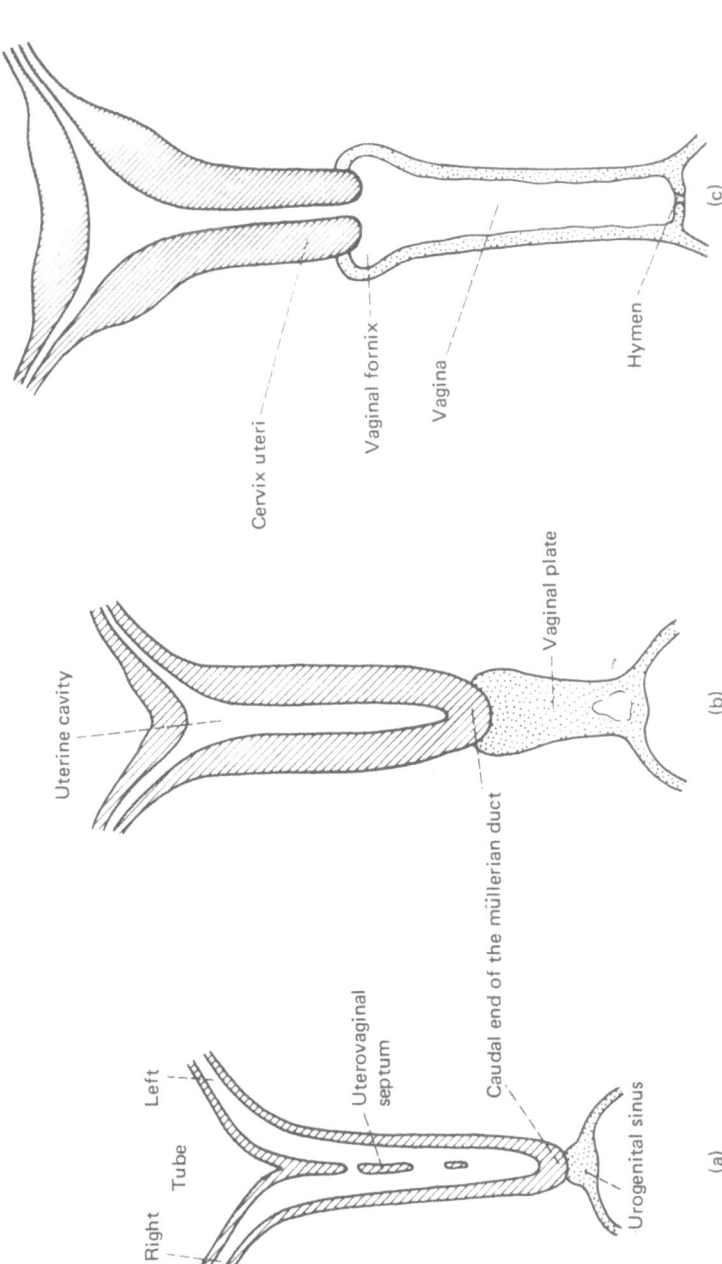

Fig. 10. Development of the uterus and vagina. (a) By the ninth week of development the uterovaginal septum is resorbed. (b) At the end of the third month the uterine cavity is formed. Between the uterus and the urogenital sinus the vaginal plate is developed. (c) At birth the uterus has an arcuate appearance; the cervix is twice as long as the uterus. Vagina, hymen, and vaginal tissues are definitively formed. (After Langman, 1969).

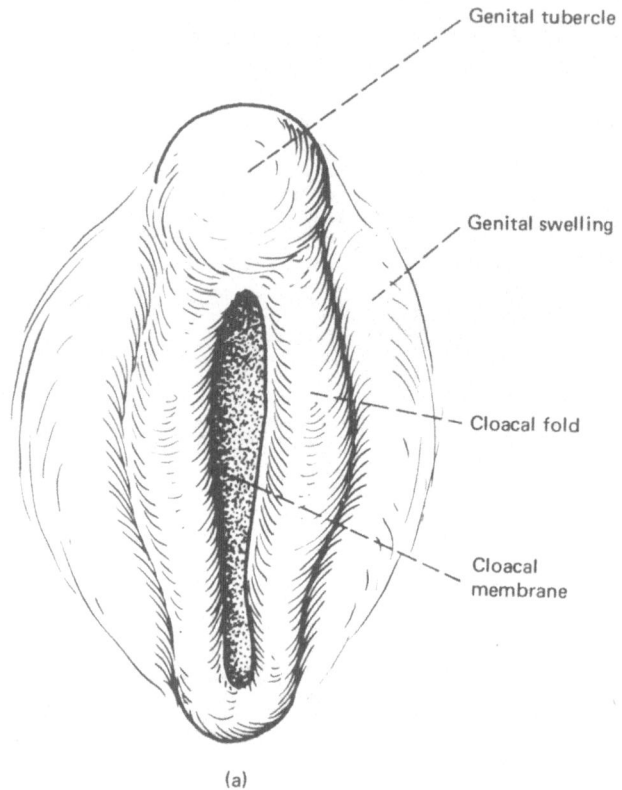

Genital tubercle

Genital swelling

Cloacal fold

Cloacal
membrane

(a)

Fig. 11. Development of the female external genitalia. (a) The undifferentiated stage at 6 weeks' fetal development. (b) The external genitalia at birth.

cent to the vagina may form Gartner's duct. They are of clinical significance, since tumors and cysts may arise from these structures.

The vagina develops from an epithelial rim below the fused müllerian ducts and the urogenital sinus. It is not yet clear whether the vaginal plate derives primarily from the müllerian ducts (mesoderm) or primarily from the urogenital sinus (endoderm).

The vagina is formed as a tube in the fifth month of fetal development. The cranial parts span the caudal parts of the cervix in a wing-shaped manner and become the anterior and posterior vaults of the vagina. Malformations in this area may result in absence of the vagina (vaginal aplasia) or partial atresia, and septate or subseptate vagina.

The lumen of the vagina is separated from the urogenital sinus by the hymen (Fig. 10). It consists of epithelium of the sinus and a thin

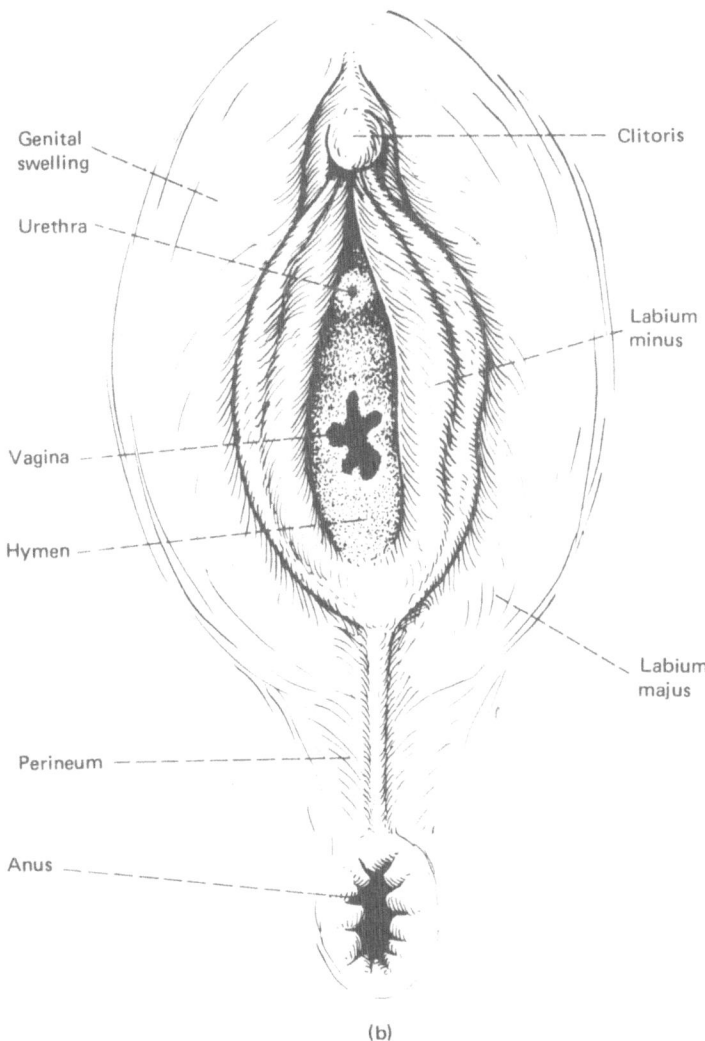

Genital swelling

Urethra

Vagina

Hymen

Perineum

Anus

Clitoris

Labium minus

Labium majus

(b)

mesodermal layer. The final form of the hymen depends on the manner of perforation into the vagina.

The Development and Differentiation of the External Genitalia. The differentiation of the external genitalia occurs later than that of the internal genitalia; they develop primarily form these three indifferent structures:

(1) the lower part of the urogenital sinus
(2) the genital tubercle
(3) the lateral genital ridges

The urogenital sinus, derived from the cloacal membrane, forms the connection between the internal organs and the body surface. The genital tubercle can be detected in its primitive form during the fifth week after conception. It develops into a cylindrical phallus and forms the lateral folds that border the lower aspect of the urogenital sinus. The indifferent stage lasts until approximately 10 weeks after conception (Fig. 11a).

After this time, differentiation occurs specifically in either the male or the female direction. In the female, the primitive phallus is transformed into the clitoris, including glans and prepuce. The genital folds become the labia minora on both sides. The labia majora derive from swellings lateral to the genital folds (lateral genital folds) that flank the clitoris and extend dorsally to the anus forming the so-called posterior commissure. The lower part of the urogenital sinus remaining in the female becomes the vestibule (Fig. 11b.) Formation of the female external genitalia therefore involves less change in structure than does the male and more closely resembles the indifferent state of development.

Sex of the embryo can be ascertained macroscopically first at 10 weeks after conception. Female embryos are identified by fusion of the lateral genital folds (posterior commissure).

Control Mechanism for the Differentiation of Secondary Sex Organs. The differentiation of sex-specific structures of the internal and external organs and the regression of the contralateral sex ducts begin only after the gonads have differentiated. There is, therefore, a long "neutral" phase in development of accessory sex organs.

Bilateral castration of male rabbit embryos (Jost 1947) revealed regression of the wolffian ducts, whereas, the müllerian ducts persisted and the urogenital sinus differentiated in the female direction. Castration of female fetuses resulted in regression of wolffian ducts, but the müllerian ducts and the urogenital sinus differentiated into female structures even though the ovaries were missing. On the basis of experiments with androgenic hormones, it became evident that the differentiation of the external organs into male or female structures is dependent on the hormones of the gonad. Differentiation of the gonads into testes provides androgens and a still unidentified factor (factor X) in interstitial cells (Leydig cells). Androgens and factor X are responsible for the stabilization of wolffian ducts and the regression of the müllerian ducts, with normal male development. Differentiation of the primitive gonad into ovaries, however, results in failure of the production of both androgens and factor X. The wolffian ducts are, therefore, not stabilized and the müllerian ducts remain, with differentiation proceeding in the female direction (Figs. 12 and 13). Androgens are also required for the development of male external sex organs. In their absence differentiation into female sex organs occurs.

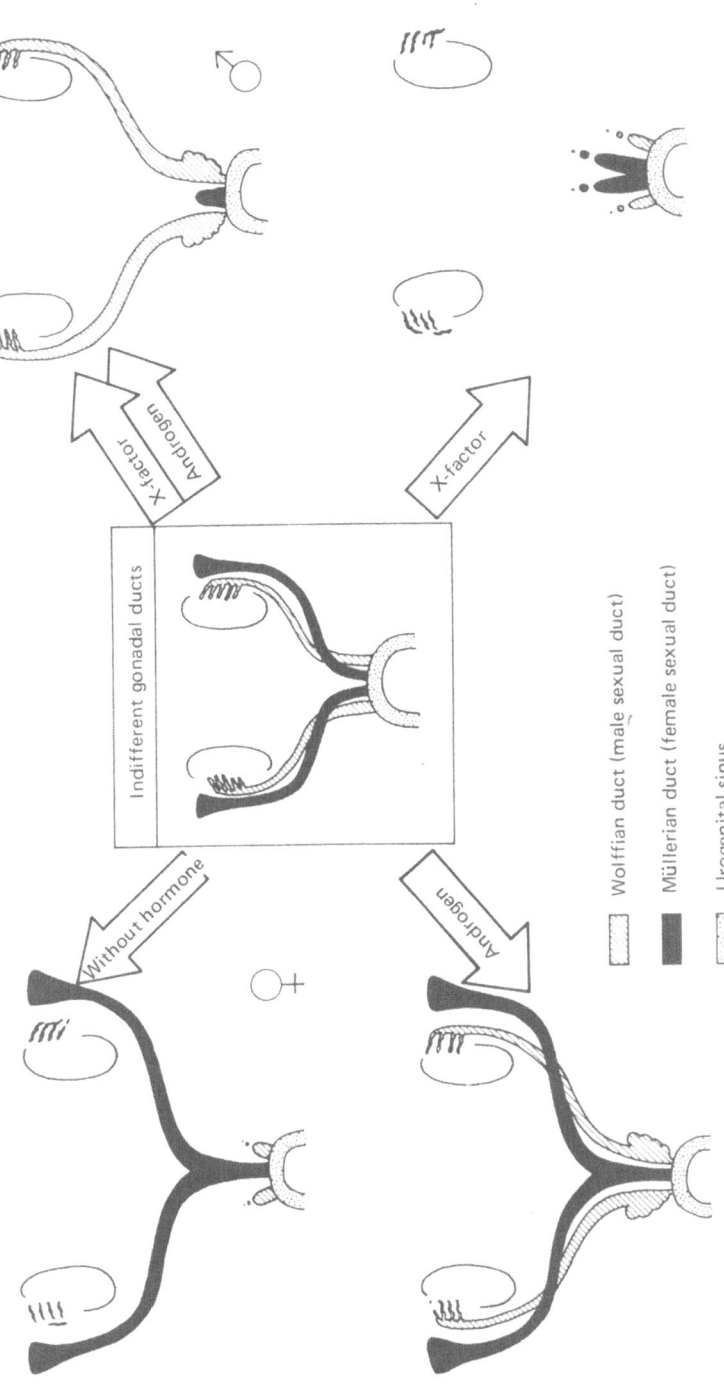

Fig. 12. Hormonal regulation of differentiation of the ductal systems. *Center:* Undifferentiated gonads with the anlagen of wolffian and müllerian ducts. *Top right:* If androgen and factor X are elaborated, the müllerian ducts are suppressed and the wolffian ducts differentiate. Normal male ducts are thus developed. *Upper left:* If both androgen and factor X are lacking, wolffian ducts are suppressed and müllerian ducts are stabilized, resulting in a female duct system. *Lower right:* Lack of androgens in the presence of factor X results in suppression of müllerian ducts but failure of stabilization of wolffian duct; both ductal systems remain rudimentary. *Lower left:* Lack of factor X in the presence of androgens results in stabilization of the wolffian ducts, but failure of suppression of the müllerian ducts; both ductal systems are thus present (after Neumann, 1967).

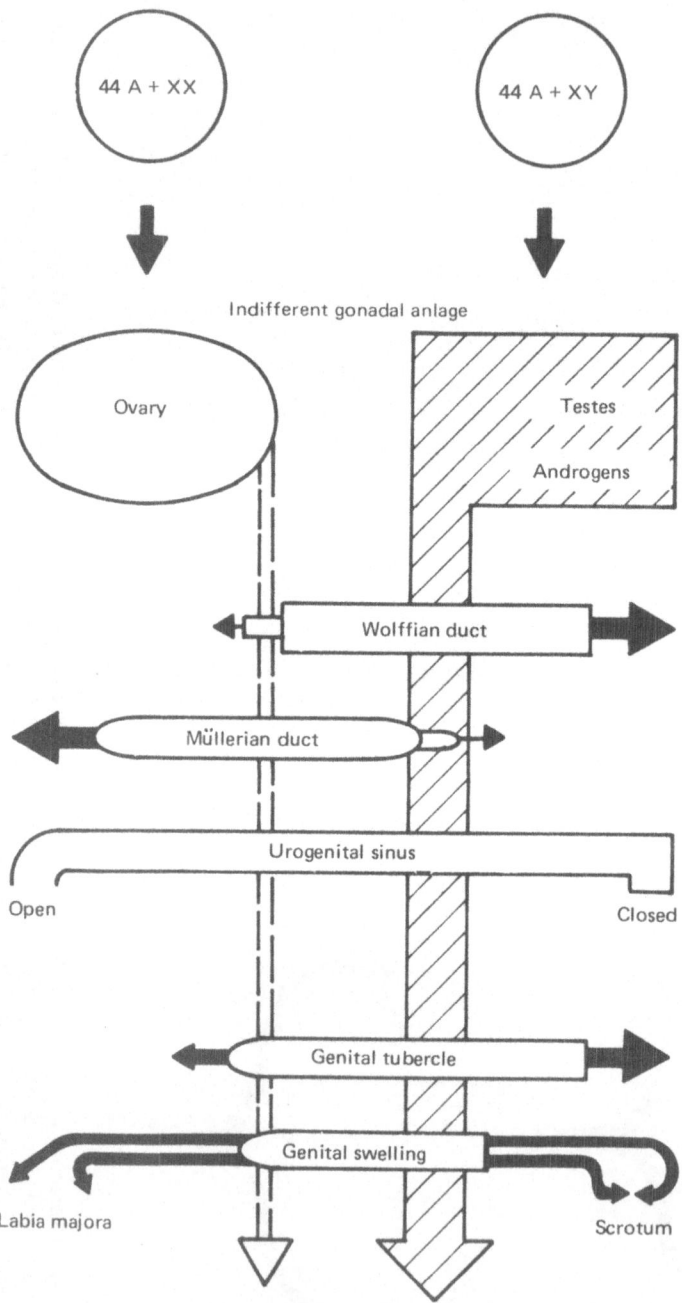

Fig. 13. Genetic and hormonal factors in the normal differentiation of duct systems and external genitalia. (Modified after Tuchmann-Duplessis, 1970.)

Selected Reading

Tuchmann-Duplessis, J., Herbert, Q. S., Haegel, S., and Pierre, T.: *Organogenèse*. 2nd ed. Paris: Masson (1970).

Gossen, P. E.: Giemsa banding patterns of human chromosomes. Clin Gen. 3:169, 1972.

Langman, J.: *Medical Embryology*, Baltimore: Williams and Williams (1969).

Ohno, S.: *Sex Chromosomes and Sex-linked Genes*. Berlin-Heidelberg-New York: Springer-Verlag (1967).

CHAPTER 2

Functional Anatomy and Histology of the Female Genitalia

The Female Pelvis

The pelvis is composed of four bones: the sacrum, the coccyx, and the two innominate bones, which consist of the ilium, the ischium, and the pubis. The innominate bones are joined by the sacrum at the sacroiliac notches and anteriorly at the pubic symphysis.

The pelvic architecture represents a compromise between its two primary functions. First, as a result of the erect posture of man, the pelvis must support the upper part of the body. Second, the pelvis must provide room for the passage of the fetus.

The Pelvic Floor. The pelvic floor comprises a three-layered fibromuscular system that extends laterally and caudally (Fig. 14). It contains the *pelvic diaphragm,* formed by the levator ani, which originates on the pelvic wall and together with the coccygeus muscle forms a muscular plate that slopes downward toward the midline (Figs. 14, 15). In the middle is a triangular passage with its base at the pubic ramus.

The urogenital diaphragm consists of a firm plate of connective tissue, which extends between the angle formed by the two pubic bones. The transversus perinei muscle strengthens this plate and provides elasticity. The urethra passes through this structure.

The lowest part of the pelvic floor consists of the voluntary sphincter ani and bulbocavernosus muscle. The superficial transversus perinei muscle and the ischiocavernosus muscles support the pelvic floor diagonally and transversely and are involuntary.

The pelvic floor is stressed during the third trimester and at delivery. Laceration and contusion, especially in the anterior part of the levator, disturb the integrity of the muscular supports and may lead to vaginal relaxation (page 249). Avoidance of injuries to the pelvic floor by episiotomy and low forceps is, therefore, part of good preventive obstetrics.

The Ligamentous System of the Pelvis. A system of fibromuscular tissue extends from the pelvic wall to the uterus. Three such structures join the uterus near the internal os: The *cardinal ligaments* originate on the lateral pelvic wall; the *uterosacral ligaments* originate on the

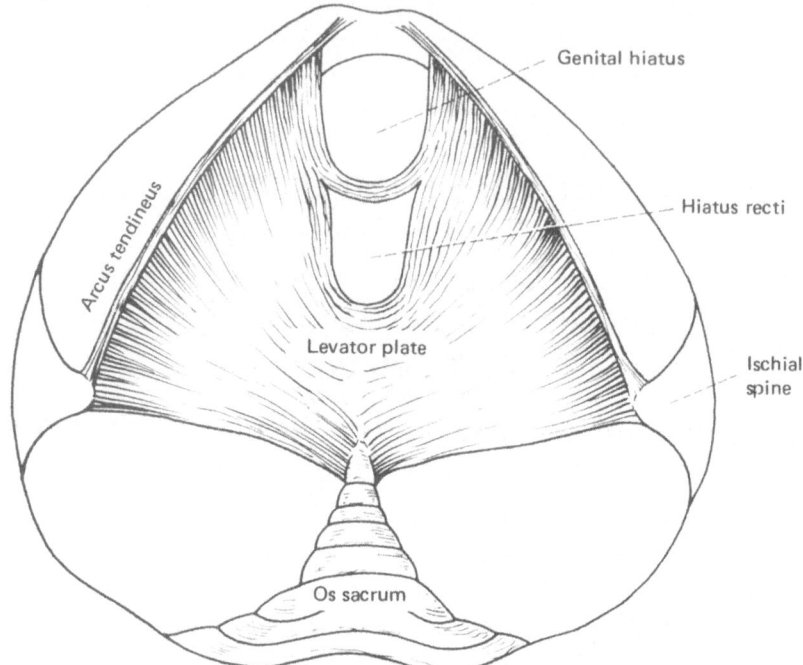

Fig. 14. Pelvic diaphragm (course of the levator ani).

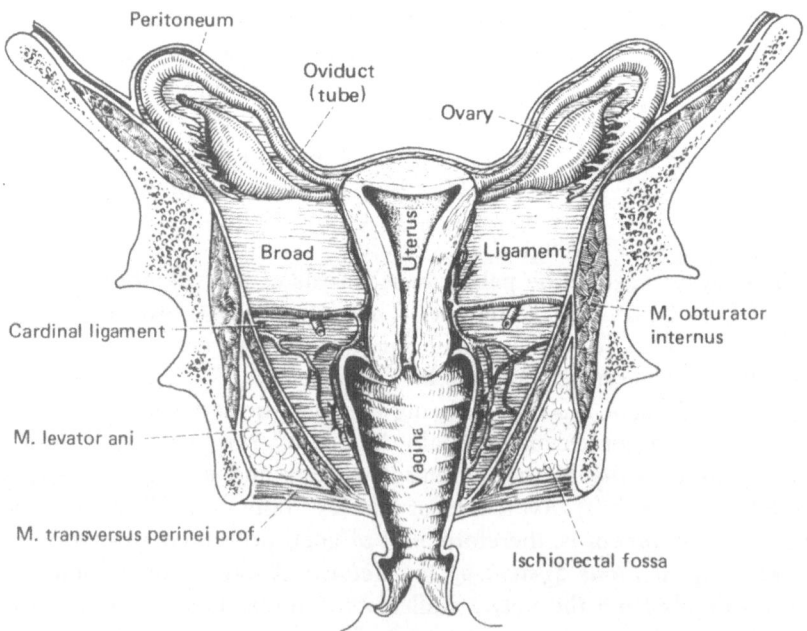

Fig. 15. Anatomy of the female pelvis (frontal section).

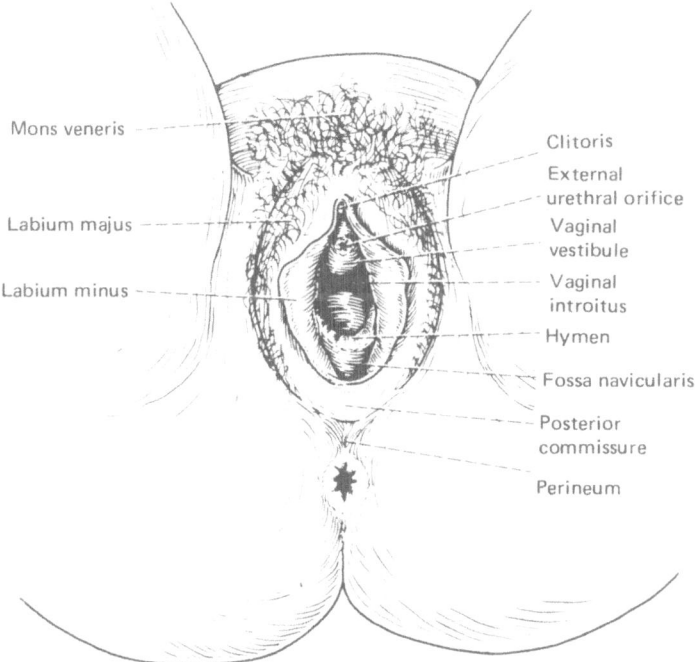

Fig. 16. Adult female external genitalia.

sacrum and surround the rectum on both sides; and the *pubovesical ligaments* arise from the perivesical and periurethral tissues.

The *round ligaments* originate near the cornua of the uterus anteriorly, extending into the inguinal canal and finally into the labia majora.

The *broad ligaments* extend from the sides of the corpus uteri to the pelvic wall (Fig. 15). The ureters and uterine arteries pass between the folds of the broad ligaments. Myomas extending from the uterus between the leaves of the broad ligament are designated intraligamentous. They often distort anatomical structures and must be removed with care to avoid damage to the adjacent ureters.

The *ovarian ligaments* are short structures extending from the uterine cornua to the ovaries.

The *infundibulopelvic ligaments* (suspensory ligaments of the ovary) extend from the fimbriated ends of the fallopian tubes to the pelvic wall and contain the ovarian vessels.

The External Genitalia

The external genitalia form a functional unit (Fig. 16). They are target organs for estrogenic hormones. During sexual activity the various parts are controlled by complex neural activity. As the result of decreasing hormonal activity, the external genitalia undergo changes with age. After menopause, because of lack of estrogens, the external genitalia become atrophic and predisposed to a variety of lesions.

The vulva is framed by the labia majora and the mons veneris, which consists of the hair and subcutaneous fat around the symphysis. The normal pubic hairline in the woman forms a transverse line across the lower abdomen.

The labia majora extend on either side from the mons veneris and merge posteriorly into the perineum, forming the posterior commissure. Occasionally, hernias along the course of the round ligaments extend into the labia majora and must be differentiated from Bartholin cysts (page 217). The labia majora are covered by squamous epithelium containing hair follicles and sebaceous glands with underlying fat. The labia majora are the homologues of the scrotum. The number of hair follicles decreases toward the inner part of the introitus.

The *labia minora* divide superiorly into two lamellae. Fusion of the lower two forms the frenulum of the clitoris, whereas the upper two merge into the prepuce. Inferiorly, the labia minora unite as small ridges to form the fourchette. The fourchette may be stretched or completely destroyed in multiparas. The labia minora consist of connective tissue that contains little fat but a rich supply of nerves and blood vessels. The vestibule is the remnant of the urogenital sinus of the embryo. It is bordered anteriorly by the clitoris, laterally by the labia minora, and posteriorly by the fourchette.

Embryologically, the *clitoris* is the homologue of the penis. Two corpora cavernosa merge below the symphysis into a short cylindrical body, the corpus clitoridis, covered by the prepuce of the clitoris. Glans and prepuce have a rich supply of nerve fibers and are highly vascularized by a plexus of enlarged veins. They are embryologically the remnants of the corpus cavernosum urethrae of the male.

The Glands of the Vestibule

All glandular structures in the subcutaneous tissue and the connective tissue of the external genitalia empty through their ducts into the vestibule.

The greater vestibular, or Bartholin's, glands are located below the bulbus vestibuli beneath the bulbocavernosus muscle. Their secretions are carried by ducts that open at the base of the vestibule on either side. They secrete a whitish-gray mucus during sexual excitement (page 77). Infection of the duct of the gland (frequently with gonococci) results in a Bartholin abscess (page 217).

The lesser vestibular glands are small alveolar structures distributed throughout the entire wall of the vestibule. The largest of these mucus-producing glands open through two to four ducts between the urethra and the vaginal introitus.

Paraurethral, or Skene's, ducts are small structures that open lateral to the urethra and may harbor microorganisms. They are commonly involved in acute gonococcal infections.

After menopause the turgor and elasticity of the labia minora and the clitoris decrease and the glands and corpora cavernosa undergo atrophy. The involution, however, does not necessarily interfere with the ability to perform intercourse.

The external os of the urethra is located in the anterior part of the vestibule. The secretion of the sebaceous glands in this area produces a layer that protects the vulnerable tissues against urine.

The vaginal introitus, the border between the external and internal genitalia, is located in the lower portion of the vestibule. The hymen is a plate consisting of connective tissue with a rich vascular but scant nerve supply. The openings of the hymen vary in shape and size.

The hymen ruptures as a rule during the first coitus (defloration), tearing at one or several points, usually posteriorly. The remaining tags border the introitus. They may be further injured during childbirth and remain as myrtiform caruncles.

Vagina

The vagina extends from the external genitalia to the center of the pelvis as a musculomembranous tube covered by a noncornifying stratified squamous epithelium which varies greatly under hormonal influences (Fig. 17 and Table 1).

The lower part of the cervix, the portio vaginalis, is a conelike projection into the vagina, forming anterior, posterior, and lateral fornices. The posterior fornix extends lower in the pelvis and is therefore longer than the anterior fornix. It is separated by only a thin layer of connective tissue from the pouch of Douglas (cul-de-sac). Collections of blood in the peritoneum may bulge into the posterior fornix and abscesses of the cul-de-sac may perforate spontaneously through it. The close relation of the posterior fornix to the cul-de-sac makes the pouch of Douglas quite accessible for diagnostic procedures such as culdocentesis or culdoscopy (page 187). The posterior wall of the vagina is longer than the anterior wall, which is separated from the urethra and bladder only by the vesicovaginal septum.

If the anterior wall of the vagina is weakened, the urethra may prolapse into it and the posterior urethrovesical angle may be increased, producing a urethrocele and stress incontinence (page 251).

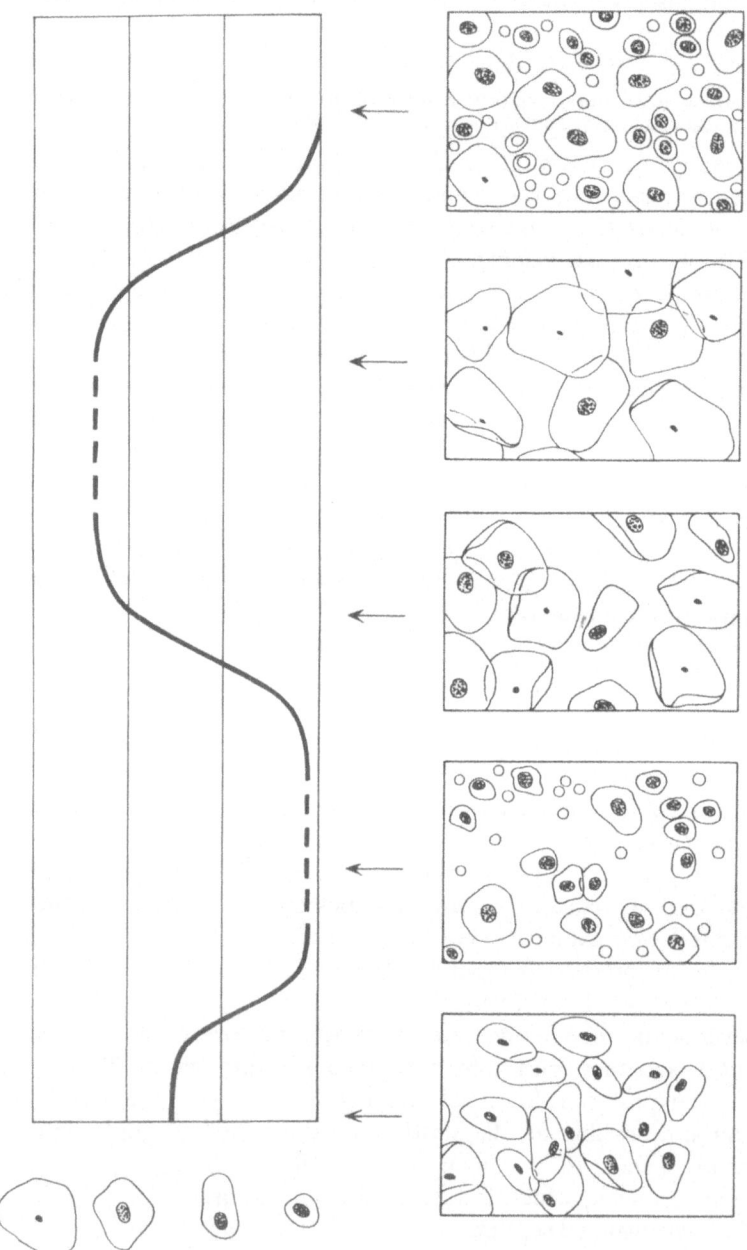

Fig. 17. Vaginal cytologic picture in various stages of life. *Top:* Graphic representation of the maturation of the vaginal epithelium. *Bottom:* Left to right: Epithelial maturation at birth, atrophic cell picture in childhood, beginning of estrogenic influence in puberty, complete maturation in the reproductive age, regression is senility.

Table 1. Most Important Characteristics of the Vaginal Epithelial Cells

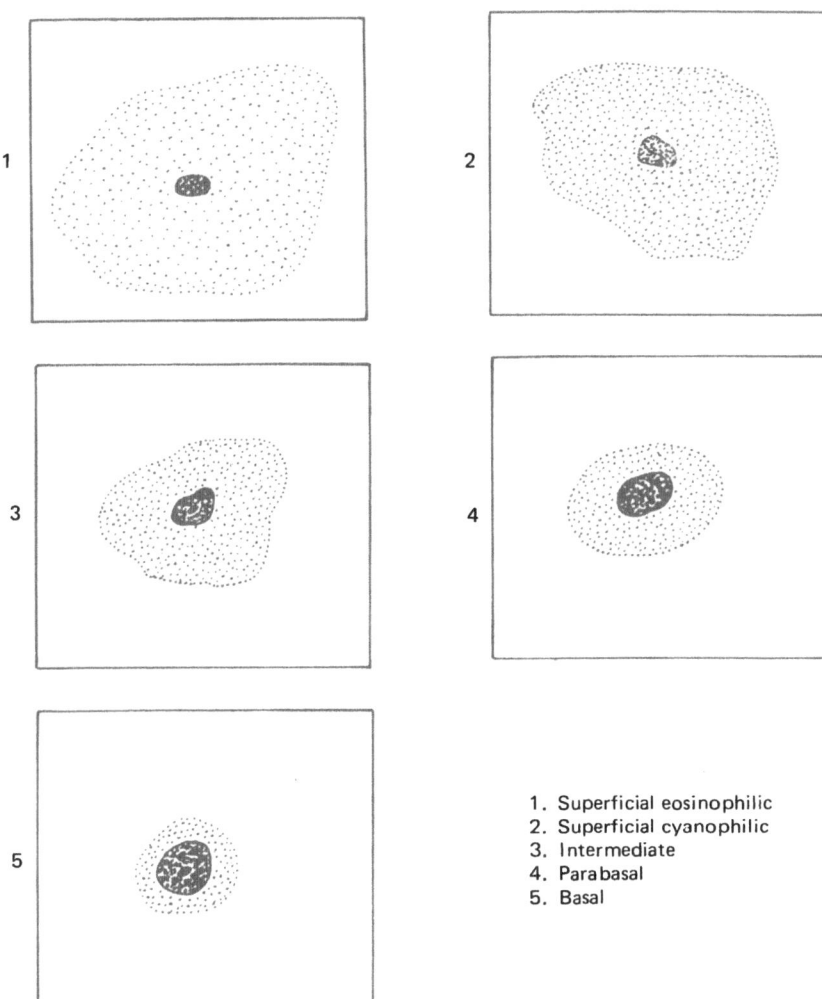

1. Superficial eosinophilic
2. Superficial cyanophilic
3. Intermediate
4. Parabasal
5. Basal

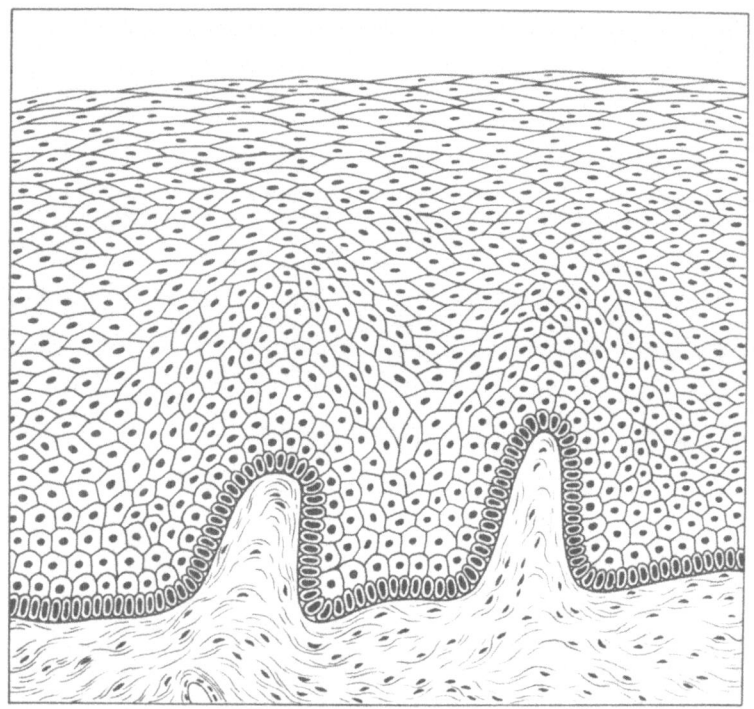

Fig. 18. Vaginal epithelium in reproductive age. Well-developed epithelium under the influence of estrogens; the individual layers (basal, parabasal, intermediate, and superficial) are easy to recognize. The same layers are found on the surface of the portio.

The anterior and posterior vaginal walls lie in contact and form an H-shaped space, which can be greatly expanded. The subepithelial connective tissue is rich in venous channels and elastic fibers that form transverse ridges, or rugae, which are occasionally obliterated after repeated childbirth or menopause.

The squamous epithelium is a target tissue for sex hormones. In the newborn the vaginal epithelium is histologically similar to that of the adult because of the estrogen stimulation during pregnancy (Fig. 18). As a result of a reduction in estrogen stimulation, the thickness of the vaginal mucosa decreases during childhood, when it appears similar to that of the postmenopausal vagina. At puberty the vaginal mucosa begins to assume its adult appearance (Figs. 19, 20, 21).

Under the influence of estrogens, the epithelium differentiates into basal, parabasal, intermediate, and superficial layers. Maturation of the vaginal epithelium is a sensitive index of estrogenic activity. For instance, estradiol in a dose of 6 micrograms produces cytologically

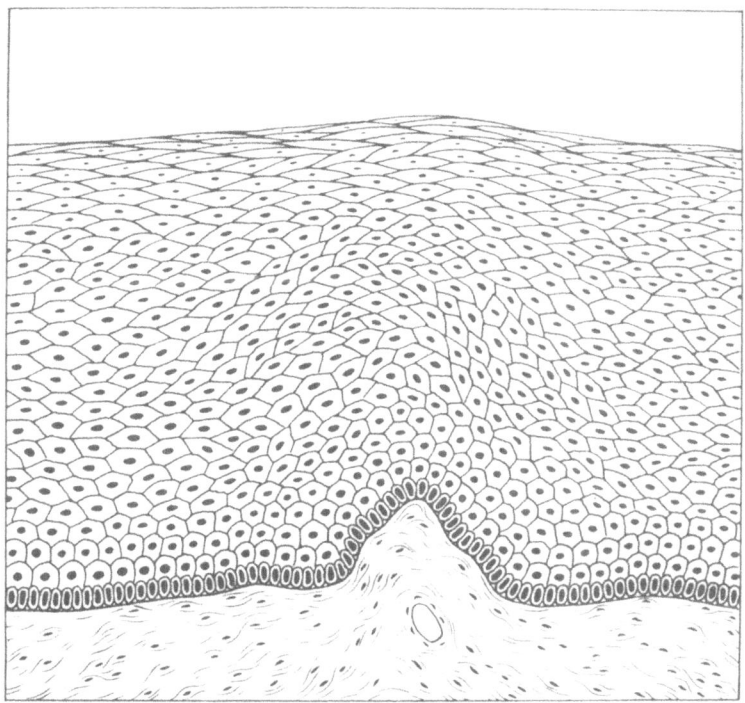

Fig. 19. Vaginal epithelium in the newborn. Under the influence of the placental estrogens the epithelium attains almost the same thickness and development as in the mature woman.

recognizable proliferative changes in the atrophic vagina. The threshold for the endometrium is 30 times higher.

In hormonal diagnostic cytology, a smear is obtained from the lateral wall of the vagina and stained by Papanicolaou's method. The hormonal activity can be estimated from the proportions of the various types of cells.

Maturation of the vaginal epithelium fails to occur in the absence of estrogens, and only basal and parabasal cells are seen. This pattern is considered "atrophic" and is characteristic of childhood and old age. Under pathological conditions during the reproductive age, this pattern may be found when the ovaries produce insufficient estrogen, either because of a lack of stimulation by the pituitary-hypothalmic system or a functional disorder of the ovary itself. Complete maturation is characterized by a predominance of superficial cells, most apparent at the time of ovulation.

The ratio of estrogen to progesterone may be suggested in the vaginal smear by the proportions of the various types of superficial cells. The so-called karyopyknotic index is the ratio of superficial cells with

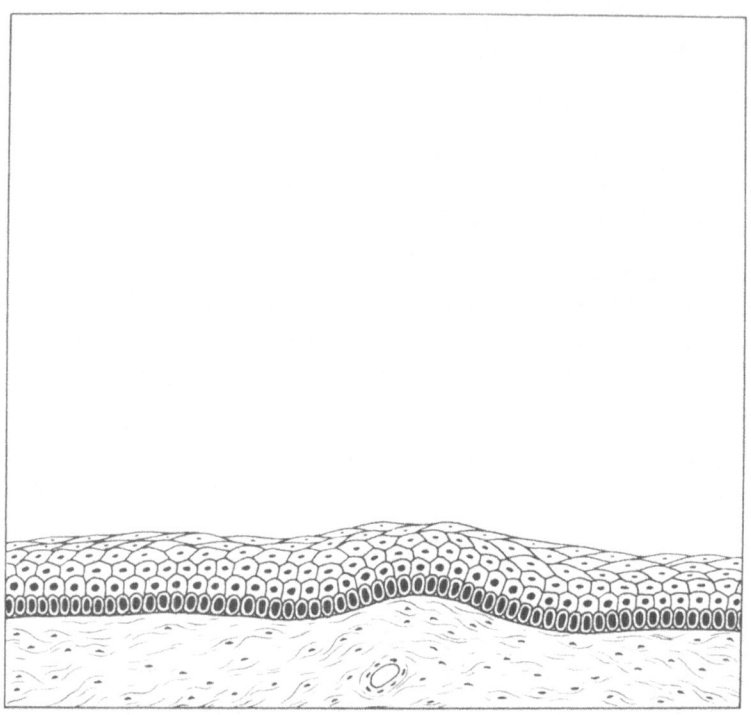

Fig. 20. Vaginal epithelium in childhood. Before the onset of ovarian function the epithelium is low.

pyknotic nuclei to those with vesicular nuclei. The relation between basophilic and eosinophilic superficial cells is called the eosinophilic index.

Progesterone activity is reflected in a decrease of the karyopyknotic and eosinophilic indices and an increase in cell clusters with folded margins. The physiology of the vagina is discussed on page 220.

Uterus

The uterus is located above the vagina in the center of the pelvis (Fig. 22). During pregnancy, it serves for reception, retention, and nutrition of the conceptus. The cavity of the uterus is lined by the endometrium, which is cyclically prepared by steroid hormones for nidation.

Lateral to the cervix is the condensed connective tissue known as the parametrium.

The external os of the cervix is normally located in the center of the portio vaginalis. It is small and round in the nullipara, but in the multipara may form a transverse slit that divides the cervix in anterior and posterior lips. As a result of cervical lacerations, the external os

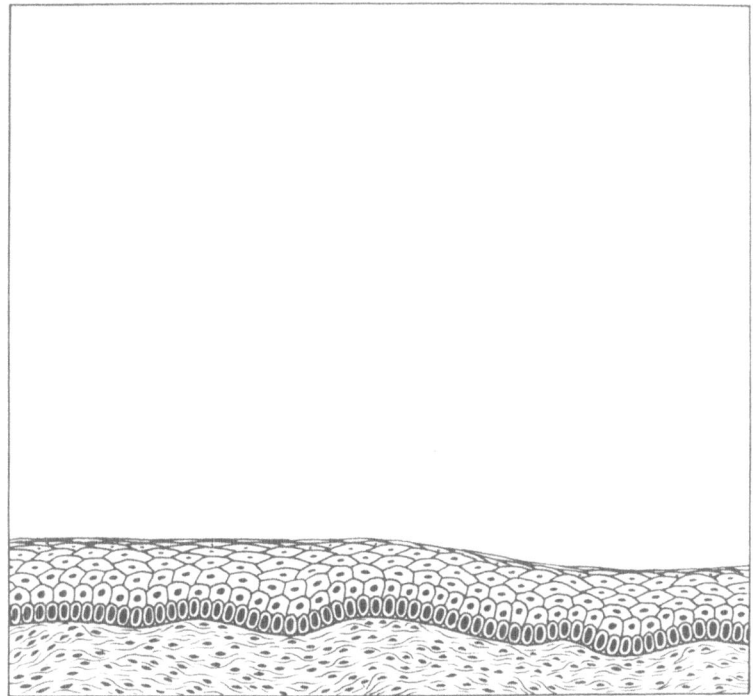

Fig. 21. Vaginal epithelium in old age. After cessation of ovarian function, the epithelium atrophies and the subepithelial connective tissue appears cellular and thick.

may be found at the end of a deep scar that extends into the lateral fornices.

A noncornified, stratified squamous epithelium normally covers the portio vaginalis. The endocervix, however, is lined by a single layer of mucus-producing columnar epithelium. The high pH of the mucus (between 7 to 8) serves as a barrier to infections. The hormone-dependent changes in physical and chemical characteristics of the mucus have great influence on the penetration of sperm. The racemose glands are probably clefts produced in the epithelial lining.

The boundary between squamous and columnar glandular epithelium is called the squamocolumnar junction, which is clinically important because most cervical cancers develop there (page 291). The location of the junction varies with age and parity. The squamocolumnar junction is located inside the cervical canal during childhood and in the postmenopausal woman, whereas during reproductive life it often extends to the surface of the portio vaginalis (Fig. 23). This change probably reflects the influence of estrogens. The clinical term "erosion" for areas

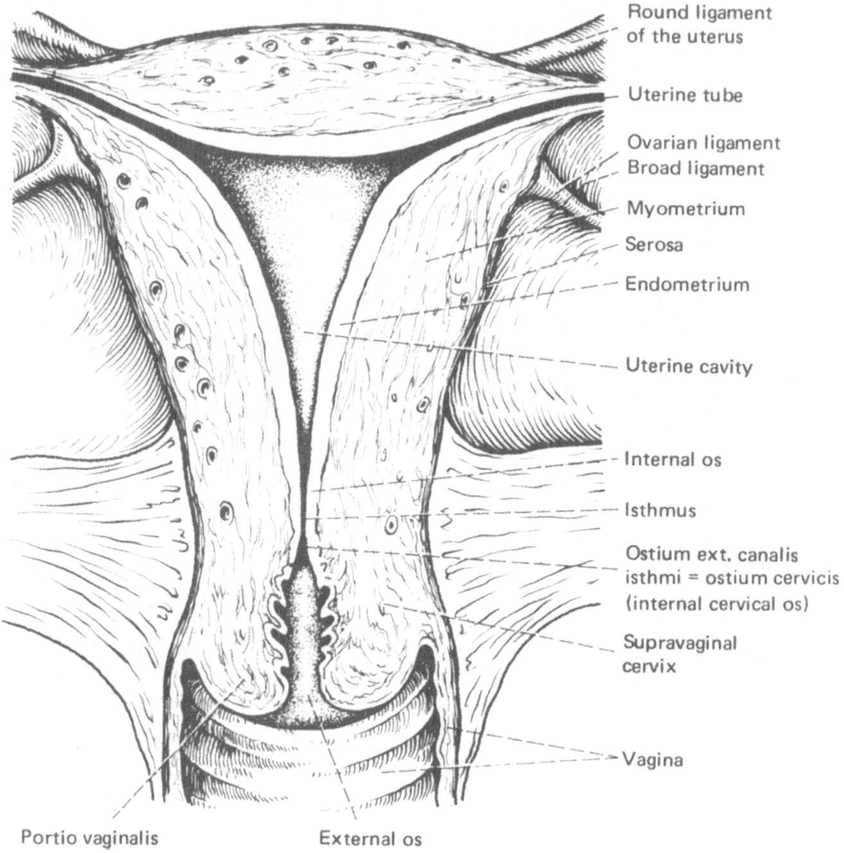

Round ligament
of the uterus

Uterine tube

Ovarian ligament
Broad ligament

Myometrium

Serosa

Endometrium

Uterine cavity

Internal os

Isthmus

Ostium ext. canalis
isthmi = ostium cervicis
(internal cervical os)

Supravaginal
cervix

Vagina

Portio vaginalis   External os

Fig. 22. Anatomy of the uterus (frontal section).

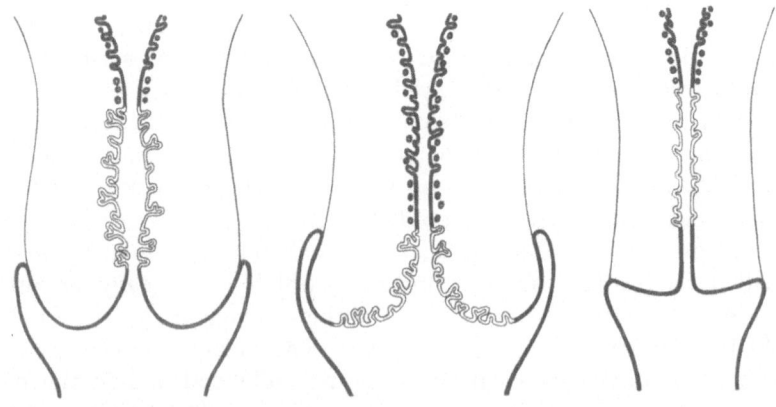

Fig. 23. Changes in the squamocolumnar junction (arrow) in various stages of life. *Left:* childhood; *Center:* reproductive years; *Right:* old age.

of eversion of the endocervix onto the portio is not accurate, for there is no loss of surface epithelium. A true erosion can be seen under colposcopic magnification (page 105) in only 1% of cervices examined. The correct term for the gross replacement of squamous epithelium by columnar epithelium on the portio is *ectopy* or *ectropion*.

The isthmus uteri is considered by some gynecologists to be a functional entity between cervix and corpus uteri. The internal os of the cervical canal is located in this area. The borderline between endocervical and endometrial epithelia is not sharp. There is a small zone in the cervix that is lined by an epithelium similar to that of the endometrium. The isthmus becomes part of the lower segment of the uterus during pregnancy. During labor, however, it is not part of the contractile muscular portion but forms, with the cervix, the passive segment of the uterus.

The *corpus uteri* is covered by peritoneum (serosa) (Fig. 22). The folds of peritoneum known as the broad ligaments are analogous to mesenteries of the intestine. The peritoneum of the uterus (perimetrium) continues anteriorly onto the bladder (bladder flap). The wall of the uterus comprises mainly smooth muscle (myometrium) with interspersed connective tissue. The blood supply of the uterus is derived from the uterine and ovarian arteries (page 42).

The *endometrium* directly abuts the myometrium for there is no submucosa. During the reproductive years the endometrium undergoes the hormonally induced changes of the menstrual cycle. The cyclic changes are necessary for the nidation and early development of the fertilized eggs. Histologically, the cyclic changes involve the epithelium and the stroma of the functional layers and the spiral arteries. The basal layer undergoes relatively little change and serves in part to regenerate endometrium during the next cycle (Fig. 24). Coiled arteries arise from the radial arteries and supply the middle and upper thirds of the endometrium. The walls of these arteries respond to the hormonal changes of the menstrual cycle.

The major histologic changes (Table 2) during the menstrual cycle are as follows:

(1) Mitoses in the epithelium and stroma indicate the proliferative activity (present in the preovulatory and very early postovulatory phases) of the cycle

(2) Pseudostratification of glandular nuclei indicates the proliferative and very early secretory phases (day 8 to 17) of the cycle.

(3) Basal vacuolation is the earliest histologic evidence of ovulation (present primarily from day 16 to 18).

(4) Secretion into the lumina of the glands is maximal during days 19 to 22, but is found also in the earlier phases of the cycle.

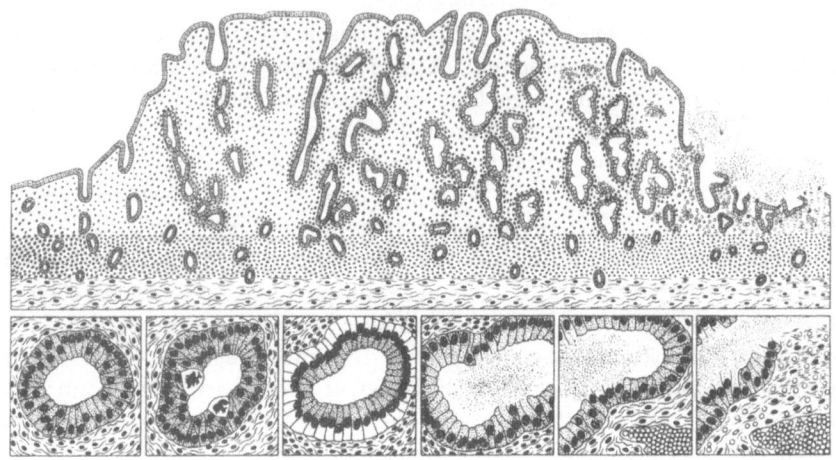

Fig. 24. Endometrium during the phases of the menstrual cycle. *Top:* Endometrium with basal and functional layers. From left to right: Proliferative phase, secretary phase, and menstrual phase. *Below:* Glandular epithelium in the phases of the menstrual cycle (see description in Table 2).

(5) Stromal edema, which favors implantation, is maximal at around day 22.

(6) A pseudodecidual reaction of the stroma, beginning around the arterioles, is found from day 23 to 28.

(7) Leukocytic infiltration marks imminent onset of menstruation.

The *fallopian tubes* extend from the uterine cornua to the ovaries. They are covered by peritoneum (perisalpinx) that is continuous with the upper margin of the broad ligament. The tubes provide a passage from the uterine cavity to the general peritoneal cavity. The tube consists of three layers: an outer serosa, a muscularis, and a mucosa (endosalpinx). The mucosa consists of a single layer of columnar cells, some ciliated and others secretory (Fig. 25). The cyclic changes are much less evident than those in the endometrium.

The current produced by the cilia is in the direction of the uterus. The ampulla is the wide lateral part of the tube that widens into the infundibulum, with its characteristic fibriated extremities. The narrow proximal portion of the tube is the isthmus. A short section of the tube is intrauterine or interstitial.

The ovary is contracted during ovulation by the fimbriated extremities of the tube. The fimbriae help to divert the ovum into the fallopian tube, or oviduct. The inner circular and outer longitudinal muscular layers permit peristalsis, which moves the ovum into the uterus. The plicae of the endosalpinx may be seen on cross section to be most complex in the infundibulum, becoming progressively simpler as the isthmus is

Table 2. Structural Changes in Endometrial Glands and Stroma
During the Menstrual Cycle (28 Days)

Phase of the cycle	Day	Glands	Stroma
Early proliferative	5–7	Short, circular cross sections, epithelium low, nuclei near base, occasional mitoses.	Spindle cells with large nuclei, few mitoses.
Midproliferative	8–10	Longer, slight tortuosity, early pseudostratification of the nuclei, many mitoses.	Variable stromal edema, many mitoses.
Late proliferative	11–14	Marked tortuosity (corkscrew glands), marked pseudostratification.	Less edema, many mitoses.
Early secretory	15–18	Widening of glandular lumen. Nuclei basal (day 16) and formation of subnuclear vacuole, which passes nucleus on day 18. Rare mitoses.	Little edema. Rare mitoses.
Midsecretory	19–23	Sawtooth glands. Nuclei returning basal. Secretion in lumen maximal. No mitoses.	Edema maximal at day 22. Earliest suggestion of predecidual cells around arteries at day 23.
Late secretory	24–28	No further development of glands. "Secretory exhaustion." Luminal surface fragmented.	Predecidual reaction maximal. Infiltration of leukocytes and erythrocytes.
Menstruation	1–4	Breakdown of epithelium from menstrual slough.	Hemorrhage into stroma and menstrual slough.

reached (Figs. 26, 27). The diameter of the ampulla ranges from 4 mm proximally to 10 mm distally. The isthmus of the tube has a diameter of 2 to 3 mm. The interstitial portions, buried in the uterine cornua, have the smallest lumens (Fig. 28). Inflammatory reactions interfere with the function of the tube in several respects. The fimbriae may be obliterated and the lumen may be occluded. Adhesions, furthermore, may impede coordinated peristalsis. The result may be trapping and implantation of the fertilized egg in the tube to form an ectopic pregnancy.

The blood supply of the fallopian tubes is provided by branches of the ovarian artery entering through the mesosalpinx.

The *ovaries* are almond-shaped organs attached to the posterior fold of the broad ligament (mesovarium) and lying free in the peritoneal

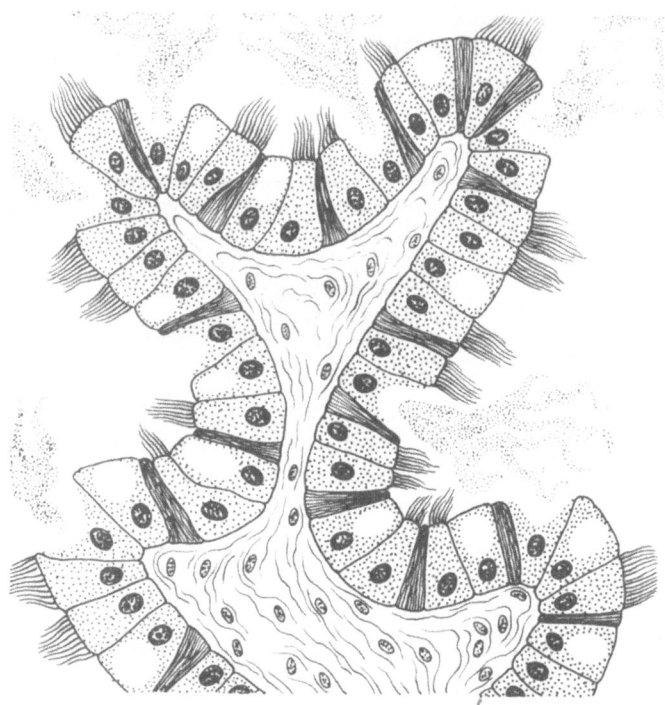

Fig. 25. Histologic appearance of oviductal mucosa. The single-layered columnar epithelium comprises ciliated cells and secretory cells.

cavity (Fig. 29). The ovary has two principal functions: the production of sex hormones and the release of ova.

During the reproductive years, the ovary weighs 7 to 10 g and measures between 2 to 5 cm in length and 1.5 to 3 cm in breadth. The surface is normally more or less corrugated. The ovary of the prepuberal girl is considerably smaller and it has a smoother surface. The ovary after menopause undergoes atrophy. It is also small but its surface is scarred and irregular.

The ovary at first is covered by a single layer of cuboidal cells, the so-called germinal epithelium. The tunica albuginea just beneath the germinal epithelium consists of dense white connective tissue. The underlying ovarian cortex contains the graafian follicles and corpora lutea, the atretic follicles and corpora albicantia, and the thecae internae and thecae externae. The inner medulla contains the blood vessels, the nerve supply (sympathetic and parasympathetic), and the lymphatics of the ovary. A rudimentary rete is occasionally found in the region of the mesovarium (page 10). Large polygonal cells are sometimes located in the hilus (hilus cells). The ovaries at birth contain the woman's entire

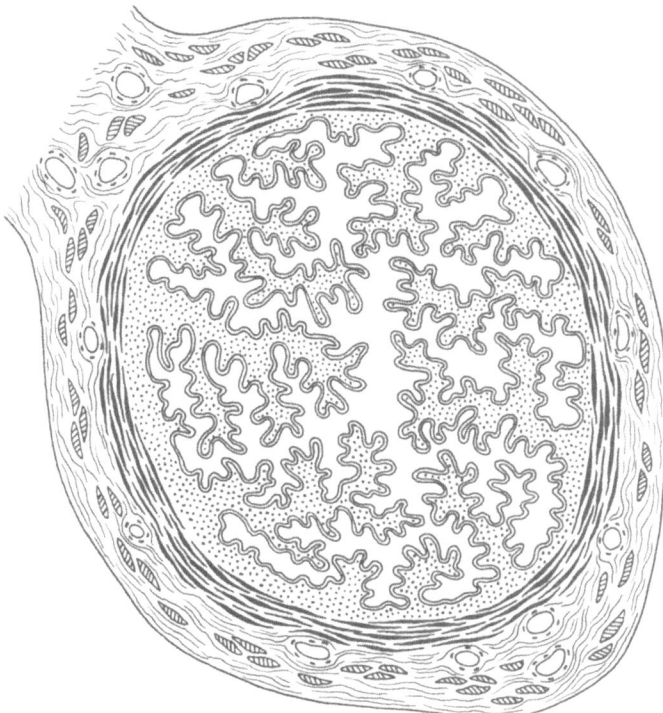

Fig. 26. Histologic appearance of the ampullary portion of the tube. The mucosa is thrown into many folds. The inner circular and outer longitudinal muscular layers are evident.

supply of ova (page 14), numbering hundreds of thousands. Fewer than 500 oocytes, however, mature during the reproductive years to be released from the ovary at ovulation. The majority undergo atrophy.

The primordial follicle consists of a single layer of cuboidal follicle cells, which later form several layers of so-called granulosa cells. These cells in turn are surrounded by a layer of cells that form the theca interna. The theca externa is a layer of essentially typical connective tissue. At this stage of development, a secondary follicle (up to 0.2 mm in diameter) results, containing an ovum that has reached its final size of 100 to 130 μ in diameter.

With further growth of the granulosa cells, fluid-containing spaces develop in them, enlarging to form an antrum filled with *liquor folliculi*. The ovum is enclosed in a mound of granulosa cells (*cumulus oophorus*) that project into the antrum. The crown of cells on top of the ovum is the *corona radiata*. The structure is now called a *tertiary follicle*, which measures between 0.5 mm and 1.0 cm in diameter (Fig. 29).

All three developmental states can be seen in the ovary before

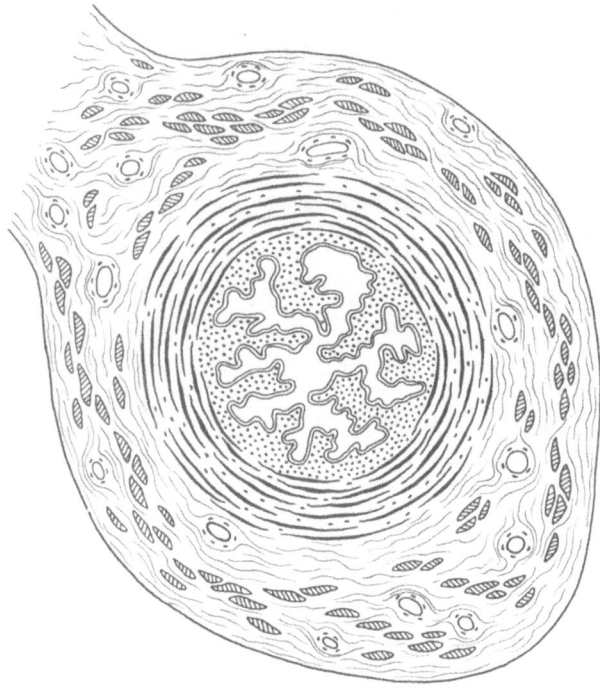

Fig. 27. Histologic appearance of the isthmic portion of the tube. The mucosa is thrown into fewer folds and the lumen in general is smaller. The two muscular layers are more clearly defined than in the ampullary portion.

puberty. During reproductive years, one tertiary follicle matures each month under the influence of sexual hormones to form the *graafian follicle*.

During ovulation, the cumulus oophorus, including the oocyte, separates from the follicle wall. The precise mechanism by which the egg escapes from the ovary remains unclear. In any case, the oocyte with its surrounding granulosa cells (corona radiata) is taken up by the fallopian tube. The ovum undergoes the first meiotic division before ovulation and starts the second division, which is carried out to the metaphase. The second meiotic division is completed only upon fertilization. Unfertilized eggs become atretic and degenerate.

Corpus luteum: The rupture of the graafian follicle is followed by collapse and folding of its walls (Fig. 29). Blood escapes into the follicular lumen. Endothelial sprouts from the invaginated theca interna grow and fibrin fills the space. A layer of *granulosa lutein* cells is quickly formed. These cells contain lipochrome pigments and produce progesterone. Estrogens are produced in the theca interna (theca lutein cells).

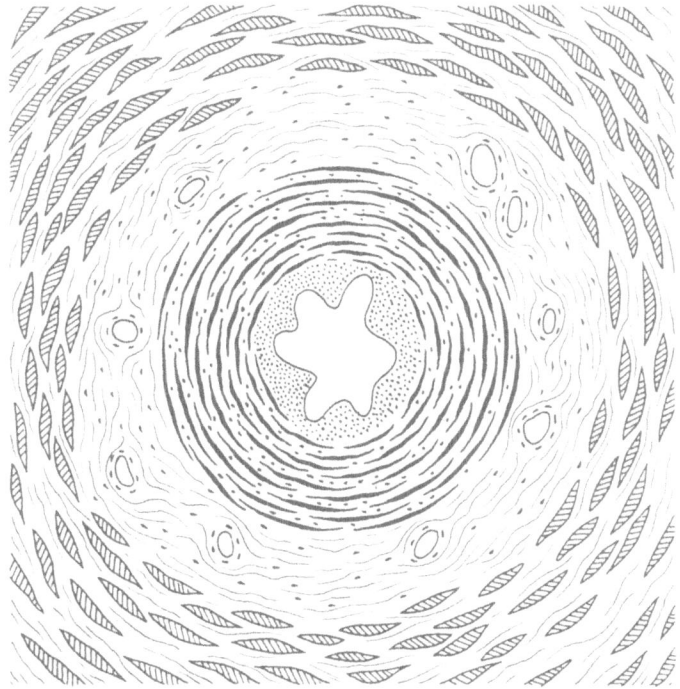

Fig. 28. Histologic picture of the interstitial portion of the tube. The tubal mucosa has very few plicae; the lumen in this area is narrowest. The tube is surrounded by a well-developed layer of circular muscle, which is incorporated into the myometrium.

The term corpus luteum refers to the yellow pigment in the structure. If the ovum is not fertilized a corpus luteum of menstruation results. If fertilization and nidation take place the corpus luteum matures, becomes increasingly vascular, and develops into the corpus luteum of pregnancy.

Failure of fertilization results in regression of the corpus luteum after approximately 10 days. The lutein cells degenerate and a scar-like corpus albicans results.

Follicular atresia: Only a small number of follicles reach maturity. The majority become atretic and degenerate. Only the tertiary follicles ever develop hyperplasia of the theca interna, which produces estrogens that affect the endocrine changes during the menstrual cycle.

Innervation of the Genitalia

Contractility of the uterus does not depend primarily on nerve supply, for hemiplegic patients may still have normal uterine contractions.

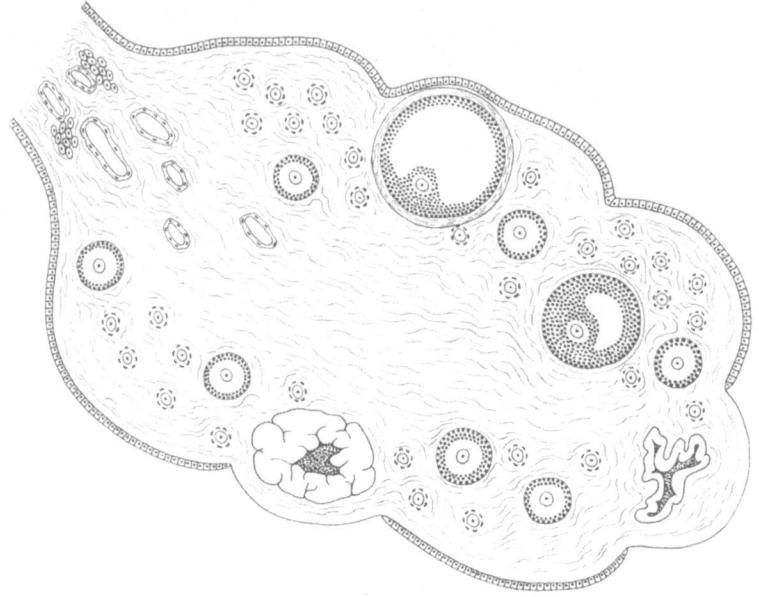

Fig. 29. Structure of the ovary. On the periphery the so-called germinal epithelium is still visible for the most part. Beneath is the fibrillar, acellular tunica albuginea. Internally is the cellular cortex, which contains follicles of various developmental stages. A corpus luteum is seen at 7 o'clock and a corpus albicans at 4 o'clock. The center represents the medulla. In the hilar region polygonal hilus cells are found in addition to the vessels.

The cervical plexus of Frankenhäuser is located lateral to the lower portion of the rectum. It receives sympathetic motor fibers originating from T7 and above that pass through the aortic plexus to the superior and inferior hypogastric plexus. The parasympathetic sensory nerve supply derives from the second, third, and fourth sacral nerves, which form the pelvic nerve, which also terminates in Frankenhäuser's plexus. Both sympathetic and parasympathetic nerves supply the uterus, bladder, and upper part of the vagina.

Surgical removal of the hypogastric plexus (*presacral neurectomy*) may relieve pelvic pain originating in the uterus.

Blood Supply of the Genitalia

The uterus, fallopian tubes, and the ovaries are supplied by two major arteries.

The ovarian artery is a direct branch of the aorta. It reaches the uterus through the infundibulopelvic ligament. One branch anastomoses with the ascending branch of the uterine artery. Removing the ovaries (oophorectomy) requires the interruption of their blood supply by ligation of the infundibulopelvic ligaments.

The fallopian tube is excised by placing a clamp parallel to it on the mesosalpinx. The interstitial portion of the tube is removed by incising the cornual end of the uterus.

The uterine artery is a branch of the hypogastric artery. It enters the uterus at the junction of the corpus with the cervix, after passing through the broad ligament and crossing the ureter approximately 2 to 3 cm lateral to the uterine wall. The uterine artery divides into ascending and descending branches. The lower branch supplies the cervix and upper vagina. The ascending branch forms an anastomosis with the ovarian artery near the cornu. Two smaller branches supply the fallopian tube and the fundus of the uterus.

Total hysterectomy requires bilateral clamping of both ovarian and uterine arteries. Care must be taken to place the clamp very close to the uterus to avoid damage to the ureters, which lie just lateral to the vessels. Especially in pregnancy the ureters cross the uterine arteries very near to the uterus.

Since the hypogastric artery gives rise to the uterine artery and several other branches, its ligation often achieves relative hemostasis even when bleeding points in the pelvis cannot be identified.

The upper vagina is supplied by the descending branches of the uterine artery. The middle third is supplied by the inferior vesical arteries and the lower third by the middle hemorrhoidal and internal pudendal arteries. The venous drainage is through the hypogastric vessels. The labia majora and minora are supplied by the internal pudendal arteries and a small branch of the obturator arteries. The dorsal artery of the clitoris is a terminal branch of the internal pudendal artery. The venous drainage is through the vesicovaginal plexus and the inferior hemorrhoidal veins.

Selected Reading

Bailliere,: *Atlas of Female Anatomy.* 7th Ed. Baltimore: Williams and Wilkins (1969).

Boving, B. G.: Anatomy of Reproduction *in Obstetrics.* 13th Ed. Philadelphia: W. B. Saunders, (1965). J. P. Greenhill (ed.)

CHAPTER 3

Endocrine Control of
Female Reproductive Functions

Steroid Hormones

The steroid hormones are of either ovarian or adrenal origin and usually are divided into estrogens, progestins, androgens, and corticosteroids. All steroid hormones share a common structural nucleus which consists of a four-ring structure known as cyclopentanoperhydrophenanthrene. It consists of three rings of six carbon atoms each (phenanthrene) and one ring of five carbon atoms (cyclopentane) (Fig. 30). Compounds derived from this basic structure are known as steroids. If one or more oxo groups are present, the compound is called a steroid. The rings are designated by letters (A to D) and the carbon atoms are numbered 1 to 17 (C_1–C_{17}). The estrogens contain a C_{18} methyl group attached at C_{13}. Androgens and pregnane derivatives (gestagens and corticosteroids) contain a C_{19} methyl group attached at C_{10}. If a side chain is present at C_{17}, the carbon atoms are numbered C_{20}–C_{27} (Fig. 31).

There are five basic steroid nuclei: *estrane* (18 carbon atoms), which is the source of the natural estrogens; androstane and etiocholane, which give rise to almost all of the natural androgens (both these nuclei differ in the stereo position of the C_5-substituted hydrogen); and pregnane and allopregnane (21 carbon atoms), which are the source of progesterone, its derivatives, and the corticoids (Fig. 32–33).

All common steroids derive from these basic steroid nuclei. The biological activity of these hormones depends on their structural configuration. Structural changes are indicated by specific suffixes and symbols.

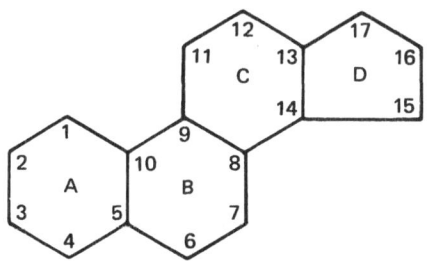

Fig. 30. The cyclopentanoperhydrophenanthrene nucleus.

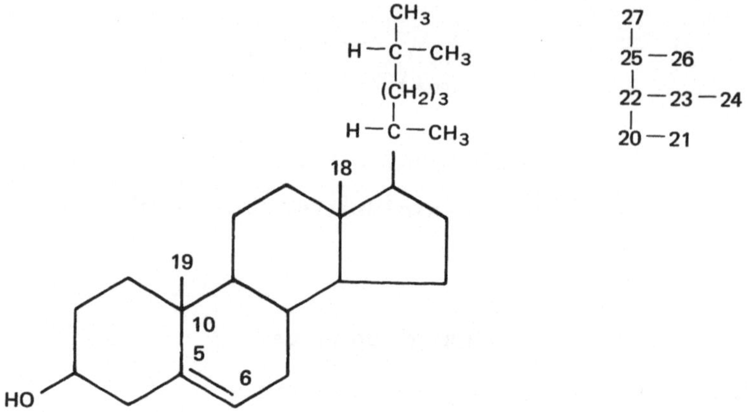

Cholesterol

Fig. 31. Cholesterol.

C_{18} C_{19} C_{21}

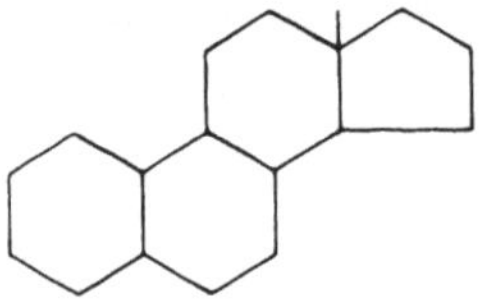

Estrane

Estrone
17β-Estradiol
Estriol
16-Epiestriol
16α-Hydroxyestrone
16β-Hydroxyestrone
16-Oxoestradiol
2-Hydroxyestrone
2-Methoxyestrone

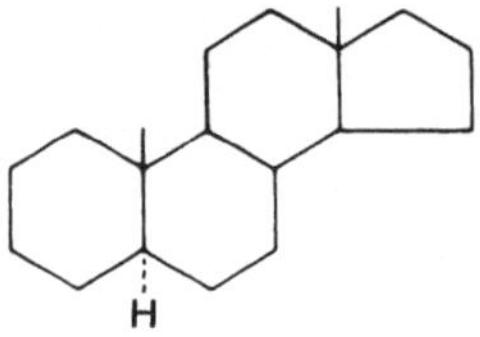

Androstane (5β)

Testosterone
Androstendione
Androsterone
Dehydroepiandrosterone
Adrenosterone

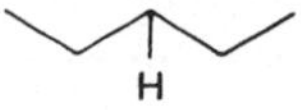

Etian

Etiocholanolone

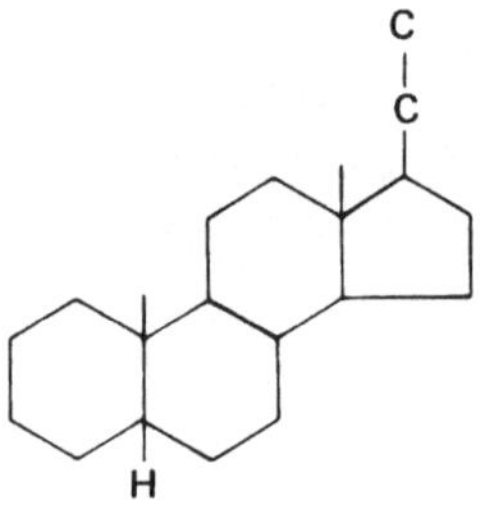

Pregnane (5β)

Progesterone
Progesterol-20α
Progesterol-20β
Pregnanolone
Pregnandione
Pregnandiol
Cortisone
Cortisol
Corticosterone
Aldosterone

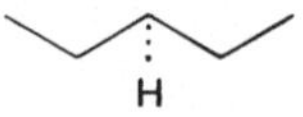

Allopregnane (5α)

Allopregnandiol

Basic steroid nuclei

Fig. 32.

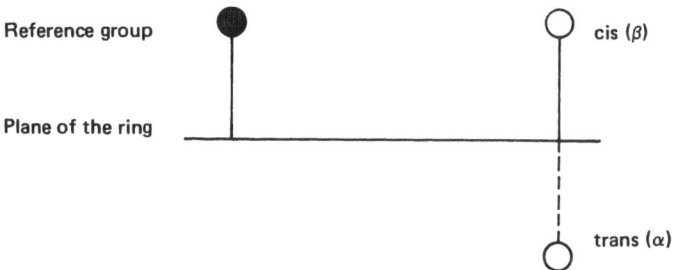

Fig. 33. Comparison of 5α and 5β configurations. Basic steroid nuclei and their derivatives.

For instance, if between adjacent carbon atoms one hydrogen atom each is lost, meaning desaturation, a second line is inserted in the structural formula, designating a double bond and an unsaturated ring. When one or more double bonds are present in a ring, the *-ane* ending is replaced by the suffix *-ene*. The Greek letter delta, Δ, also indicates the presence of a double bond, its location being indicated by a superscript number. When a hydroxyl (—OH) group replaces a hydrogen, the suffix *-ol* is used together with the carbon atom number. The suffix *-one* indicates substitution by an oxo (keto) group. Hydroxyl, oxo (keto), and methoxy groups can be affixed to C_3, C_6, C_7, C_{11}, C_{16}, C_{17}, C_{18}, C_{19}, and C_{21} of the steroid molecule. A steroid containing 21 carbon atoms is referred to as a C_{21} steroid.

Thus, all steroidal estrogens derive from estrane, which is characterized by the lack of a methyl group at C_{10}, by an aromatized A ring, and by a hydroxyl group at C_3.

The C_{19} steroids such as testosterone and androstenedione lack the side chain at C_{17}. The C_{21} steroids carry a two-carbon side chain at C_{17}. The main steroids of this group are progesterone and 20-dihydropregn-4-ene-3-one (progesterol-20 α and -20β).

Biosynthesis. Steroid hormones are synthesized in a similar fashion in all endocrine organs with the exception of the placenta. They are formed from acetate or cholesterol into pregnenolone and progesterone. Progesterone was believed to be the steroid from which androgens as well as estrogens arise. Recently it has been observed that progesterone can be bypassed. The basic compound is pregnenolone, which has little-known biologic activity. Ovaries, testes, and adrenal differ in the concentration of enzymes significant in the formation of hydroxyl groups and double bonds; 17-hydroxyl groups are added in the ovary, the side chain is split off the progesterone, and finally through androgens the pathway ends in estrogens. In the corpus luteum the pathway is partly arrested at the step of progesterone. C_{11} hydroxyl formation is a significant step in the adrenal (Fig. 34).

The liver is the basic organ of metabolism and breakdown. Steroids

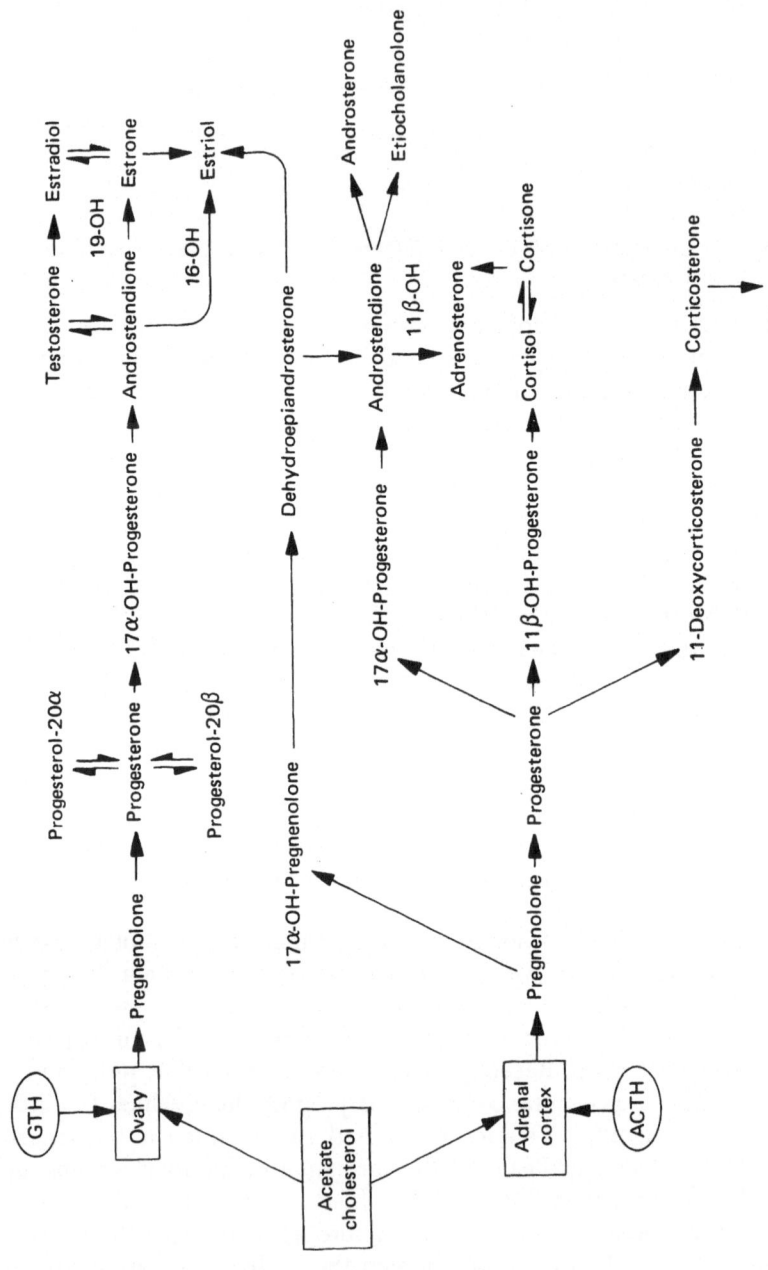

Fig. 34. Biogenesis of steroid hormones in the ovary and adrenals.

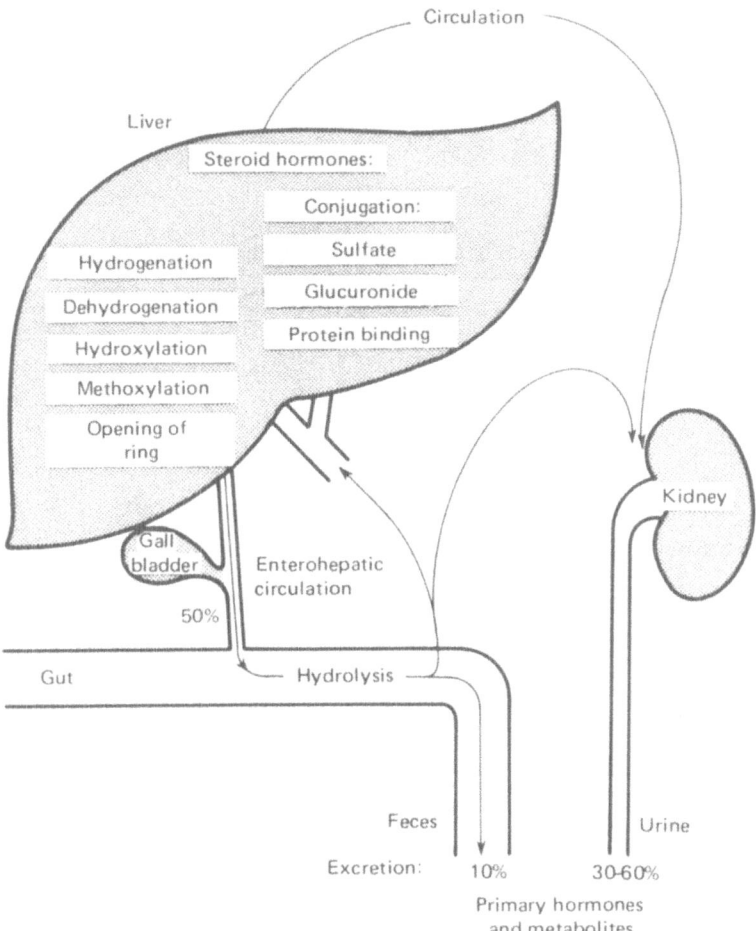

Fig. 35. Steroid hormone metabolism in the liver and enterohepatic circulation and excretion in urine and bile.

are metabolized and solubilized for excretion by adding hydroxyl groups and double bonds, methyl groups, oxidation and reduction, and conjugation with sulfate or glucuronic acid. During this pathway steroids pass serveral times through the enterohepatic circulation (liver–gall bladder–gut–liver). They are excreted by the kidney by glomerular filtration and tubular secretion. Metabolites are conjugated to form sulfates and are partly reabsorbed in the tubules. A larger quantity of steroids is excreted through fecal elimination (Fig. 35).

Estrogen

The steroids that are most active in order of potency are estradiol, estrone, and estriol (Fig. 36). There are at least 30 naturally occurring

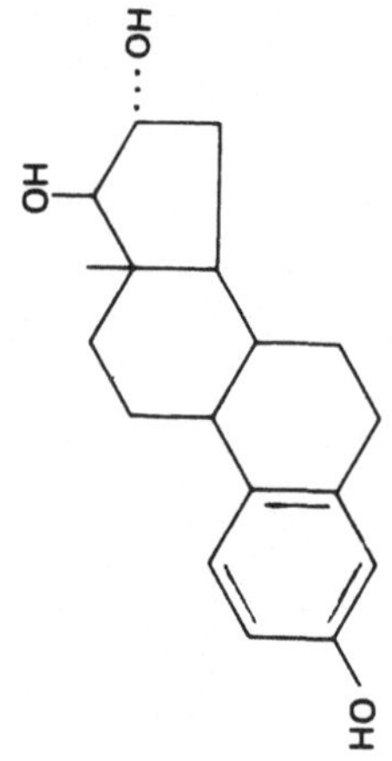

Estriol

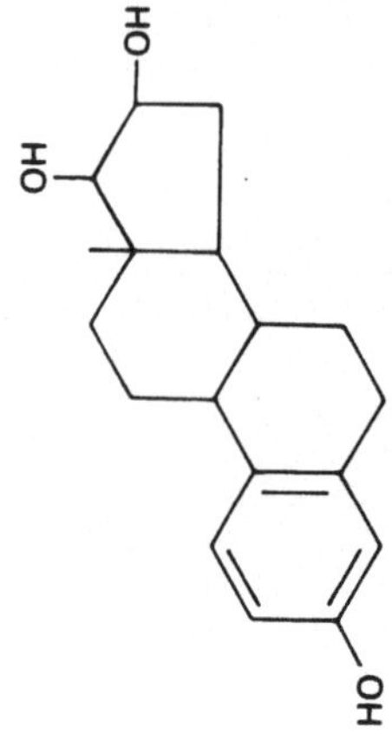

16-Epiestriol

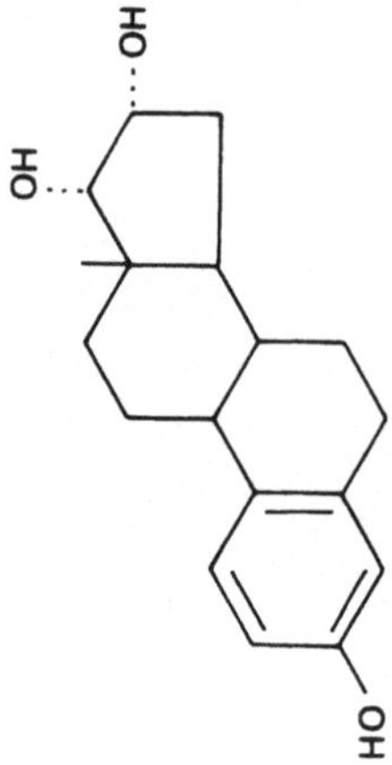

17-Epiestriol

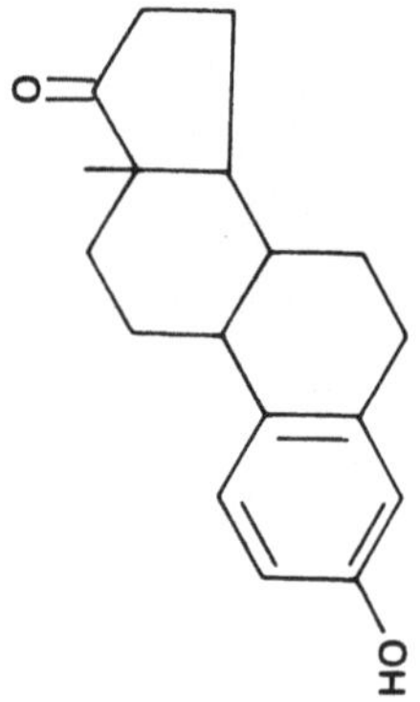

Estrone

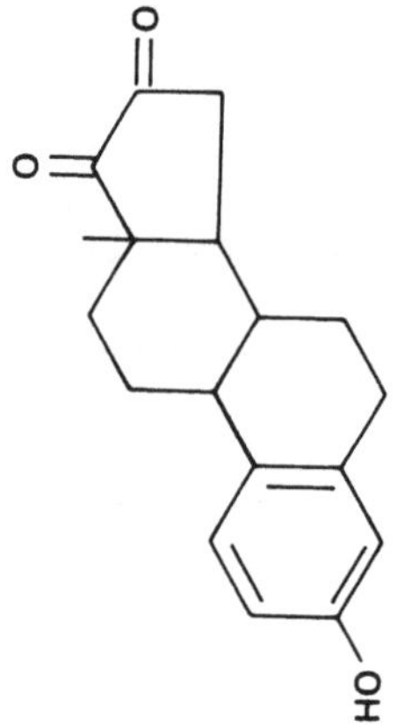

16-Oxoestrone

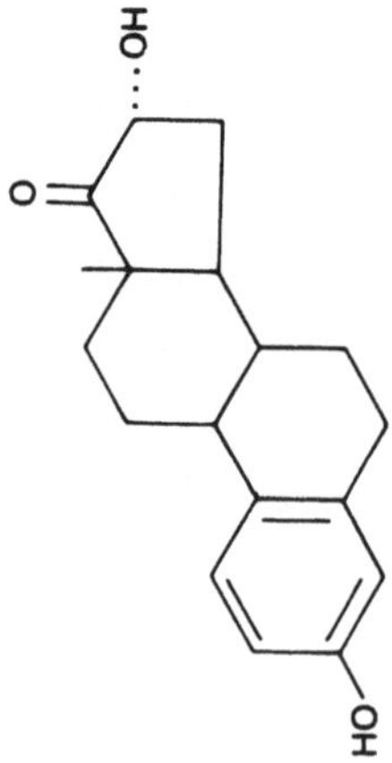

16α-Hydroxyestrone

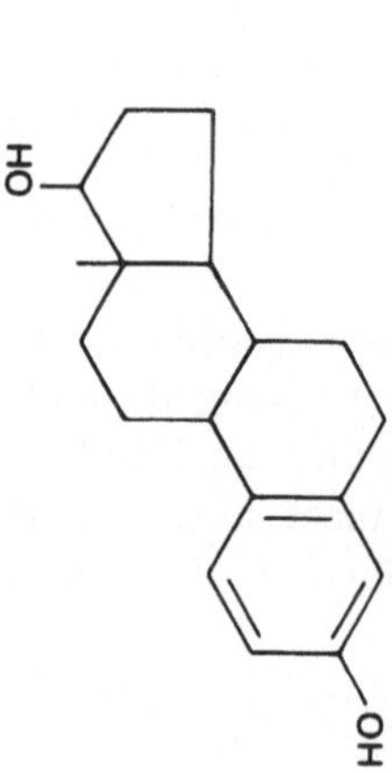

17β-Estradiol

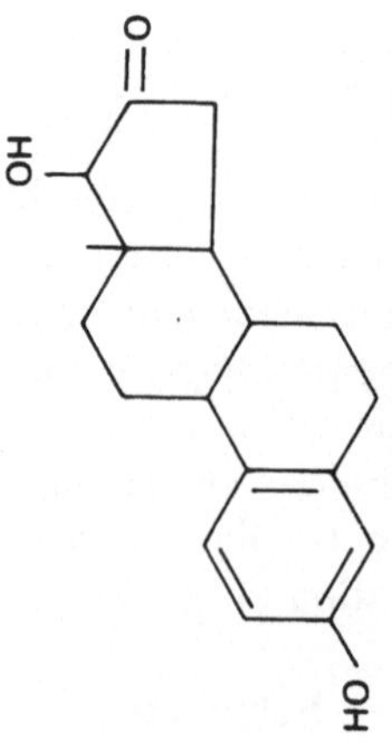

16-Oxo-17β-estradiol

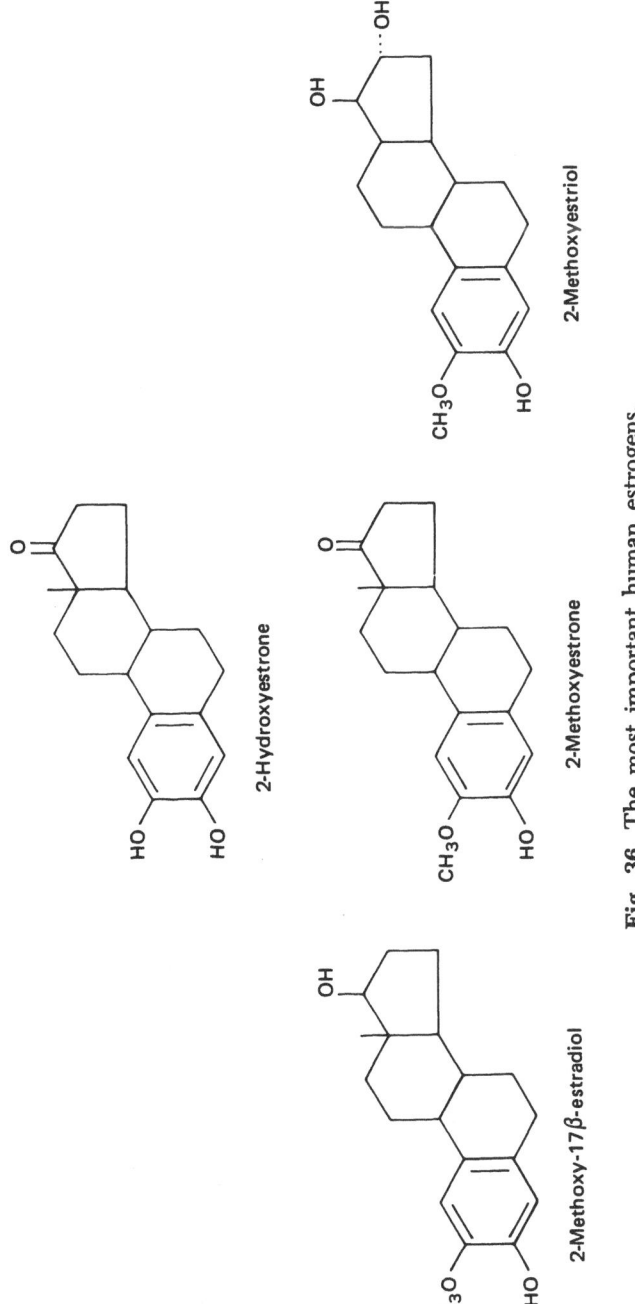

Fig. 36. The most important human estrogens.

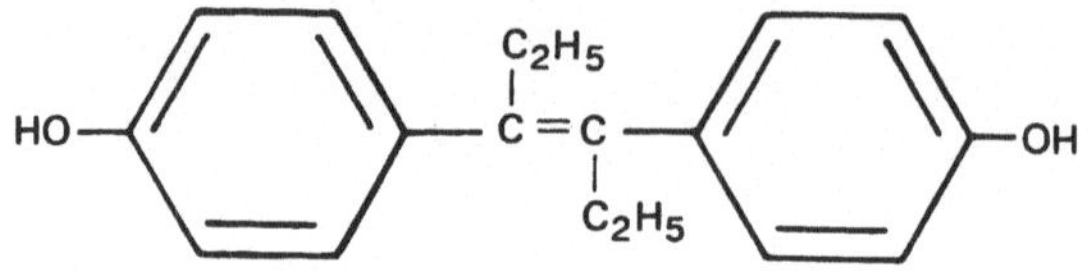

Fig. 37. Diethylstilbestrol.

estrogens in the human body. The term "estrogen" is used for every chemical structure that produces estrus in the rat or mouse as indicated by cornification of the vaginal epithelium. Therefore, some chemicals are included in the term "estrogens" that are not even steroids, for instance, the stilbenes, which are connected by a carbon bridge with two ethyl groups (Fig. 37).

Biogenesis. Estrogens can be formed from acetate and from cholesterol, pregnenolone, progesterone, and finally, by aromatization of the ring A from C_{18} (nor) and C_{19} steroids.

Estrogens are excreted in the urine as water-soluble conjugates. Estriol is conjugated primarily to glucuronic acid and is excreted as estriol glucosiduronate. A large amount of estrone and a smaller quantity of estriol are excreted as sulfates.

Progesterone

Progesterone is present in the corpus luteum and also in the adrenal glands as the most important intermediate product. Dehydration to C_{20} in α or β-position results in 20-dihydropregn-4-en-3-on (progesterol 20 α and 20 β). Pregnanediol is the excretion product of progesterone resulting from reduction of progesterone, pregnandione, and pregnanolone (Fig. 38). Pregnanediol is inactive in regard to gestagen effects.

Biogensis. Progesterone is formed in the granulosa lutein cells of the corpus luteum, in the zona reticularis of the adrenal, and the trophoblastic syncytium. As already mentioned, progesterone is the basic steroid for the biosynthesis of a number of hormones such as estrogens, androgens, and corticosteroids. Progesterone is synthesized from acetate, cholesterol, and pregnenolone. Only in the corpus luteum and the placenta is progesterone formed as the unmetabolized active hormone.

Metabolism. Although the major metabolite of progesterone is pregnanediol, progesterone is excreted in smaller amounts as pregnanedione, pregnanolone, and allopregnanediol. The plasma concentration of progesterone and the urinary concentration of pregnanediol is increased in the secretory phase of the cycle and in pregnancy. Small amounts of pregnanediol, however, can be detected in urine of male children and menopausal women.

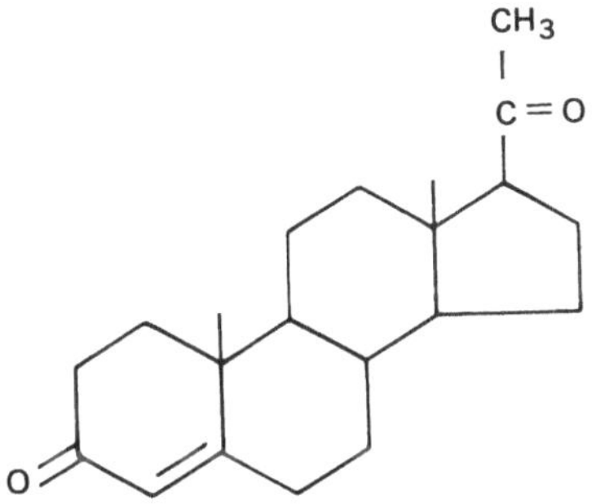

Progesterone
(Pregn-4-en-3.20-dione)

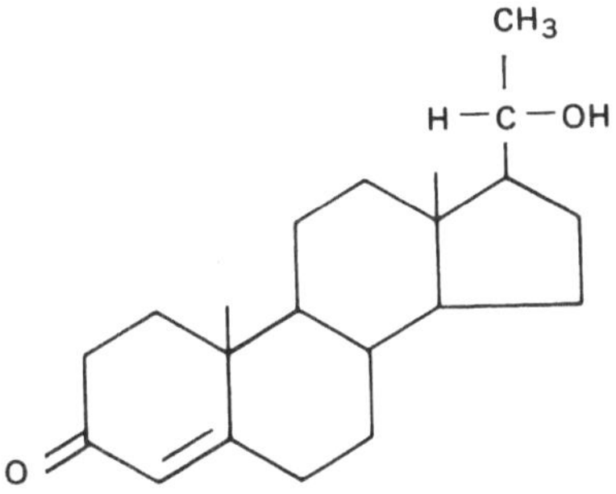

Progesterol-20α
(20α-Hydroxypregn-4-en-3-one)

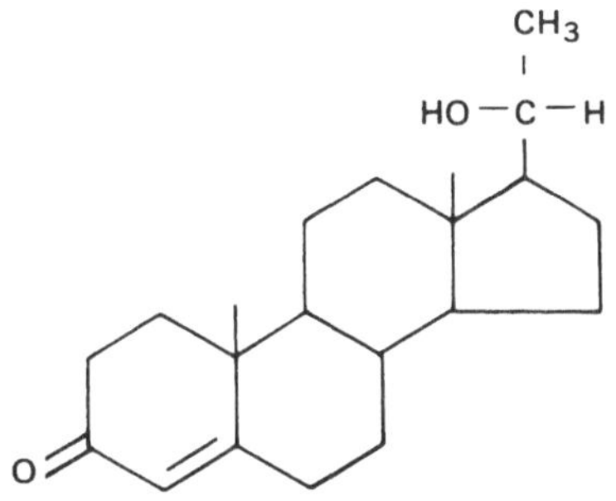

Progesterol-20β
(20β-Hydroxypregn-4-en-3-one)

Fig. 38. Progesterone and its 20-dihydro derivatives.

Androgens

Androgens belong to the C_{19} steroids (Fig. 39). Testosterone is biologically the most active androgen and is characterized by a hydroxyl group at the C_{17} position and a Δ^4-3-oxo configuration of ring A. It can be detected in small amounts also in the female. The similar Δ^4-androsten-3, 17-dione is found in ovaries, testes, and adrenals. Androstendione belongs to the 17-ketosteroids and is formed by oxidation of the C_{17} hydroxyl group of testosterone. A variety of 17-ketosteroids are found in urine as metabolites of androgens. In the female, the larger part results from androgen production of the adrenal. In the male, about one-third is derived from the testes, whereas the ovary synthesizes androgen in only small amounts.

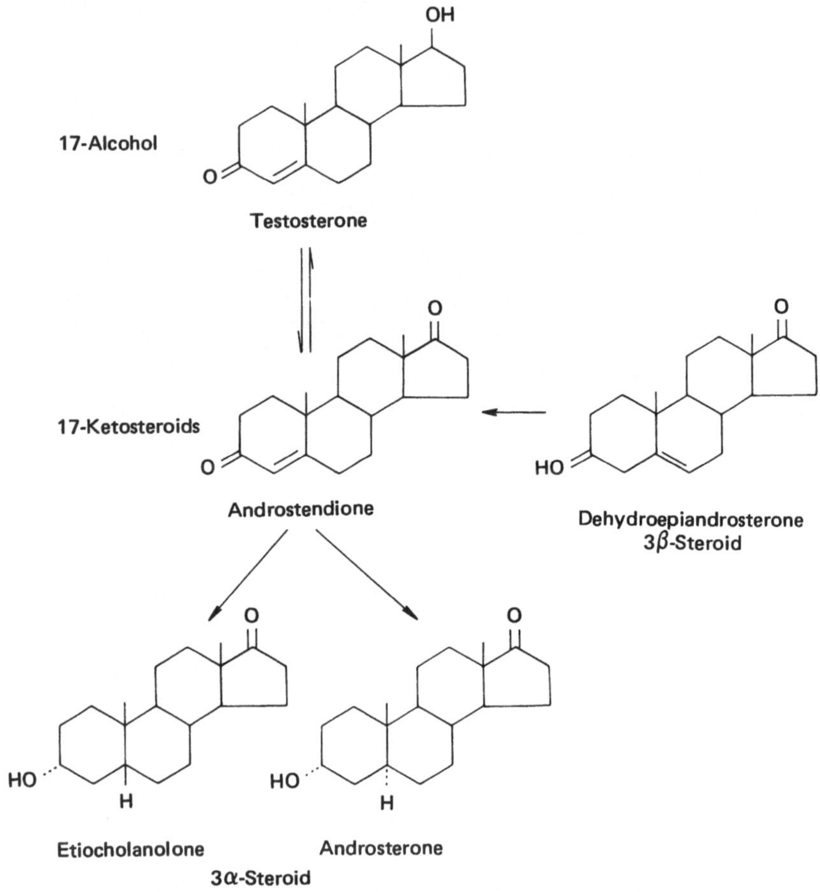

Fig. 39. The most important androgens and their metabolites.

Protein Hormones

All protein hormones have a basic structure of amino acids linked by peptide bonds (R—NH—OH—R'). FSH and LH contain sugar residues in addition to peptide chains and are therefore termed glycoproteins. The molecular weight of the protein hormones varies from 1000 in certain polypeptides (oxytocin and vasopressin) to about 30,000 or 40,000 (LH and FSH).

The chemical structures of vasopressin and of oxytocin are known and they are both available in synthetic form. To date the primary structure of the protein hormones has been partly (LH, FSH, prolactin) or completely (ACTH, GH, insulin, glucagon) elucidated.

Gonadotropins. The three gonadotropins are listed below:

(1) *Follicle Stimulating Hormone (FSH)*. A specific (basophilic) cell population in the anterior pituitary lobe is thought to produce LH and FSH. The so-called Δ_2 type cell is believed to secrete FSH. This hormone induces follicular maturation up to the stage of maturity of the graafian follicle and causes estrogen secretion.

(2) *Luteinizing Hormone (LH)*. It cellular origin is the Δ_1 type cyanophil. Together with FSH, the LH completes follicular maturation, causes ovulation, high estrogen secretion, and corpus luteum formation, together with the luteal production of estrogens and progestins. The gonadotropins act synergistically and possibly in part interchangeably during the various phases of the reproductive cycle.

(3) *Prolactin.* It is generally assumed that prolactin is produced by the eosinophilic alpha cell of the anterior pituitary lobe. In some species it causes maturation of the corpus luteum, its maintenance, and the stimulation of its hormone secretion. The significance of prolactin in the human reproductive cycle remains unknown. Its existence in the human body, however, has recently been proved and the galactopoietic effect has been shown during pregnancy and post partum.

Hypothalamic Releasing Hormones

Hypothalamic releasing hormones have been identified for all anterior pituitary tropic hormones. Hypothalamic inhibiting hormones also exist for the regulation of prolactin and MSH. All the releasers identified thus far are low-molecular-weight peptides containing between three and ten amino acids.

Mechanism of Action of Hormones. Steroid hormones do not interact with all cells of the organism in a similar fashion, but precisely with cells of special target organs. These cells possess hormonal receptors that are able not only to bind steroid hormones but also to transport them and increase their concentration. Most available information concerns the estrogens.

Estrogen receptors bind the steroid hormone to the cell. The hormone is then able to act on the genetic material of the cell nucleus. The genetic information is accumulated in the chromosomal deoxyribonucleic acid (DNA). Repressors, which are small protein molecules, block, in the resting period, the liberation and "readability." The hormone binds the repressor and therefore liberates the chromosomal DNA. Stimulation of RNA polymerase induces the formation of messenger ribonucleic acid (mRNA) and, therefore, transcription (see page 5). The newly formed mRNA adheres to the ribosomes of the endoplasmic reticulum (transla-

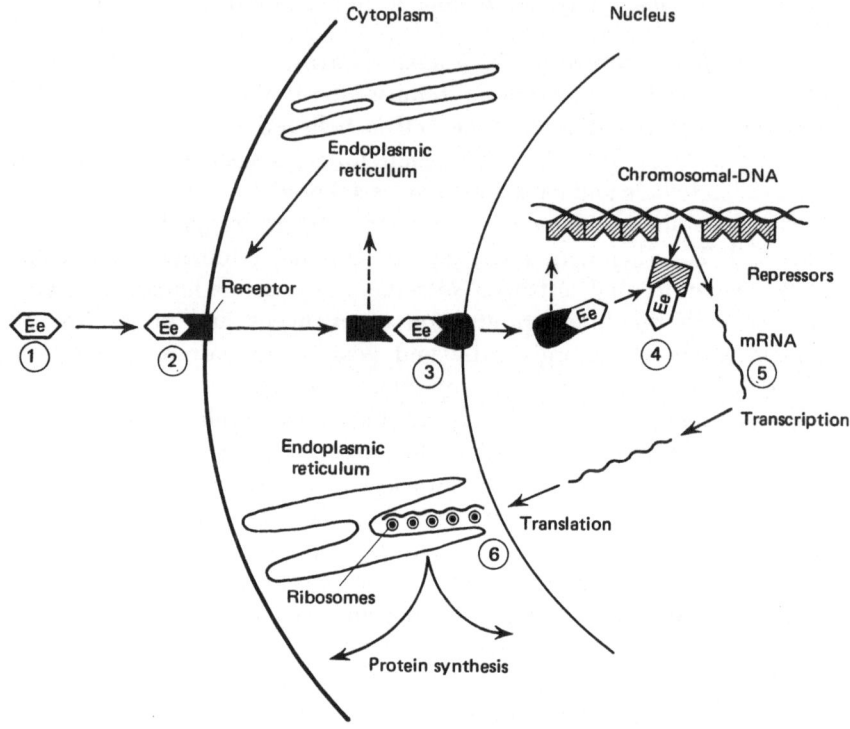

Fig. 40. Current (partly hypothetical) concepts of the mechanism of action of steroid hormones in the cell, using estrogen as an example: (1) estrogen molecule moves toward the cell; (2) estrogen is bound to the receptor of the cell wall and transported through the cytoplasm; (3) estrogen is bound by the receptor of the nuclear membrane and transported through the nucleus; (4) estrogen-receptor complex releases the repressors of the chromosomal DNA; (5) formation of messenger RNA and transcription; (6) attachment of the ribosomes of the endoplasmic reticulum and synthesis of proteins.

tion), leading to synthesis of specific proteins. Estrogen-stimulated growth of the uterus may occur in this fashion (Fig. 40).

This information is not yet available for progestational hormones and androgens. However, it can be assumed that the action is similar.

The mechanism of action of gonadotropic hormones is based on a general increase in metabolism of the ovary. There is stimulation of the enzymes that are significant in the steroid production from acetate to cholesterol or the epimerization and 3 β-hydroxylation of Δ5-3 oxosteroids, as in the case of pregnenolone to progesterone.

Releasing hormones are believed to act by specific depolarization in the cells of the anterior lobe of the pituitary. The depolarization of the membrane is associated with a reception of the calcium ion, which in turn releases the stored granules. This release of the granules or another as yet unidentified mechanism activates the synthesis of gonadotropins.

Dopamine is the synaptic carrier substance that stimulates the liberation of the FSH- and LH-releasing hormones from the tuberoinfundibular neurons to the portal vascular system.

Hypothalamic-Pituitary Axis

Central nervous and endocrine principles interact in the regulation of the human reproductive cycle. Ovarian function is thus controlled by the hypothalamic-pituitary axis. Certain morphologically defined areas of the ventral and basal median hypothalamus (arcuate, ventromedian, and the anterior periventricular nucleus, as well as the suprachiasmic and retrochiasmic regions) register the concentration of certain sex steroids (sensor). On the basis of this information, neuroendocrine impulses are emitted to the gonadotrope of the anterior pituitary (hypophysiotropic impulse system) (Fig. 41). The ovaries, via their steroid secretion, are thus integrated in a feedback system of a hypothalamic sensor with a hypophysial and a gonadotropic impulse and a gonadal effector. The reference point is the peripheral steroid concentration. For instance, low estrogen levels increase gonadotropin release, whereas elevated circulating estrogens result in an inhibition of gonadotropin secretion. A positive feedback results when a steroid hormone stimulates gonadotropic hormone secretion. A negative feedback results when a steroid hormone inhibits the gonadotropic impulse.

The transmission of the steroid-induced neuroendocrine impulse is provided by a neurohumoral pathway. A capillary portal system links the median eminence and pars tuberalis to the sinusoids of the anterior pituitary lobe.

The neurohumoral substances carrying the information from the hypothalamus to the pituitary are termed releasing hormones (or releasing factors). Thus far, one decapeptide has been shown to be an LH- and FSH-releasing hormone in the human and a tripeptide to be a TSH- and prolactin-releasing hormone. The existence of releasing hormones and one (prolactin) inhibiting hormone was demonstrated for all human tropic hormones. The chemical structure of these human releasing hormones is not yet known, but there is no evidence for species specificity of these hormones in mammals. The human gonadotropin releasing hormones are probably decapeptides. Compounds with LH–FSH releasing activity are used in clinical trials. The releasing hormones not only cause release but also synthesis of their respective tropins. Production of the

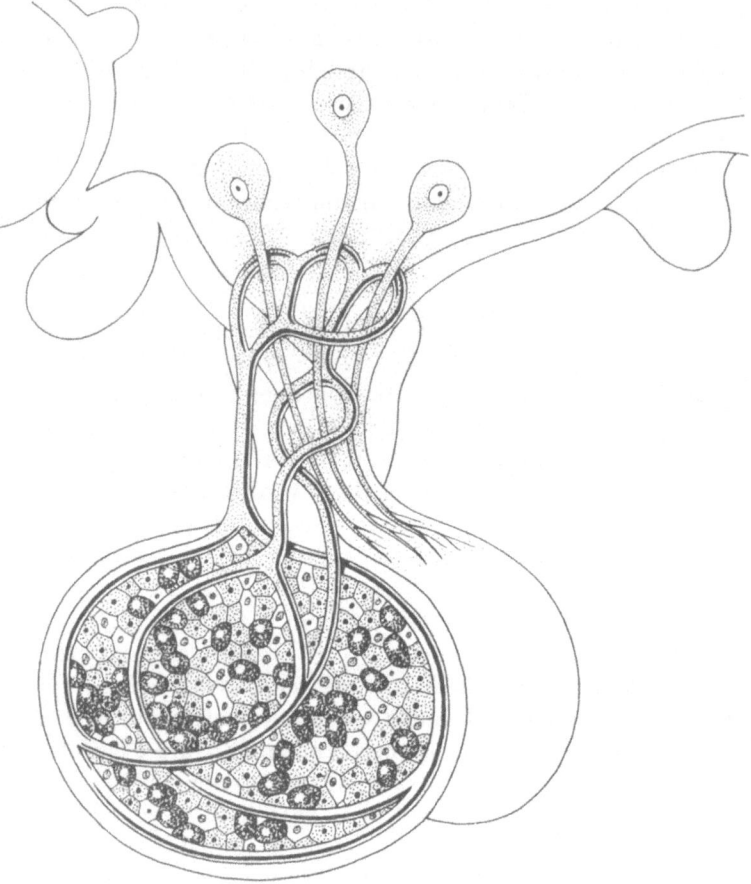

Fig. 41. The hypothalamic-pituitary system (hypophysiotropic regions, neu-rohumoral carrier system, and anterior pituitary lobe). *Top:* Hypothalamic FSH- and LH-releasing regions and ACTH-inhibiting region. *Center:* Neuro-vascular transmission of the releasing hormone by the special vessels of the median eminence of the anterior pituitary lobe.

gonadotropin-releasing hormones is located mainly in the median basal hypothalamus and in the suprachiasmic and preoptic areas.

The median basal hypothalamus is responsible for the tonic gonado-tropin release (tonic center). The suprachiasmic and preoptic areas pro-vide the cyclic impulse (cyclic center), which exists in females only. In males, only a tonic center is active. The two types of regulatory centers originally exist in both sexes. Differentiation into the male (tonic) and female (cyclic) patterns occurs during the late fetal and neonatal period as a result of gonadal testosterone produced by the male fetus.

In response to the releasing hormone stimulus the gonadotropins are secreted.

The Ovary

Morphology: At the beginning of each cycle, gonadotropic hormones stimulate growth of the follicles, one of which differentiates and matures. The rest of the follicles regress and become atretic or remain undeveloped. The maturing ovum increases in size from day 12 to 15 from 100 to 140 μ in diameter. The follicle grows from 2 to 12 mm at the time of ovulation. The follicle cells multiply and form 12 to 16 layers. At this stage, they are called the granulosa cells. A capillary net forms in the theca interna, which has differentiated from the external theca of connective tissue, facilitating the follicle's path to the surface. Follicular fluid is formed by dissolution of some granulosa cells and by secretion of the theca cells. Estrone and estradiol can be demonstrated in the follicular fluid as well as the theca and granulosa layers. The production of estrogens initiates the feedback mechanism already described (page 57). Just before rupture of the follicle, progesterone begins to form as a result of LH stimulation. Progesterone, progesterol-20α, and progesterol-20β have been demonstrated in the mature follicle. The follicle ruptures around day 14 to 15 in the normal cycle and liberates the egg. In most instances there is slight bleeding in the center of the follicle. Estrogen production decreases slightly following ovulation and increases around day 22 to a second peak that is somewhat lower than the preovulatory peak.

The granulosa cells increase in size after ovulation, and after storage of a lipoid yellow substance they develop into granulosa lutein cells. Connective tissue cells and capillary sprouts grow into the granulosa cell layer from the theca interna. The corpus luteum as an endocrine gland is thus formed. The synthesis of progestational hormones decreases slightly after ovulation and then increases continuously to a maximum at day 22. Simultaneously with regression of the corpus luteum, progesterone synthesis declines steeply (Figs. 42 and 43).

Endocrine Effects. The endocrine structures of the ovary produce, under the influence of the gonadotropic hormones, estrogens (estrone and estradiol). Progesterone, progesterol-20α, and progesterol-20β (dihydroxyprogesterone) are found in the corpus luteum along with some estriol.

Estrogens are produced primarily in the theca interna, but granulosa cells are also capable of synthesis. Estrone and estradiol are metabolized and interconverted in the ovary. Progesterone is produced in the granulosa layer. Theca interna and granulosa are closely interrelated in steroid biogenesis.

It is assumed that through negative feedback, by rather small amounts of estrogen, FSH increases up to midcycle, when it is inhibited by high estrogen concentrations, and then decreases in the second half

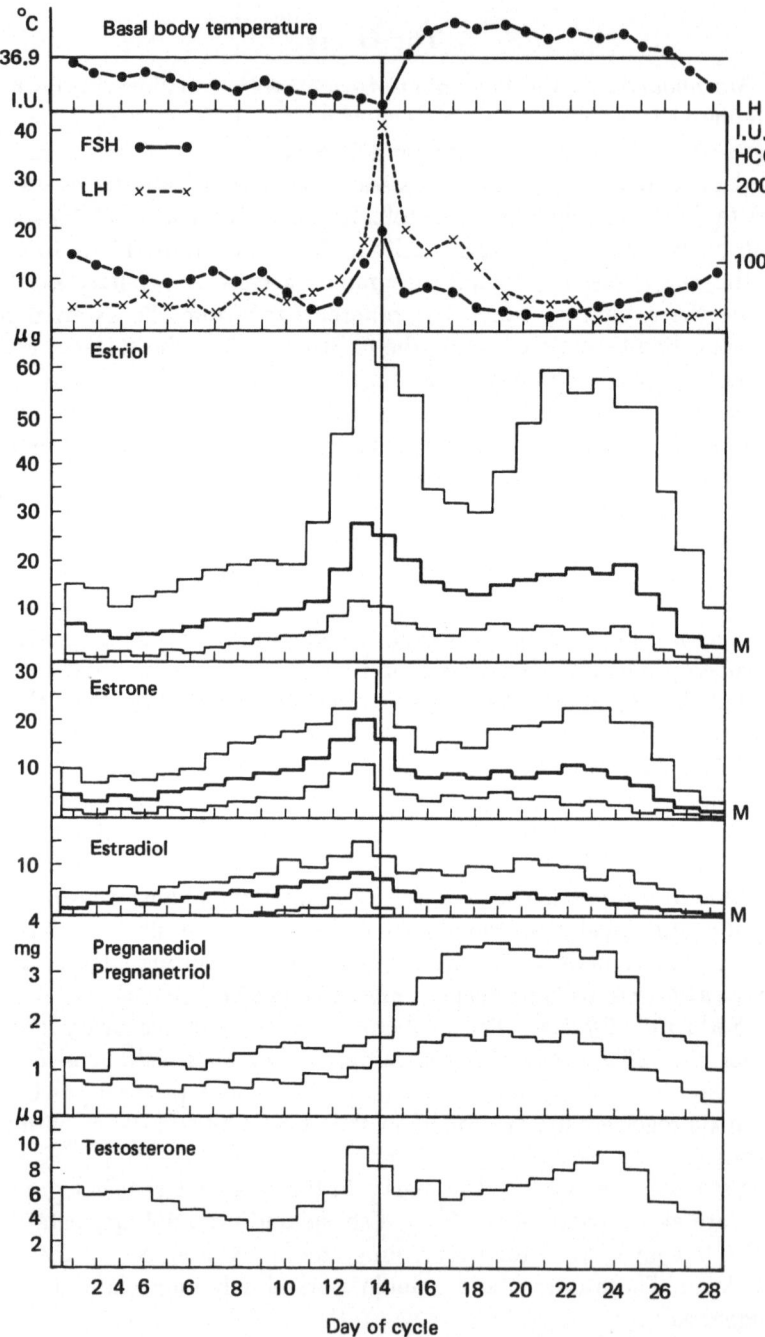

Fig. 42. Hormone excretion during the menstrual cycle.

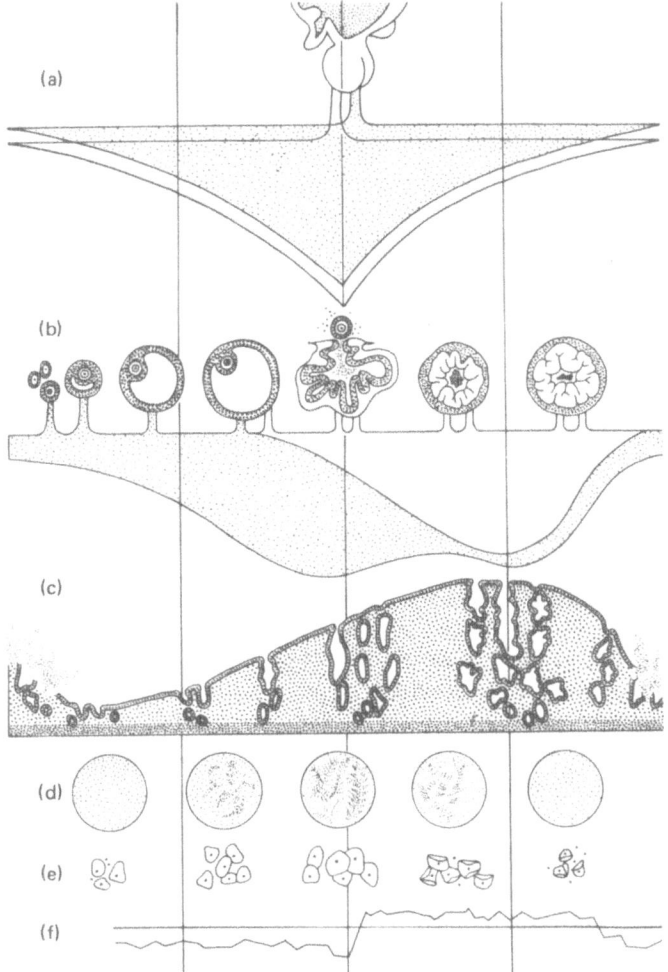

Fig. 43. Effect of anterior pituitary hormone on ovary, endometrium, and vagina. (a) Anterior pituitary secretion of FSH and LH; (b) secretion of estrogen and progesterone by ovary with development of follicle, ovulation, and formation of corpus luteum; (c) response of the endometrium: menstrual, proliferative, secretory, and menstrual phases; (d) cervical mucus with fern pattern at midcycle; (e) vaginal cytology; (f) basal body temperature with postovulatory rise.

of the cycle. Small concentration of estrogens also stimulate LH production. The steep increase in estrogen production and secretion around midcycle is associated with an LH peak (LH surge), which results in ovulation. The high estrogen level in association with progesterone inhibits LH production and secretion, resulting in a decrease of LH in plasma and urine in the second stage of the cycle. Hormone production of the corpus luteum may continue without further stimulation. There seems to be an increase of LTH (prolactin) in the second phase of the cycle, but the biologic function is still unclear.

Hormone Excretion. The excretion of total estrogen (estrone, estradiol, and estriol) varies between 20 to 100 μg per 24 hours, with a peak at midcycle (Fig. 42). Estrone and estradiol are excreted in about equal amounts. The excretion of estriol is about equal to the sum of the other two estrogens.

The main excretion product of progesterone and progesterol-20 α is pregnanediol-3α-20 α (see page 52). During the first half of the cycle, pregnanediol values per 24-hour urine are below 2 mg. After ovulation, the excretion increases to between 2.5 and 6 mg per 24-hour urine. Similarly there is an increase in 17α-hydroxyprogesterone in plasma and a rise in its excretion product (pregnanetriol) in the urine.

Furthermore, there is a variation in testosterone excretion during the cycle. The value is somewhat higher in the second stage of the cycle than in the first.

Hormonal effects on the Target Organs: Fallopian Tubes. The ciliated cells of the tubal mucosa increase in size under the influence of estrogens and are maximally developed at the time of ovulation. The secretory cells increase in size before menstruation. Glycogen is present in these cells during the entire cycle but predominantly in the menstrual phase. There is some increased motility of the musculature and ciliated cells under the influence of high levels of estrogen and a decrease during the corpus luteum phase.

Endometrium. The shedding of the functional layer of the endometrium takes approximately 3 days. Current work suggests that not all of the zona functionalis is lost during menstruation. Immediately after menstruation the regeneration begins under the influence of estrogens. The endometrial layer grows up to 5 mm in height and the stroma becomes thicker (page 35).

Unless conception occurs, regression of the corpus luteum takes place because of the decrease in LH and the absence of stimulation by human chorionic gonadotropin (HCG). Normal menstruation can be explained as a result of the combined withdrawal of estrogen and progesterone.

Menstrual Blood

Menstrual blood is incoagulable because of a lack of fibrinogen. In addition, various coagulation moieties are missing (factors II, V, and VIII) or present in lower concentrations (platelets). The erythrocyte concentration is reduced in menstrual blood and there is frequently hemolysis as well. The lack of fibrinogen is primarily a result of high proteolytic activity, especially fibrinogenolysis, and to a lesser extent of intravascular coagulation. The normal amount of blood lost during a menstrual period varies normally from 50 to 80 ml but may be as much as 150 to 200 ml (hypermenorrhea). The heavier flow is usually associated with "clots." These clots, however, are not fibrin clots but combinations of erythrocytes, mucoid substances, glycoproteins, and glycogen. They are now believed to form in the vagina rather than the uterine cavity. Fibrin clots in the uterine cavity may be present after curettage, since the functional layer, which contains the proteolytic activity in the form of plasminogen activator, is mechanically removed. Clotting forms as a result of release of tissue thromboplastin.

The Cervix

With increasing levels of estrogens, the external os begins to dilate (up to 4.5 mm in the nulliparous patient). Cervical secretion increases and becomes clear and watery. Mucus, placed between two glass slides at ovulation time, can be stretched up to 20 cm before breaking. This phenomenon is called Spinnbarkeit. When mucus is allowed to dry on a glass slide, a crystallization pattern can be observed with low magnification. The typical estrogenic pattern is referred to as a fern (Fig. 44). This crystallization begins at day 8 of the cycle, reaches a maximum at midcycle, and disappears about 8 to 10 days after ovulation.

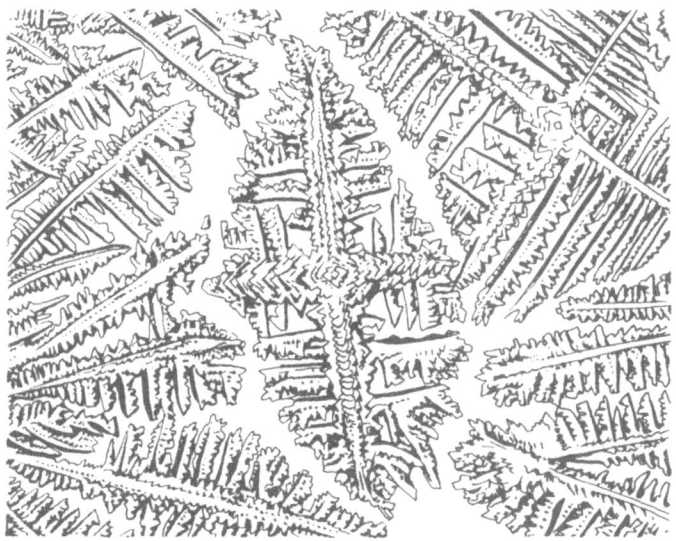

Fig. 44. Fern pattern in cervical mucus. This is a sign of a predominant estrogen effect.

Crystallization is a result mainly of an increase in sodium chloride. Certain proteins in cervical mucus are found in only low concentrations at the time of ovulation. They include plasminogen activator, albumin, gamma globulin, lysozyme, and α-1-trypsin inhibitor. These changes in biochemical and physical characteristics make the mucus optimally penetrable by sperm. Progesterone reverses these effects: The amount of mucus decreases; the mucus has a whitish appearance because of incorporation of cells (mainly polymorphonuclear leucocytes); Spinnbarkeit decreases to 1 to 2 cm; and the mucus resists penetration by sperm (Fig. 45).

The Vagina

The vaginal epithelium proliferates under the influence of estrogenic hormones; the cellular changes are diagrammed on page 29. The vaginal epithelium is an even more sensitive index of steriodal hormonal effects than is the endometrium.

The Breasts

Estrogenic hormones stimulate the growth and proliferation of epithelial cells of the lactiferous ducts. Progesterone in association with estrogens stimulates proliferation and secretion of the alveoli.

Basal Body Temperature

The rectal temperature in the follicular phase of the cycle varies between 36.2 and 36.7°C, or 97.2 and 98°F. Increased production of progesterone after ovulation results in an upward shift in basal body temperature. Although the basal body temperature is only an indirect reflection of the hormonal concentrations, it is still the most widely used method for recognizing the time of ovulation (page 185).

Metabolic Function

Estrogens decrease the level of cholesterol for a certain period of time. Although the concentration of serum lipid increases, the ratio of β- to α-lipoproteins decreases. Estrogens also stimulate the retention of calcium. Potassium increases extracellularly and sodium intracellularly. Both extracellular water and plasma volume are increased Estrogens also increase the binding of corticosteroid and thyroid hormones. There occurred increased incorporation of mucopolysaccharides in the perivascular connective tissue, pigmentation, and changes in glucose metabolism. Progestational hormones in general oppose many estrogenic meta-

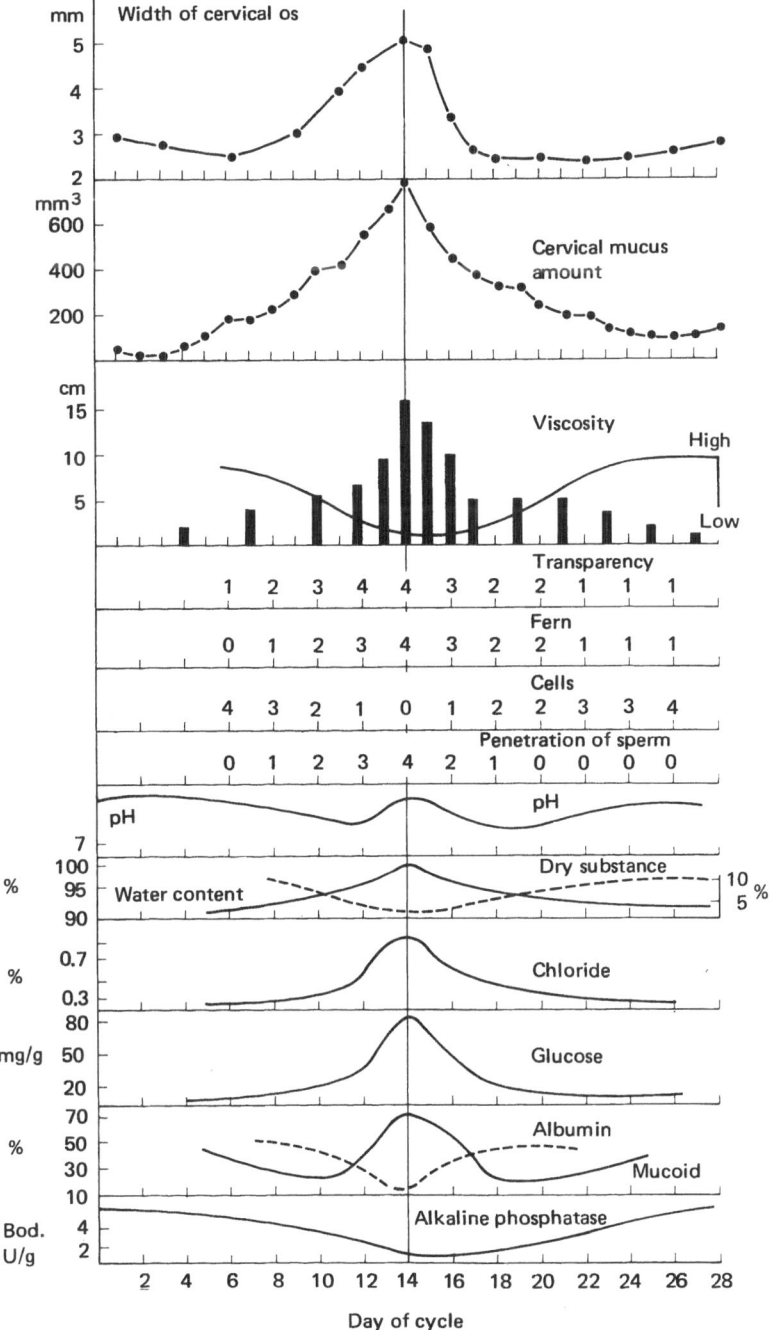

Fig. 45. Cyclical estrogen- and progestin-dependent changes in the cervical os and mucus.

bolic effects. They stimulate excretion of urea and sodium but may, nevertheless, lead to water retention.

Psychological Influence

Many women undergo emotional changes during the menstrual cycle. In particular, depression, irritability and antisocial behavior are more commonly observed in the premenstrual phase of the cycle (see page 152 on premenstrual syndrome).

CHAPTER 4

Physiology of the Stages of the Woman's Life

Childhood

The gonads and external genitalia in both boys and girls enter a period of rapid growth at age 7.

Estrogens, which are present in both sexes up to the age of 7, increase rapidly thereafter in the girl. It is possible that the major fraction of these estrogens is produced as androgens in the adrenal and metabolized by enterohepatic pathways to estrogens. Production of estrogen by the ovary begins at about the age of 11. Concomitant with this estrogen production, the sexual organs begin to mature. The corpus uteri in particular increases in size more rapidly than does the cervix, and the ratio of corpus to cervix, which is 1:2 in the child, changes to 2:1 in the adult (Fig. 46). Gonadotropins are present in small amounts in childhood but they can be measured in 24-hour urine samples $1\frac{1}{2}$ years before menarche.

Puberty

Puberty is the period of life in which the gonads begin to assume adult function. The secondary sexual characteristics become evident and the psychological female pattern develops. Puberty begins between the ages of 8 and 9 and lasts for approximately 4 to 8 years, ending with the acquisition of reproductive capacity shortly before the end of the period of growth. Some authors define puberty as ending with menarche, and the time after menarche as adolescence.

Puberty occurs in the girl about 2 years earlier than in the boy and is dependent upon hereditary and racial factors as well as social factors and nutrition. In all Western societies an acceleration has been observed. That is, menarche seems to be occurring earlier than in preceding decades. The reason is not clear but it is assumed to be related to nutritional and other environmental factors. The mean age of menarche is $12\frac{1}{2}$ years, with a bell-shaped distribution.

Bone development is related to ovarian function and puberty can be predicted by radiologic examination. During puberty the extremities

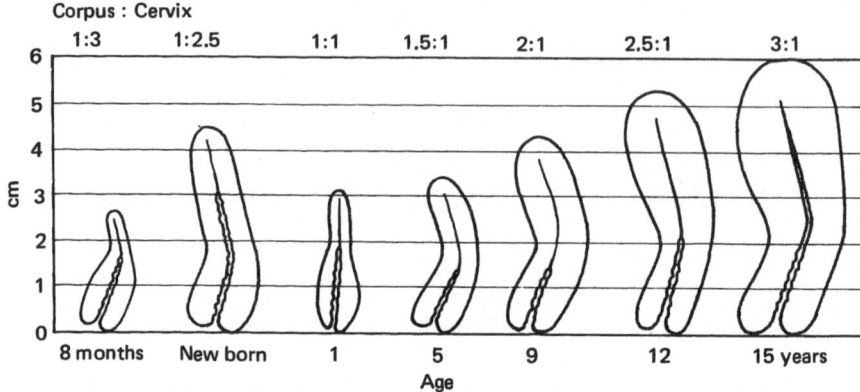

Fig. 46. Growth of the uterus and the relations of size of corpus to cervix during the first 15 years of life.

are relatively long, for they grow faster than the trunk and terminate their growth earlier. The bony pelvis begins to grow in the girl between the ages of 8 to 10, whereas the shoulder circumference grows more prominently in boys. In girls fatty tissue begins to develop in the breasts and thighs at this time.

The ovaries begin to increase in size and weight during puberty. Primary follicles become larger and the granulosa layer increases.

The first menstrual period is usually prolonged, since bleeding occurs from an endometrium that has been growing for a long while, and the menstrual cycle is frequently prolonged. Whereas the menarche is anovulatory, the number of ovulatory cycles begins to increase slowly. Only about 30% of the cycles are ovulatory at age 12 and about 50% at age 16. Fertility, therefore, is usually not attained earlier than 2 years after menarche. The corpus luteum, furthermore, is frequently inadequate during the earliest ovulatory cycles.

Estrogens are produced not only in the follicles but also in the interstitial cells of the ovary. During this time the vagina begins to mature, the cervix produces mucus, and because of the increase in size of the corpus, the uterus is usually anteflexed (page 245). The breasts increase in size and the nipples begin to form (Fig. 47). Not infrequently one breast develops faster than the other. After development of the breast, pubic hair appears around the introitus. The first sesamoid bone can then be visualized radiologically. Axillary hair begins to grow approximately 1 year later, or 2 years before menarche (Fig. 48). During this period facial acne may develop as a result of relative excess of androgens. Closure of the epiphyses occurs between the ages of 16 and 17 and general growth ceases at that point.

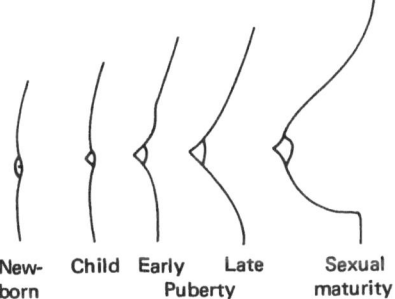

Fig. 47. Development of the female breast from birth to adulthood.

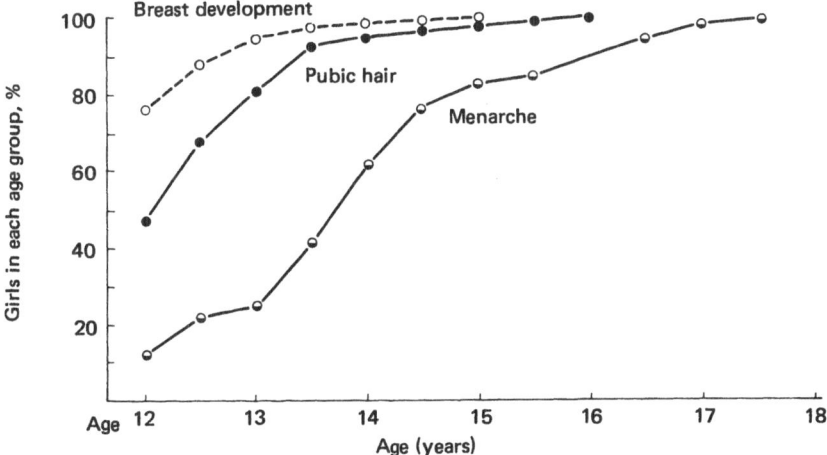

Fig. 48. Onset of the most important signs of puberty.

Excretion of estrogen begins to fluctuate before puberty. Excretion of 17-ketosteroids and testosterone increases, but less strikingly in girls than in boys. Excretion of pregnanediol is low; it becomes cyclic when biphasic cycles occur, with ovulation and corpus luteum functions. Adult values are reached at age 13. Production of gonadotropins begins to increase at ages 11 to 12 and shows a slight peak $1\frac{1}{2}$ years before menarche. Gonadotropins are increasing in all well-developed girls by the age of 14.

Reproductive Period

The reproductive period is defined as that portion of the life of a woman in which she is able to reproduce. It is characterized by morphological, functional, and psychological maturity. It begins with the

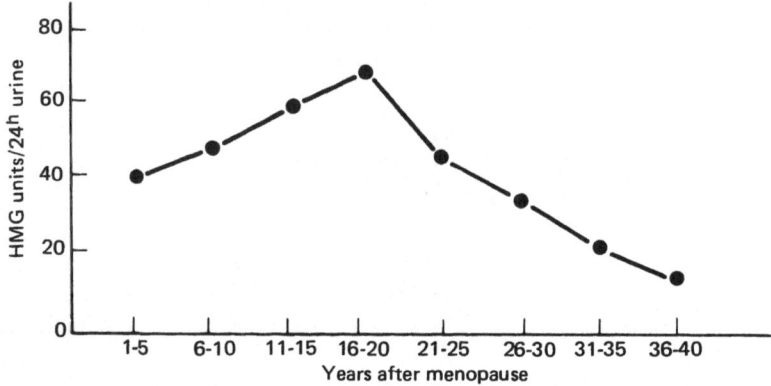

Fig. 49. Gonadotropin excretion (in units of HMG) in the urine during climacteric and old age. Rise up to about 15 years after menopause, followed by a fall. The values remain higher than in the reproductive years.

development of secondary and tertiary sexual characteristics and ends with the premenopause. Criteria are ovulatory cycles and the potential for pregnancy.

Climacteric—Menopause—Old Age

The climacteric is that stage in a woman's life in which the gonads cease their generative function and decrease their hormonal production. The start of the climacteric is somewhere between the ages of 45 and 50. Response of the follicle to gonadotropins decreases in the preclimacteric phase (Fig. 49). Ovulation and corpus luteum formation are absent, perhaps as a result of regressive changes secondary to sclerosis of ovarian vessels, increase of connective tissue, and decrease of follicles. The ovary decreases in weight but the pituitary-hypothalamic axis remains intact.

Menopause is defined as the last uterine bleeding induced by ovarian function. It normally occurs between the ages of 45 and 55. This definition is, of course, retrospective, usually made one year after the last uterine bleeding. The menopause is followed by the postmenopause. Old age, in endocrine terms, is reached by age 65.

Since puberty and menarche seem to be occurring earlier and since the climacteric and the menopause seem to be delayed in recent decades, the reproductive phase is apparently lengthened.

Hormone Production

Pregnanediol excretion in the second half of the cycle is absent or markedly reduced as a result of failure of ovulation and of formation

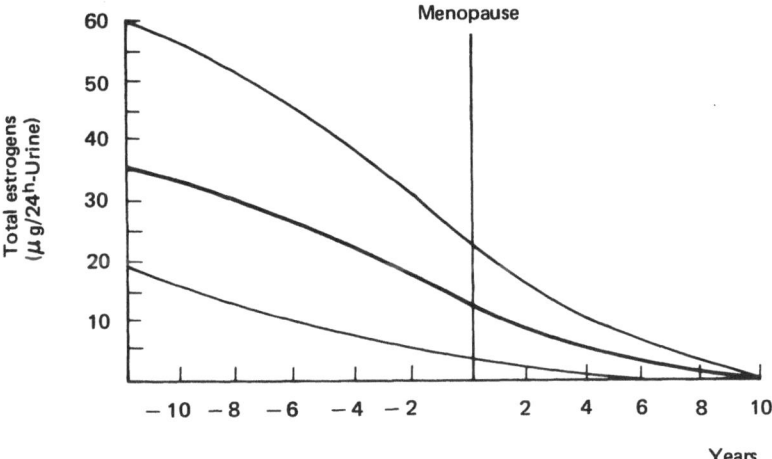

Fig. 50. Excretion of total estrogens in the urine during the climacteric and old age.

of the corpus luteum. Furthermore, the average production of estrogen is reduced, compared with that in the reproductive period, and the ovulation peak is less pronounced. Concentrations of 5 to 20 μg per 24 hours are found in the urine in the late postmenopause. The endometrium fails to respond to concentrations below 10 μg. This value corresponds to a total production of 90 to 100 μg of total estrogen per day, which is produced in small amounts by the interstitial cells of the ovary and to a greater extent by the adrenal (Fig. 50). The adrenal, however, most likely produces C_{19} steroids, which are in turn metabolized by the enterohepatic pathway to estrogens.

As a result of a negative feedback from the decreasing ovarian function, gonadotropin release increases continuously during the menopause. After castration by oophorectomy or irradiation, the gonadotropic hormone production increases rapidly. It reaches a maximum 5 to 6 weeks after castration, a value approximately 5- to 20-fold that observed during the normal cycle. The concentration increases further and reaches a maximum after 10 to 15 years.

The increase is related mostly to FSH; LH release is only slightly higher than at the time of ovulation in the normal biphasic cycle. The FSH increase frequently produces follicle cysts and an increase in ovarian stroma. For a period of 3 to 5 years after menopause the ovaries may be stimulated by the excessive release of FSH, resulting in an increase in estrogenic hormones. Then the response of the ovaries ceases completely. The ovary produces only small amounts of androgens.

Table 3. Special Endocrinological Assays

Quantitative Information	Indication	Discussed in text on page listed
Gonadotropins		
FSH	Endocrine disturbances	146
LH	Infertility	181
(LTH)	Induction of ovulation	179
	Inadequate corpus luteum	144
Steroid hormones		
Estrogens	Endocrine disturbances	146
Pregnanediol	Infertility	181
Androgens		
(17-ketosteroids, some-	Hirsutism	53
times fractionated)	Virility	53
especially dehydroiso-	Intersexuality	172
androsterone and		
testosterone		
17-Hydroxycorticoids		47
Pregnanetriol		52

Table 4. Tests of Endocrine Function

Methods	Indicatian	Discussed in text on page listed
Progestin test	Secondary amenorrhea	146
Estrogen test	With negative progestin test	149
Gonadotropin test	Test of possibility of induction of ovulation	179
ACTH stimulation test	Suspicion of insufficiency of adrenal cortex	151
Dexamethasone test	Differential diagnosis of	
Dexamethasone-HCG test	hirsutism and virilization	148

Table 5. Special Cytogenetic Methods in Endocrine Disturbances

Method	Indication	Technique
Sex Chromatin	Primary amenorrhea	Smear from buccal mucosa or vagina
	Intersexuality	Drumsticks in blood smear
Chromosomal analysis	Primary amenorrhea with stigmata of	Blood culture
	Turner's syndrome, with negative or doubtful sex chromatin findings	Tissue culture

Table 6. Pregnancy Tests

Methods	Information available from	Notes
Basal body temperature	Elevation longer than 16 days	Not accurate
Hormone test (estrogen-progesterone combinations)	Fifth week after menses	Many preparations available
Immunological tests	Fifth week after menses	
(HCG) Pregnosticon	Fifth week after menses	Uses sheep erythrocytes
Gravindex	Fifth week after menses	Uses latex particles

Table 7. Contraindications to Estrogen Therapy

Absolute contraindications
  Liver diseases; severe hepatitis and seguelae
  Enzyme disorders (Rotor syndrome, Dubin-Johnson syndrome)
  Porphyria
  Thromboembolic disorders

Relative contraindications (individual decision)
  Myomas
  Endometriosis
  Fibrocystic disease of the breast
  Carcinoma of the corpus
  Carcinoma of the breast
  Diabetes

Selected Reading

Hormones

Butt, W. R.: *Hormone Chemistry.* London: Van Nostrand (1967). Salhanick, H. A., Kipnis, D. M., and Vande Wiele, R. L.: *Metabolic Effects of Gonadal Hormones and Contraceptive Steroids.* New York: Plenum (1969).

Estrogens

Diczfalusy, E., and Lauritzen, Ch.: *Estrogens in the Human.* Berlin-Göttingen-Heidelberg: Springer-Verlag (1961).

Gonadotropins

Bell, E. T., and Loraine, J. A. (eds.): *Recent Research of Gonadotrophic Hormones.* Edinburgh and London: Livingstone (1967).

Mechanism of Hormones

Mueller, G. C.: *Oestrogen Action and Genetic Expression in the Uterus.* In: *Biogenesis and Action of Steroid Hormones.* R. I. Dorfman, K. Yamasaki, and M. Dorfman, eds. Los Altos: Geron-X-Inc. (1968).

Hypothalamus-Hypophysis

Harris, G. W.: *Neural Control of the Pituitary Gland.* London: Edward Arnold (1955).

Martini, L., and Ganong, W. F.: *Neuroendocrinology.* New York and London: Academic (1967).

McCann, S. M., Dharswal, A. P. S. and Porter, J. C.: Regulation of the adenohypohysis. *Amer. Rev. Physiol.* **30**: 589, 1968.

Ovary

Zuckermann, S.: *The Ovary.* New York and London: Academic (1962).

Cycle

Loraine, J. A., and Bell, E. T.: *The Clinical Application of Hormone Assay.* Edinburgh and London: Livingstone (1965).

CHAPTER 5

Menstrual Hygiene

Vaginal tampons are the most frequently used means of absorbing menstrual blood. Adolescents should be advised about their proper use by choosing the appropriate size, depending upon the opening of the hymen. With normal flow the tampons should be changed at least once every 12 hours.

Forgotten tampons within a few hours produce a malodorous discharge and possible local infection. An ascending infection is rare, however.

There is disagreement about the desirability of vaginal douching. Douching is not harmful if an acidic solution is used of pH approximately 4.0, but it is not generally necessary because of the self-cleansing mechanism of the vagina. In no circumstances should disinfectants such as Lysol or Phisohex be used. Vaginal douching is advisable if the patient is annoyed by physiologic mucorrhea at midcycle. Douching as a mode of contraception is associated with a high failure rate.

The so-called feminine hygiene sprays offered by the cosmetic industry are but slight modifications of antiperspirants and are therefore not desirable for use within the vagina. They may produce allergic reactions and are not recommended. A history of their use should be elicited in cases of vulvitis of apparently unknown cause.

There is a close correlation between primary dysmenorrhea (page 152) and inadequate sexual education. The importance of an early explanation of the physiology of menstruation is thus obvious.

Sports during Menstruation

Light sports are permissible during menses. Calisthenics may even be beneficial in cases of dysmenorrhea. Avoidance of all sports because of pain during menstruation may lead to a permanent pattern of incapacitation each month.

Tampons allow the patient to swim during menses, since water does not normally enter the vagina. The same applies to bath water.

Intercourse during Menstruation

For many decades intercourse during menstruation was believed to lead to infections. If infections recur after menses, however, they are related not to intercourse, but perhaps to the temporary increase in pH at that time (page 221). Religious taboos and personal preferences determine whether couples will engage in intercourse during menses.

CHAPTER 6

Sexual Physiology

Consultation and advice in sexual matters is an integral part of gynecological practice. The obstetrician-gynecologist is also responsible for a large share of education in this area.

The Physiologic Sexual Response

Sexuality is basically an expression of interpersonal relationship.

Human sexual response is basically similar in men and women. Sexual stimulation in both sexes is accompanied by vascular engorgement and muscle tension.

The sexual response is divided into four phases, according to Masters and Johnson: (1) excitement, (2) plateau, (3) orgasm, and (4) resolution.

The excitement phase results from either somatic or psychogenic stimulation. It is the longest of the four stages and can be delayed voluntarily or interrupted at any time. Continuation of stimulation and increase in sexual excitement leads to the plateau phase. The plateau phase progresses spontaneously and involuntarily to orgasm. This stage lasts only a few seconds. The orgasmic phase is followed by resolution. With continuing stimulation women may experience another orgasm at any time during the resolution phase. In contrast, the male undergoes a refractory period during which orgasm cannot be reached. The refractory phase ordinarily lasts from at least 5 to 30 minutes (Figs. 51, 52).

Female Sexual Response

As a result of congestion in the lower third of the vagina and increase in tone of the bulbocavernosus muscle, the labia majora thin and are elevated. The labia minora become edematous and enlarge and protrude through the thinned labia majora. The labia minora and the introitus assume a vivid color during the plateau phase.

Bartholin's glands produce a few drops of a mucoid secretion during the excitement phase. The amount, however, is too small to provide significant lubrication of the vagina and may be barely enough to

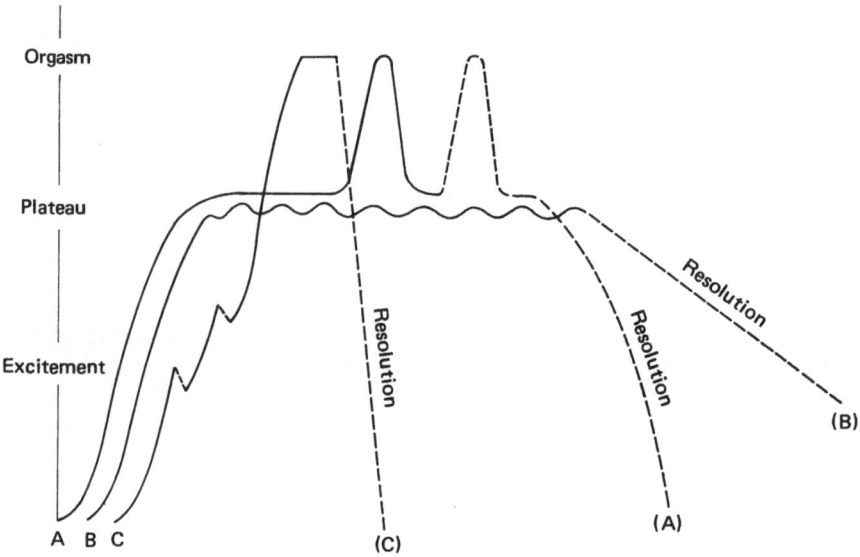

Fig. 51. Cycle of female sexual response. (A) Reaction pattern with single or multiple orgasms; (B) reaction pattern without orgasm; (C) reaction pattern without distinct plateau (from Masters and Johnson, 1967).

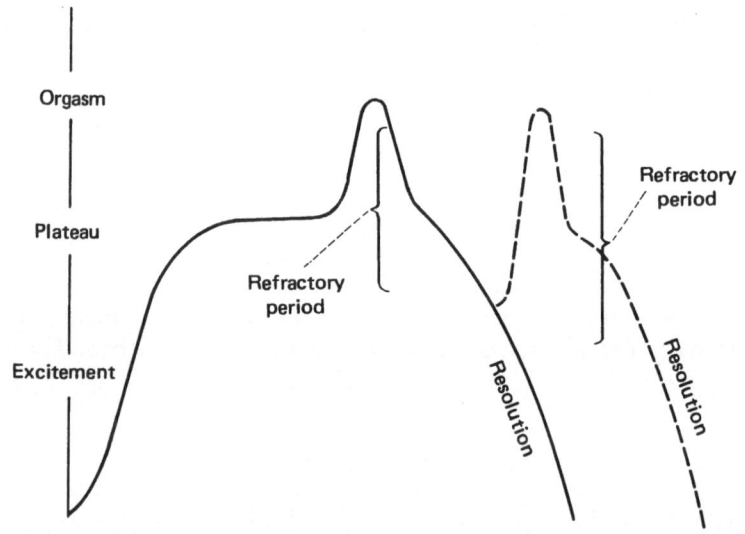

Fig. 52. Cycle of male sexual response (from Masters and Johnson, 1967).

moisten the introitus. Lubrication of the vagina is provided by a transudate (page 79).

The clitoris is a highly sensitive receptor of neurogenic impulses. The organ is a primary site of sexual excitement. The glans of the clitoris increases only slightly through congestion. In contrast, the corpus, with its corpora cavernosa, swells during the excitement phase independent of direct stimulation, but there is no erection corresponding to that of the penis. During the plateau phase there is retraction of the clitoris under the crura, shortening of the suspensory ligament of the clitoris, and contraction of the ischiocavernosus muscle. The clitoris returns to its normal position after decrease in sexual stimulation.

The vaginal mucosa begins to change during the excitement phase. Ten to 20 seconds after stimulation there is a transudation. This "sweating phenomenon" provides lubrication of the vagina in early sexual response. The transudation is an interesting phenomenon that is apparently unique. It is explained by vasodilation and congestion of the vaginal venous plexus (page 221), which also gives an intense livid color to the vaginal mucosa. This transudate is also present after hysterectomy and even after ovariectomy and is therefore independent of ovarian function. It can be seen even after creation of an artificial vagina by means of a skin graft. The biochemical composition is as yet unknown. The physiologic pH of the vagina is only slightly increased from 4 to 4.5. The neutralization required for the survival of the sperm is provided by the seminal plasma. This buffering capacity of the seminal plasma provides for motility of the sperm for up to 6 hours.

During the excitment and plateau phases the upper two-thirds of the vagina increases in length and diameter. The lower third of the vagina increases in size during the excitement phase to only a minor degree, but with progression to the plateau phase there is a pronounced congestion, producing a so-called orgasmic platform. It is this area which undergoes rhythmical contraction at orgasm. Three to five contractions occur 0.8 second apart, later decreasing in intensity. The intensity is dependent on the experience of the individual and it may differ even from orgasm to orgasm in the same individual.

Orgasms are experienced without precise location deep in the pelvis, in the crura of the clitoris, in the vagina, and in the uterus. The increased muscle tone reaches its peak with convulsions of the entire musculature during orgasm. The phenomena described return to normal during the resolution phase in the reverse order of their development. The only reaction of the cervix uteri during sexual stimulation is a slight opening of the external os during the resolution phase, possibly favoring migration of sperm, but there is no evidence of a sucking effect of the portio vaginalis.

The uterus is elevated in the late excitement and early plateau phases, presumably as a result of congestion of the entire pelvis. During orgasm the uterus undergoes contractions that start in the fundus. The uterus increases considerably in size as a result of congestion. Occasionally, congestion and contractions are felt as pain. Uterine elevation and enlargement return to normal in a few minutes during the resolution phase, but during multiple orgasms an enlargement of the uterus may persist for hours.

Physiologic Response of the Sexual Organs in the Male

The first genital reaction of the male after sexual stimulation is erection of the penis. The penis increases in length and diameter during the excitement phase. The excitement phase can be interrupted at any time in the male also. Distraction by other stimuli may lead to loss of erection despite continuation of sexual stimulation.

During the plateau phase the penile circumference increases at the coronal ridge and the glans acquires a dark bluish color as the result of venous filling. During orgasm, ejaculation by rhythmic expulsive contractions along the entire length of the penis and urethra are achieved by contractions of the sphincter urethrae, bulbocavernosus, ischiocavernosus, and transverse perineal muscles. Contractions begin at intervals of 0.8 second and they decrease in intensity and frequency after the first five to six contractions. The male experiences orgasm in the penis and less markedly in the prostate. Turgescence and increase in muscle tone with convulsions of the entire musculature follow the pattern of female response. During the resolution phase the penis decreases to about half its erect size. Detumescence of the penis may be delayed if it remains in the vagina.

Scrotum and testes react in the different phases in a characteristic sequence, depending on congestion and muscular tone. The scrotal skin undergoes tension and thickening, which disappear during the plateau phase as a result of shortening of the spermatic cords. The size of the testes increases by 50%.

Extragenital Responses during Sexual Stimulation

The extragenital reactions are also the result of congestion and increase in vascular tone.

(1) The female breast increases because of vasodilation during the various phases. The venous channels become visible because of their bluish color, most pronounced during the plateau phase and more intense in the nullipara than in the multipara. Erection of the nipples, which is the rule in the female, occasionally occurs in males as well.

(2) "Sex flush." As a result of capillary dilatation a maculopapular rash develops late in the excitement phase and spreads rapidly over both breasts, chest, flanks, and back. The flush disappears in the resolution phase in the reverse order of its appearance. The sex flush is observed in the male in approximately 25% of cases.

(3) Myotonia. At the end of the excitement phase there is some voluntary and involuntary muscle tension, which progresses during the plateau phase to semispastic contractions of facial, abdominal, and intercostal muscles. At orgasm, there is complete loss of control and spasms of various muscles are often pronounced. This reaction is similar in the male.

(4) Bladder, urethra, and rectum. The congestion of adjacent organs may be augmented by the mechanical irritation during intercourse. Urinary urgency is noted frequently in the female in the resolution phase. Contractions of the sphincter ani and gluteus maximus increase in the plateau phase.

(5) Additional extragenital reactions include tachycardia and hyperventilation, which begin in the plateau phase and continue during orgasm. The systolic blood pressure rises. Widespread perspiration may be observed in the resolution phase. These reactions are identical in the female and the male.

Normal Sexual Behavior

The frequency of coitus varies greatly. In normal young couples intercourse may occur between once in 2 weeks to several times a day. The frequency of coitus generally decreases with age. Libido in the female increases between the early twenties and the mid-thirties and reaches a plateau at about age 45 that may continue into the menopause. In the male the greatest libido is in the late teens with a decrease often occurring in the late twenties. Strenuous physical work and intense intellectual activity may decrease the libido.

The male is, as a rule, more easily excited than the female and reaches orgasm earlier. Simultaneous orgasms in both partners are the exception rather than the rule. Ignorance of this fact is often related to complaints of sexual incompatibility.

The female may be more easily excited sexually without significant physical contact. The fact that she is desired by her partner may in itself give satisfaction. If female orgasm is delayed, it may be elicited by digital stimulation, but the partner should remember that the clitoris retracts during the plateau phase (page 79). Proper technique of digital stimulation is therefore important. The concept of two different orgasms, a vaginal and a clitoral, must now be regarded as a fallacy. Physiologically there is no difference in response. The failure to achieve orgasm after penetration of the penis into the vagina, which has virtually no sensory nerve supply, may no longer by considered a manifestation of sexual immaturity.

Another fallacy is the idea that a small penis is less effective in causing vaginal orgasm. First, the differences in size of the erect penis are considerably smaller than those of the flaccid penis. Second, the penis comes in contact with the clitoris in only a few coital positions. After hysterectomy, even radical hysterectomy, the shortened vagina is sufficient for coitus. The loss of the uterus has no effect on orgasm except possibly psychologically. This fact should be explained by the gynecologist in detail to the patient and her husband before operation.

The distensibility of the vagina, however, is reduced after irradiation and the vaginal mucosa is more vulnerable as a result of relative avascularity (page 313). This fact may bear on the selection of treatment of cervical cancer in younger women, in whom radical surgery in suitable patients may be preferable to primary irradiation (page 312) for this reason.

Circumcision does not reduce sexual sensation nor can the circumcised male delay ejaculation beyond the usual time.

Three common coital positions are of clinical interest: (1) female-supine position, (2) female-superior position, and (3) knee-chest position.

The female-supine position may be preferable in a woman with an anteverted uterus and a problem of infertility. The seminal pool is deposited in the posterior fornix of the vagina. The reflux of semen can be delayed by elevating the pelvis with a firm pillow. This position may not be suitable late in pregnancy because of the protuberant abdomen. Since the male is required to perform more actively in this position, it may not be the position of choice for men with cardiovascular disease.

In the female-superior position many women have a feeling of greater satisfaction. The base of the penis comes in contact with the clitoris. The female partner has a greater ability to keep the semierect penis in the vagina after ejaculation. This position may be preferable for partners who are obese and for men with cardiovascular problems. The position is not preferable for patients with infertility problems, since the semen may leak rapidly out of the vagina.

The knee-chest position is advisable immediately after operations on the perineum (episiotomy, posterior colporrhaphy) and it may help to overcome difficulties at defloration (rigid hymen) because the direction of the penis avoids pressure on the perineum. The position is also preferable for patients with infertility problems and a retroverted uterus, since the seminal plasma pools in the anterior fornix.

Sexual Response in the Elderly

Increasing age does not necessarily reduce libido or orgasm in the female. On the contrary, knowledge that pregnancy is no longer possible may improve sexual response. The involution of tissues as a result of estrogen lack, however, may create problems. Congestion and muscle tone decrease; the dilatation and transudation of the vagina are reduced; and the vaginal mucosa becomes susceptible to trauma. The phases of the cycle of sexual response may be delayed.

Theoretically, there is little reduction in sexual potency in the male, but he also may experience a decrease in congestion and tonus and, therefore, a delayed reaction.

In general, however, sexual potency is often prolonged into old age. The mechanical problems resulting from involution may sometimes be treated successfully by hormonal stimulation (page 160).

Sexual Compatibility

The majority of cases of incompatibility are psychogenic. Only rarely do the difficulties have an organic basis.

Sexual Disturbances in the Female

Dyspareunia is pain on intercourse. Organic causes include atrophy and narrowing of the introitus as a result of scars. Severe vaginitis also may cause discomfort on intercourse. An additional organic cause is endometriosis of the uterosacral ligaments. Dyspareunia often results from psychogenic causes as well (page 87) often allied to the pelvic congestion syndrome (page 277). On pelvic examination the utero-sacral ligaments may be spastic and painful when in contact with the penis. Typically, the spasm disappears during examination under anesthesia, which aids in differential diagnosis (page 277).

Frigidity is lack of libido or desire for sexual activity. Frigidity is often mistaken for lack of orgasm. A detailed history will reveal, however, that sexual desire may be present without the achievement of orgasm.

Loss of libido is physiologic after heavy exercise and serious physical illness. Libido can be reduced also during intense intellectual activity. Drugs such as morphine, heroin, and LSD may decrease libido, as may chronic alcoholism.

With few exceptions, permanent frigidity is a psychosexual problem, frequently associated with subconscious repression. Many patients complain of lack of orgasm, but a carefully obtained sexual history reveals that they may achieve orgasm through digital stimulation but not intercourse. Complete lack of orgasm is extremely rare, probably a result of psychosexual disease. Explaining to the patient that there is no difference between clitoral and vaginal orgasm is helpful advice for many couples.

Nymphomania is defined as an excessive sexual desire in women, although it is difficult to quantitate. It may be a reason for promiscuity, but it bears no relation to prostitution.

Vaginismus. In this condition, the pelvic musculature is spastic, narrowing the introitus and preventing penetration by the penis. This problem is, almost without exception, psychogenic. It requires psychiatric evaluation.

Sexual Disturbances in the Male

Impotence is considered primary if an erection has never been achieved or maintained during coitus. Secondary impotence develops after at least initial satisfactory intercourse.

The cause of primary impotence is psychological. There are innumerable causes of secondary failure (Table 8). In general, failure of erection initiates a vicious cycle in which fear of failure augments the initial impotence. These patients should be managed by skilled psychiatrists.

Premature ejaculation. This problem may be defined as failure to delay ejaculation for 30 seconds after penetration. A transient form of premature ejaculation may be physiologic. It may occur, for instance, at first intercourse after long periods of abstinence. In its severe and chronic form premature ejaculation must be considered a psychosexual disorder. Alcohol in moderate amounts as well as tranquilizers can occasionally delay ejaculation. Anesthetic creams applied to the glans may have limited value. Masters and Johnson (1970) have described the treatment of this condition in detail.

Borderline between Normal and Abnormal Sexual Behavior

Data are still insufficient to define clearly the psychological and behavioral characteristics of normal and abnormal sexual patterns. According to Kinsey, anything is normal that pleases two heterosexual partners without harming them physically or psychologically. It may be appropriate to extend the definition to include homosexual acts between consenting adults in this context. Statistically, every act must be considered "normal" if it is performed by a large number of persons. This definition includes orogenital techniques as well as masturbation.

Sexual Counseling

Sexual counseling requires knowledge of psychological and physical reactions to the sex-specific differences in the cycle of sexual response, and of the range of variation in psychological and physical behavior in both sexes. This knowledge allows a differentiation of disturbances based on ignorance and insecurity from true sexual neuroses and organic

Table 8. Classification of Physical Causes of Secondary Impotence

Anatomic
Congenital deformities
Testicular fibrosis
Hydrocele

Cardiorespiratory
Angina pectoris
Myocardial infarction
Emphysema
Rheumatic fever
Coronary insufficiency
Pulmonary insufficiency

Drug Ingestion
Addictive drugs
Alcohol
Alpha-methyl-dopa
Amphetamines
Atropine
Chlordiazepoxide
Chlorprothixene
Guanethidine
Imipramine
Methantheline bromide
Monoamine oxidase inhibitors
Phenothiazines
Reserpine
Thioridazine
Nicotine (rare)
Digitalis (rare)

Endocrine
Acromegaly
Addison's disease
Adrenal neoplasms (with or without Cushing's syndrome)
Castration
Chromophobe adenoma
Craniopharyngioma
Diabetes mellitus
Eunuchoidism (including Klinefelter's syndrome)
Feminizing (interstitial-cell) testicular tumors

Infantilism
Ingestion of female hormones (estrogen)
Myxedema
Obesity
Thyrotoxicosis

Genitourinary
Perineal prostatectomy (frequently)
Prostatitis
Phimosis
Priapism
Suprapubic and transurethral prostatectomy (occasionally)
Urethritis

Hematologic
Hodgkin's disease
Leukemia, acute and chronic
Primary anemia

Infectious
Genital tuberculosis
Gonorrhea
Mumps

Neurologic
Amyotrophic lateral sclerosis
Cord tumors or transection
Electric shock therapy
Multiple sclerosis
Nutritional deficiencies
Parkinsonism
Peripheral neuropathies
Spina bifida
Sympathectomy
Tabes dorsalis
Temporal lobe lesions

Vascular
Aneurysm
Arteritis
Sclerosis
Thrombotic obstruction of aortic bifurcation

diseases. The gynecologist is the physician most frequently confronted by patients with complaints of sexual inadequacy. The gynecologist is also responsible for excluding organic causes in the female. It should be the special concern of the gynecologist to counsel young people, for it is much easier to prevent psychosexual neuroses than to treat them.

Conception control is part of counseling, since fear of an unwanted pregnancy is frequently the origin of sexual inadequacy. The gynecologist must advise patients about possible changes in sexual response after major operative procedures as well as in the prenatal and postpartum periods. It is not unusual for couples to restrict intercourse unnecessarily for fear of harming the fetus or causing prematurity. Libido in the female may be increased or decreased during pregnancy, whereas the male's libido usually remains unchanged.

Sexual counseling in the elderly will become increasingly important as life expectancy increases and as the pattern of geriatric disease changes.

Sexual counseling requires the time and patience of the physician. The first office visit usually provides the basis for a doctor–patient relationship of mutual confidence. Treatment then requires careful sexual histories obtained independently from the female and male partners. Counseling by a team comprising at least a gynecologist and a psychologist has obvious advantages. If during the initial interviews, there is evidence of a deep-seated neurosis the patient should be referred to a psychiatrist.

Selected Reading

Kinsey, A. C., Pomeroy, W. B., and Martin, C. E.: *Sexual Behavior in the Human Male.* Philadelphia and London: Saunders (1948).

Kinsey, A. C., and Gebhardt, P. H.: *Sexual Behavior in the Human Female.* Philadelphia and London: Saunders (1953).

Masters, W. H., and Johnson, V. E.: *Human Sexual Response.* Boston: Little, Brown (1966).

Masters, W. H., and Johnson, V. E.: *Human Sexual Inadequacy,* Boston Little, Brown (1970).

CHAPTER 7

Fundamentals of Psychosomatic Gynecology

Pathogenesis

Psychosomatic functions are not subject to voluntary control. They are more like reflexes. "Emotional reflexes" influence not only the vasomotor system (blushing), secretion (perspiration caused by excitement), and smooth muscle (diarrhea induced by fear), but also the endocrine organs, as for example the secretion of adrenalin in dangerous situations. Emotional reflexes can be pathologically heightened or modified. A strong emotion can precipitate loss of consciousness and cessation of menstrual flow. Vaginal bleeding induced by fear is known to occur.

Bleeding and abortion in pregnant women involved in car accidents occur rarely as a result of direct bodily injury. Fright itself has frequently been considered the only explanation for the bleeding or even the abortion. The more intense the emotions (fear, terror, frustration), the more likely are psychosomatic reactions to occur. The significant factor is not the objective extent of danger, but rather the intensity of the experienced fear. Obviously, this intensity is modified by individual variation and emotional threshold. Previous traumatic experiences, inappropriate attitude toward the environment, and psychic and physical stress are some of the predisposing factors. It is irrelevant to the initiation of the psychogenic diseases whether the patient suffers an acute emotional shock (terror) or a more chronic tension (anger, fear). In the chronic situation, more numerous factors subconsciously determine the symptom. The most striking fact is that the symptom, that is, the "illness," fulfills a more or less recognizable purpose. For example, painful menstruation is used to justify absenteeism from school; sudden vaginal bleeding permits the patient to refuse her husband's sexual advances without obviously offending him; vaginismus prevents premarital intercourse that the patient is unable emotionally to accept and may exaggerated sign or symptom referable to the genitalia, such as vaginal discharge, amenorrhea, or dyspareunia, may provide a "reason" to postpone an engagement or a wedding. The patient is more or less conscious not only of her "escape into illness" but also of her need to be protected from her own insufficiencies. Affection, protection, help, and compassion are thus exacted through the illness.

If no genuine conflict can be elicited, the patient may be manifesting a *primitive reaction.* For example, the symptom may stem from the wish to play truant to avoid a certain task or to postpone an examination.

In a *neurosis,* a serious conflict induces and perpetuates the illness, but the causal relation between the conflict and the symptom is repressed. For example, the patient may complain of infertility. The family and the husband desire an heir to the family business. The patient considers pregnancy unreasonable so long as she has to work herself, but she does not want to oppose the whole family. In fact, she recognizes their desire as rational, but objects to a pregnancy at this particular time. The result may be psychogenic sterility. Later, in more favorable circumstances, the patient conceives without medical assistance.

Transitions from such *circumstantial neuroses* to true *neuroses* are common. In the latter, the conflict and its circumstances play only an initiating role in the problem. The real "failure" is in personality structure.

For example, a 22-year-old psychosexually infantile girl, in order to please her parents, marries her father's business friend, because her older sister had just married and she had thus lost her friend and companion. The marriage results in frigidity, vaginismus, and divorce. The wedding night is a more or less incidental event that exposes the conflict between the psychologically infantile girl and the demands of her environment. The actual cause is the psychosexual immaturity of the patient (dependence on her parents and lack of adaptability to heterosexuality).

Circumstantial conflicts such as selection of a partner, marriage, sexual fulfillment, pregnancy, and climacteric, as well as any stress situation can frequently cause *"circumstantial" neuroses,* manifested by specific gynecologial symptoms. The same circumstances, however, are occasionally only the exterior motive through which a latent conflict becomes manifest. The diagnosis of circumstantial neuroses is relatively easy. The conflict is usually conscious and thus easily accessible and understandable. The prognosis in such cases is good. The therapy frequently consists in explanatory discussions. These patients, however, frequently present the physician with fictitious conflicts during the psychotherapy in order to divert him from the actual problems.

Personality neuroses should, therefore, always be handled by an experienced psychotherapist.

Symptoms. The correlation between emotion and somatic reaction (expression of symptom) is more specific for the physiologic emotional reflex than for the psychosomatic illness. The cat ruffles its fur only in case of imminent danger (anger, rage). Shame evokes redness, not pallor. Sorrow is evident through tears, and fear through perspiration and pallor. The somatic equivalent of the emotion is apparently an informative signal.

The symptoms of psychosomatic diseases provide similar signals, but they are only rarely comprehensible at first sight. The reasons are that the means of somatic expression are limited and the complex psychological mechanisms can be understood fully only by a trained psychoanalyst. There are interesting correlations between neurotic symptoms and dreams.

To understand psychosomatic diseases, it is of utmost importance to realize that the determinants act subconsciously to create the symptoms. Consequently, the patient herself cannot recognize her symptom as an expression of her conflict or her emotion.

The symptom serves a purpose. At the beginning of the illness, the symptom is supposed to achieve something that the patient can not or could not have achieved rationally.

For example, the very young girl gains her mother's sympathy by complaining about her periods, or vulvar eczema prevents sexual intercourse toward which the young patient is ambivalent.

It is not always so easy to recognize the tendency (the initiator of the illness) as in these examples. First, the motivation may be related to a currently unknown conflict. Second, the symptom can persist as a conditioned reflex, although the initiating conflict is no longer present (so-called residual symptom). For example, vaginismus that results from rape may persist even though the patient is happily married for one year and desires to have intercourse with her husband.

The example of vaginismus demonstrates the multiple causes of the symptom. It could mean (1) rejection of the unloved first sexual partner; (2) fear of pregnancy; or (3) complete rejection of sexuality for various reasons (for example, psychosexual infantilism or inadequate sex education). Only the patient herself can provide the true explanation of the symptom and thus permit the proper psychiatric investigation and treatment.

In a few cases the symptom is just as comprehensible as a physiological emotional reflex. For example, neurotic blushing is an expression of a pathologically intensified embarrassment, and frigidity implies that the man or men in general "leave the patient cold." This knowledge, of course, does nothing to explain why this indifference toward sex arose in the first place. Furthermore, amenorrhea is such a well-known sign of pregnancy that a secondary psychogenic amenorrhea can easily be interpreted as reflecting fear of pregnancy.

The condition of an organ can determine the nature of the symptom. For example, under the hormonal influences of early pregnancy, a conflict can transform physiological emesis into hyperemesis. This somatic manifestation is so important that it may become practically the only neurotic symptom of pregnancy-related conflicts. In the climacteric patient, the endocrine changes may induce the well-known vasomotor complaints of hot flashes and flushes. In the neurotic patients these complaints are exaggerated.

In addition, symptoms of a disease may become neurotically fixed or revived. For example, a concussion may lead to permanent headaches and a dysmenorrhea caused by peritonitis may persist even years after the original event. The symptom may result from identification or may follow the pattern of a previously experienced disease. The diease that determines the symptom does not necessarily have to be experienced by the patient herself. In other words, a patient who suffers from severe cancerophobia may develop symptoms similar to those that she observed while caring for her mother who had cancer. Or, a young girl may develop primary dysmenorrhea only because she has learned from her older sister that periods are painful. Usually the prototype is a loved or respected person who is close to the patient.

The symptom may serve the purpose of self-punishment. In other words the organ involved in the "sinful act" is punished. A young woman who is the past had a criminal abortion might remain sterile during her subsequent marriage. This is an example of obscure psychogenesis of a gynecologic ailment that will usually be considered to be the result of postabortal tubal inflammatory disease with occlusion responsible for the sterility.

The symptom, finally, may have no ascertainable cause. During an attempted rape, the patient might receive an essentially harmless injury to the right leg. Thereafter she begins to limp, refrains from using her leg, and develops an atrophy. At the time of the first sexual intercourse, which occurred under great psychic tension (fear of being detected), the patient might incur a genital injury from coitus. The pain thus incurred and the bleeding become associated in such a way that each successive menstruation becomes painful.

This case demonstrates how difficult it can be to uncover some of these associations, particularly since the patient usually represses such experiences and does not remember them. If the associations are exposed, important therapeutic consequences can be derived. It is obvious that, in this case, one is not dealing with a neurosis, in the narrow sense of the word, that necessitates a psychoanalytical approach. Simple suggestive methods may be sufficient.

An example of multiple factors in the cause of a symptom follows. A primigravida, delivered by cesarean section, develops premenstrual fainting spells that become more severe and more frequent with increasing durations of her very irregular periods. The pathogenesis is as follows: The patient is a newlywed who has financial problems because her selfish and domineering mother refuses to turn over the family business to her son-in-law. Primarily for this reason, but also because of the previous cesarean section, she fears another pregnancy. If her period is only slightly late, she begins to panic. Whenever she faints, her mother always cares for her and, fearing a life-threatening disease, takes her to see a doctor. First, the tendency for revenge is evident in that she puts her mother repeatedly into a state of anxiety and fear. Second, fainting is an expression of fear, which is familiar even to a layman. Third, the "fainting" can be interpreted here in two ways. The patient is "powerless" against her mother; she does not have the courage or the opportunity to change the situation. Furthermore, since her husband declines birth control, she is helplessly exposed to the possibility of another pregnancy. Not perceiving the connotations of her words, she herself states during a consultation: "I have no ground to stand on," as she faints.

This example demonstrates how important it is to interpret symptoms in order to understand and treat psychosomatic disease. If they remain unintelligible, the physician will fail in his therapy.

Diagnosis. If the physician understands the symptoms of a psychosomatic illness, he can derive important clues to diagnosis and treatment. For practical purposes, the following criteria and diagnostic possibilities should serve as guides.

(1) the patient's behavior

(2) the way in which the patient presents her complaints

(3) the time of onset of the symptoms

(4) the relation of the environment to the patient (history)

(5) suggestibility of the complaints

(6) the lack of organic findings to explain the symptom

(7) unsuccessful therapeutic attempts (frequent change of physician)

(8) symptoms and combinations of symptoms that *a priori* suggest psychosomatic disease

(9) change of symptoms during therapy

(10) psychological tests and interpretation of dreams

(1) *The patient's behavior.* In their confrontation with a physician, neurotic patients are frequently insecure, nervous, frightened, and reserved. Their answers are given hesitantly and reluctantly. Frequently, they give the impression that they have to struggle to speak about themselves and their problems. Many patients are so inhibited that they appear apathetic. This attitude is correlated with the ambivalence of these patients, who, on the one hand, desire to get rid of an annoying symptom, but on the other hand, fear to lose their protective mechanisms. Subconsciously, the patient fears that the psychological associa-

tions will be discovered. Other patients, by contrast, make a great display of their illness. They appear tired and resigned, speak in a low voice, and look sorrowful, or they overemphasize their symptom (painful expressions, limping, tremors). This response is to convince themselves and others that their suffering is organic.

The patient's attitude becomes especially peculiar if, by chance or intentionally, one touches upon the subject that initiates the illness. The patient's reaction, then, is either embarrassment, sudden rejection, animosity, or a more or less acute emotional outburst, such as crying.

(2) *The way in which the patient presents her complaints.* Neurotics have a peculiar way of describing their complaints. Frequently their story is exaggerated or especially compelling.

A patient may give a highly detailed account of the time of onset and duration of her pains. Especially informative in such cases is the answer to a question about what the patient does for her complaint (calling a physician or use of narcotics for insignificant menstrual pains).

The description of the symptoms is not based on organic signs but rather on the patient's conception of her illness. The suspicion that the patient's condition is psychosomatic arises when the symptoms do not have a common denominator (abdominal pains and fainting shortly before menstruation, lack of orgasm and hypomenorrhea, or defective gait and frigidity).

(3) *The time of onset of the symptom (onset of the disease).* Much useful information can sometimes be obtained by asking the patient about the exact onset of the illness. For instance the failure to menstruate may be related to a change in geographic location. The "change" is held to be the cause of amenorrhea. The actual cause, in fact, may be enrollment in a boarding school. A patient with dysmenorrhea and obstipation may give an exact date for the onset of symptoms. In a later discussion, it may be learned that her fiancé had left her around that time.

Frequently, while speaking with the physician, the patient may realize the correlation of such events and then respond emotionally (crying, indignation, resistance). The possibility must be considered, however, that the patient attempts to mislead the physician with a false explanation or another superficial conflict. She may simulate a physical condition to explain her inadequacy. It is very important to avoid suggesting associations, for the patient may willingly accept them to evade elucidation of the actual conflict.

(4) *The relaton of the environment to the patient (history).* The patient's history should provide useful clues to the nature of the affliction. The physician should ask himself what the patient is trying to achieve or express with her symptom. Naturally the patient's husband or relative will not be able to answer objectively questions about the authenticity

of the patient's disease, since the patient's symptom must affect her immediate environment if it is to be of any value to her. Therefore, one should ask about the measures taken by the patient (bed rest or absence from work) and about the attitude of others toward the patient (anxious relatives, exaggerated concern, and sibling preference).

(5) *Suggestibility of the complaints.* If during a thorough physical examination, which is imperative, one searches for other symptoms, such as might be explained by the gynecological findings ("Do you also have occasional headaches, stomach aches, pains in the gall bladder region?"), the patient may provide affirmative answers, although she initially had no such complaints. A pelvic examination that was previously reported as "very painful," however, becomes absolutely painless during a distracting conversation.

(6) *Lack of organic findings.* Naturally, a normal pelvic examination is suggestive of psychosomatic illness. In addition, it is important to understand that many pelvic findings, such as a retroverted uterus or even large myomata may seldom or never elicit gynecological symptoms. Therefore, true pathologic findings do not necessarily exclude neurosis.

(7) *Unsuccessful therapeutic attempts (frequent change of physician).* Should the history reveal that several thorough therapeutic attempts were unsuccessful, or that the patient had been "shopping around" for physicians, it may indicate that the patient is eluding effective therapy (for example, sterility). Temporary therapeutic success achieved with inadequate means, however, also suggests a psychogenic cause. Finally, even symptoms explained by disease can be influenced by suggestion.

(8) *Symptoms and combinations of symptoms that a priori suggest psychosomatic disease.* In obstetrics and gynecology one finds syndromes that are almost always psychogenic. They include frigidity, lack of orgasm, dyspareunia, and vaginismus (except in cases of an intact hymen, scarring, or inflammation), nymphomania, pelvic congestion (dissolution of the spasm under anesthesia), pseudocyesis, and hyperemesis gravidarum (at least in its early stage). Of primarily psychogenic origin and only seldom of organic are secondary amenorrhea without physical and endocrine findings; sterility without organic cause; dysmenorrhea (even in association with hypoplasia and retroversion of the uterus); and the combination of obstipation and dysmenorrhea. Pruritus vulvae and exaggerated climacteric complaints almost always reflect a psychogenic component. Finally, excessive complaints about insomnia, awakening at night with palpitations, reactive depressions, migraine, and phobias suggest a psychogenic cause or psychic stress in response to "organic" complaints.

Conditions that are only occasionally psychogenic include primary amenorrhea, vaginal discharge, juvenile and climacteric vaginal bleeding (anovulatory cycles), hypermenorrhea, backaches, and urinary urgency.

(9) *Change of symptom during therapy.* Occasionally, it is possible to eradicate a symptom such as dysmenorrhea or frigidity only to have another symptom appear (headache instead of dysmenorrhea, vaginismus instead of frigidity). The demonstration of an identical basis for both symptoms, for example, rejection of the partner, provides evidence for the psychogenesis of both symptoms.

(10) *Psychological tests and interpretation of dreams.* Finally, it is possible to make a presumptive diagnosis on the basis of relatively simple psychological tests and by interpretation of the patient's dreams (stereotypy of nightmares) by an expert in that field.

Not every psychogenic disease requires extensive psychotherapy. Therapeutic success can also be achieved by treating the illness organically. Psychotherapy will be required, however, as long as the initiating emotional conflict persists and whenever the patient suffers from a true neurosis.

Since the psychosomatic disease represents to the patient a temporary escape from her conflict, one should very carefully decide whether and when therapy is indicated. This decision requires thorough knowledge and experience in psychotherapy. When the gynecologist is in doubt, he should seek psychiatric consultation for his patient.

Selected Reading

Kroger, W. S.: Psychosomatic Obstetrics, Gynecology and Endocrinology. Springfield, Ill.: Thomas (1962).

CHAPTER 8

Gynecological History and Physical Examination

History

A good history is the key to any diagnosis. Compilation of symptoms in a correct sequence may provide the diagnosis even without examination. In other instances, the symptoms considered in relation to lesions found on pelvic examination will provide the diagnosis in cases where history or physical examinations alone had been inconclusive.

The cooperation of the patient is based on her confidence in the doctor.

Obtaining the History. The most frequent reasons for a visit to a gynecologist's office are the following: preventive annual check-ups, irregular uterine bleeding, dysmenorrhea, pain in the lower abdomen, vaginal discharge, contraceptive advice, diagnosis of pregnancy, infertility, and uterine prolapse.

The first question, "What brings you to the office?" provides a clue to the actual complaint and may serve as an "opener." It is advisable to allow the patient to present her problem without interruption. Her presentation is followed by specific questions designed to obtain additional information. To follow a plan of questioning is most important, although the details of sequence may vary. Detailed information on previous deliveries in a menopausal patient is less significant than in a patient coming in for a prenatal visit.

Menses. Information regarding the last two periods should be requested, preferably with exact dates, duration, and amount of flow, estimated as accurately as possible. The number of tampons or pads soaked with blood may provide some information. (A tampon completely soaked with blood absorbs approximately 30 ml.) The age at menarche (first menstrual bleeding) is important for many reasons. The menstrual period is calculated from the first day of bleeding to the first day of the next menses. The menstrual pattern is documented; irregularities in duration (polymenorrhea, oligomenorrhea, amenorrhea, and metrorrhagia) and in amount (hypomenorrhea, hypermenorrhea, and menorrhagia) are noted (page 143). Menstrual irregularities should be clarified also in regard to premenstrual or postmenstrual staining. Continuous bleeding can give diagnostic clues in many ways. It is helpful to indicate the type of bleeding in a scheme (page 142) that shows the irregularity at a glance. Therapeutic procedures can also be indicated in this scheme, for instance, a curettage or hormonal treatment.

If the patient is amenorrheic, it is important to distinguish between a primary and a secondary amenorrhea (page 146). The age at the last uterine bleeding (menopause) must be recorded. The patient is asked specifically about discharge, spotting, or any bleeding occurring 6 months or more after menopause.

Pain on menstruation requires a specific history in order to exclude certain causes. Did onset of pain and menarche coincide (primary dysmenorrhea) or did the menses become painful months or years after painless menstrual periods (secondary dysmenorrhea)? Is the pain present before the onset of menses? When is it maximal and what is the location? What drugs are taken? (The answer to this question may provide a clue to the severity of the pain; codeine prescriptions usually indicate severe pain, unless the patient is an addict). The way a patient describes her symptoms may direct attention to psychosomatic components, but the doctor should be careful not to make this diagnosis unless all organic causes have been ruled out.

Discharge. When did the the discharge begin? Amount, color, odor, and admixture of blood should be recorded. Relation to coitus, specifically on intromission, should be ascertained to rule out contact bleeding. Previous intake of antibiotics and oral contraceptives may provide clues about fungous infections.

Previous Pregnancies and Abortions. The numbers of term pregnancies, premature deliveries, abortions, and presently living children are given in a scheme of four digits, for instance, 2-1-1-3. The first digit indicates the number of term pregnancies, the second the number of premature infants, the third the number of abortions, and the last the number of infants living. The birthweight and sex of each offspring and whether the patient nursed are of interest in the younger patient. The patient should be asked whether the abortions were performed by a physician and by what means. Ectopic pregnancies are documented separately.

Previous Operations. All previous operations, especially those for gynecologic disease, should be recorded and, if known, the type of anesthesia used. When possible, it is desirable to obtain written hospital records of previous operations.

Previous Serious Illness. Previous serious physical and mental illnesses must be recorded, especially infectious disease (tuberculosis and venereal diseases), cardiovascular disease (stroke, cardiac decompensation, thromboembolic complications), urogenital diseases, diabetes, cancer, and dermatologic disorders. Information is requested also about cancer, cardiovascular disease, diabetes, and hereditary diseases in relatives.

If the history reveals indication of a particular disease or request for advice, a more detailed discussion is necessary. Several examples follow.

Contraception. The history should clarify whether a particular technique may be (1) acceptable to the patient and (2) usable. The selection of a given contraceptive depends on a variety of factors. Frequency of intercourse, religious acceptability of a method, personal preference or rejection of a certain method, agreement of the partner, and contraindications (for instance, previous thrombophlebitis, page 209).

Pain in the Lower Abdomen. The duration, severity, relation to the menstrual cycle, and associated discharge and menstrual bleeding are significant considerations, as are pain in the adjacent organs (bladder and rectum). In some patients it may be advisable to delay a detailed sexual history until the second visit.

Prolapse Complaints. A frequent complaint in women with uterine prolapse is stress urinary incontinence resulting from associated urethrocele. The type of incontinence has to be classified urgency, stress incontinence (page 252), or total incontinence.

Infertility. The care of patients with infertility problems requires patience and psychological skills. The history begins with questions regarding the primary or secondary nature of the infertility. The menstrual cycle has to be documented carefully. Previous attempts at treatment provide as important information as previous gynecologic disease (especially inflammatory disease and endometriosis). Information about the husband is obtained and a semen analysis is requested if not already available.

Consideration of the Social and Family Background. Social and psychological information can provide very important clues to diagnosis. They allow a better evaluation of the patient's personality, behavior, and adaptability to her immediate environment and to society in general. Subjective complaints without obvious organic cause may direct the attention of the examiner to psychosomatic problems. Psychological problems can, however, result from organic disease. Many patients are misinformed about the loss of the uterus and ovaries. The patient's feeling of being crippled, especially in regard to sexual function, may induce a vicious cycle of sexual inadequacy associated with feelings of inferiority. An inadequate explanation by the physician about the surgical procedure and its sequelae may initiate such psychological problems.

The discussion after pelvic examination. After pelvic examination the patient has the right to be informed in clear, nontechnical terms of the findings and how they may explain her complaints. Planned treatment and prognosis should then be discussed.

The discussion after the examination may remind the physician why the patient consulted him in the first place. Her complaints may be correlated with the pelvic findings. Not infrequently, however, the pelvic examination may reveal an unexpected lesion, for instance an ovarian tumor. This finding is now predominant in the physician's mind. If the patient requested medical advice because of vaginal discharge, the treatment of this complaint is still important to her. She may forget it temporarily because of the shocking news, but she will remember her discharge at a later time. If the doctor forgets the discharge, the patient may lose confidence in him.

The Pelvic Examination

The history is privileged information between doctor and patient. The pelvic examination, however, requires a third person, preferably a nurse. This is a very important medicolegal consideration to protect the gynecologist against any accusations by the patient of improper conduct.

The patient should empty her bladder before pelvic examination. If a clean-catch specimen is needed, the patient should be advised accordingly.

The patient is then placed on an examining table. The abdomen is inspected. Scars and hair distribution are noted. The abdomen is palpated with both hands in search of tumors and rebound tenderness. The kidneys are palpated and the liver is outlined by palpation and percussion. The groin is examined for hernias and enlarged lymph nodes. The search for umbilical, inguinal, and rectus hernias is best performed with the patient standing.

The examination of the breasts (page 113) and a brief examination of heart and lungs follow.

The stirrups of the table are extended and the patient is positioned for *pelvic examination.* The back rest of the table is raised and the patient's thighs are abducted and placed in the stirrups. The buttocks are placed at the end of the table. This positioning straightens the lumbar lordosis and relaxes the rectus muscles.

The physician sits on a stool between the legs of the patient. A sheet placed on the lower abdomen lessens the patient's feeling of exposure. The external genitalia, including perineum and anus, are carefully inspected. With the left index finger and thumb the labia minora and majora are spread apart and the introitus is inspected. Inflammatory lesions, scars, trauma, atrophic changes, and tumors are noted and recorded. The size and form of the clitoris are described in cases of intersexuality. If vaginal relaxation is present a bulging of the walls of the vagina is produced when the patient bears down.

Speculum Examination. The following instruments are required:

(1) Graves specula of several sizes and shapes

(2) several sponge sticks with cotton balls

(3) a good light source

(4) two cervical tenacula, a uterine sound, one endometrial curette, one thumb forceps, and one cervical punch biopsy forceps

(5) lubricating jelly

The speculum is chosen according to the size of the introitus and the length of the vagina. The specula are washed and sterilized after each usage.

The Graves speculum is an instrument in which the posterior blade is approximately 2 cm longer than the anterior blade (Fig. 53). The closed speculum is introduced at an oblique angle into the introitus with slight pressure. The speculum is rotated to the transverse position and slowly opened. The portio vaginalis should be easily visualized.

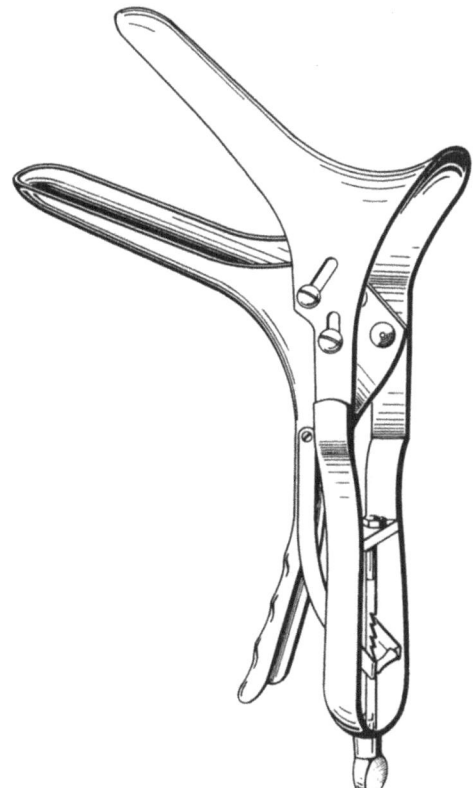

Fig. 53. Duck-bill speculum.

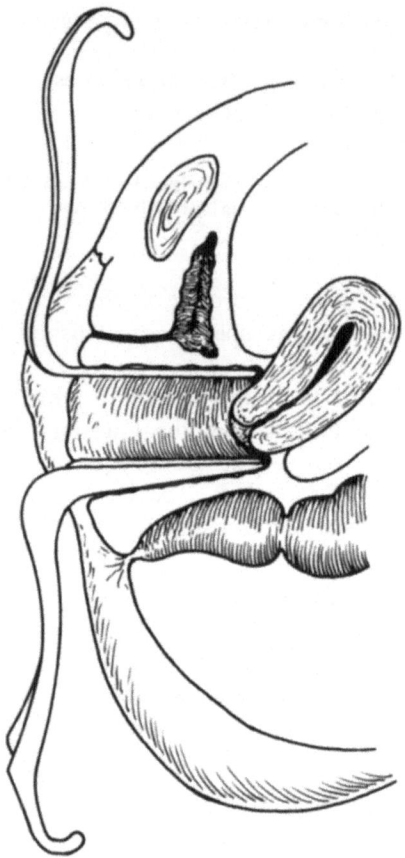

Fig. 54. Demonstration of the portio after introduction of anterior and posterior blades of speculum (Sims type).

Sometimes several attempts are necessary. The screws are adjusted appropriately to keep the speculum open. The gentle introduction of the speculum reduces, in almost all instances, trauma to the cervix. In patients who have prolapse it may be advisable to introduce a Sims speculum, using in addition a straight anterior blade (Fig. 54).

Cytologic Examination. Material for cytologic smear is routinely obtained at the first visit, regardless of the patient's age. The smear provides the following information:

(1) diagnosis of cancer and its precursors (page 279)
(2) hormonal diagnosis (page 72)
(3) microbial diagnosis (*Trichomonas, Candida*) (page 111).

Required are:

(1) a glass slide cleansed with alcohol
(2) a spray containing equal parts of 95% alcohol and ether for the fixation of the smear
(3) cotton-tipped wooden or plastic sticks.

After the speculum is inserted the entire surface of the portio vaginalis is touched with the cotton tip, which is then rolled onto the glass slide (Figs. 55, 56). The slide is immediately sprayed with alcohol/ether for fixation.

The second specimen is taken from the endocervix. The cotton tip is inserted into the cervical canal, up to the internal os, if possible,

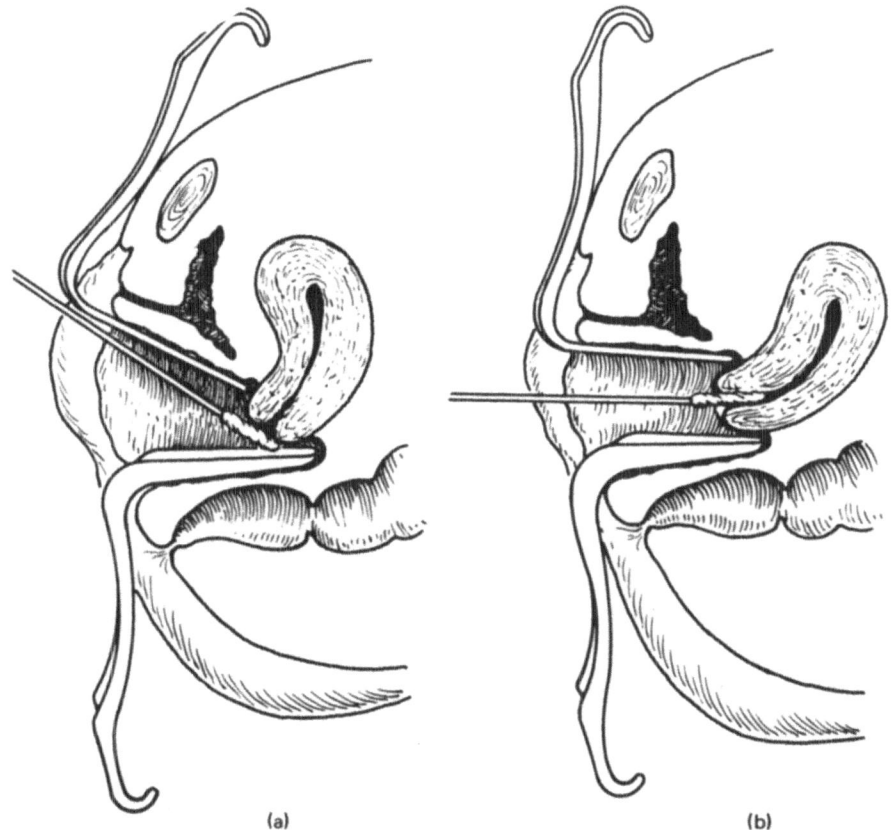

(a)   (b)

Fig. 55. Obtaining a cervical smear for diagnostic cytology: (a) smear of the portio vaginalis (ectocervix); (b) smear from the endocervix.

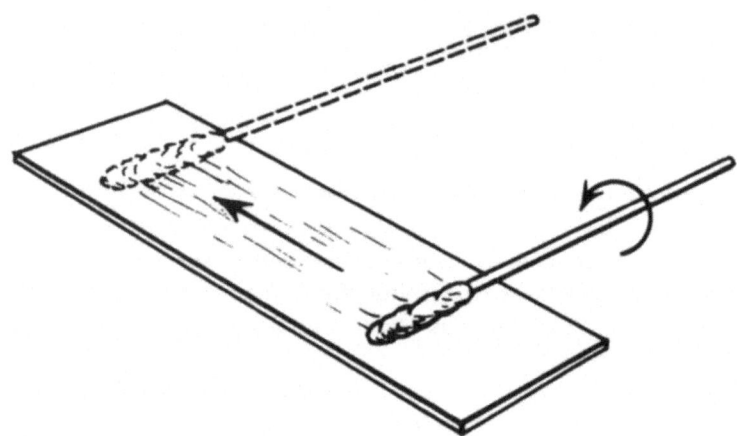

Fig. 56. Smearing the material on the slide with a cotton-tipped applicator in the direction of the arrow.

and rotated several times. After removal, the cotton tip is rolled onto a slide. Proper technique is of utmost importance. No cytologist can evaluate a poorly obtained smear. The information requested by the pathologist is recorded on a card, which comes with the slide and is submitted with the smear to the laboratory. The proper report requires special attention.

For hormonal diagnosis the smear is taken from the posterior or lateral wall of the vagina.

Colposcopy. The portio vaginalis is inspected with the colposcope (page 103).

Colposcopy is a technique by which the surface of the portio vaginalis is observed at a magnification of 12 times with a direct light source. The colposcope (Fig. 57) is fixed to the examining table and can be swung in front of the introitus. A speculum is inserted and the portio vaginalis is visualized. Colposcopy is a technique that requires training, but in experienced hands it can be performed in a few minutes.

Routine colposcopy is part of any gynecological examination in much of Europe. The technique was developed by Hinselmann in Hamburg, Germany, in 1924. After World War II, before cytologic techniques were adopted in Europe, colposcopy was developed, using new instruments, as a screening technique for malignant and premalignant lesions of the cervix. For a decade arguments raged as to which technique was preferable as a screening method. In the United States colposcopy has gained popularity increasingly over the last few years as an adjuvant to cytology. It is used to perform so-called directed biopsies (page 304) in patients with a suspicious or positive Papanicolaou smear.

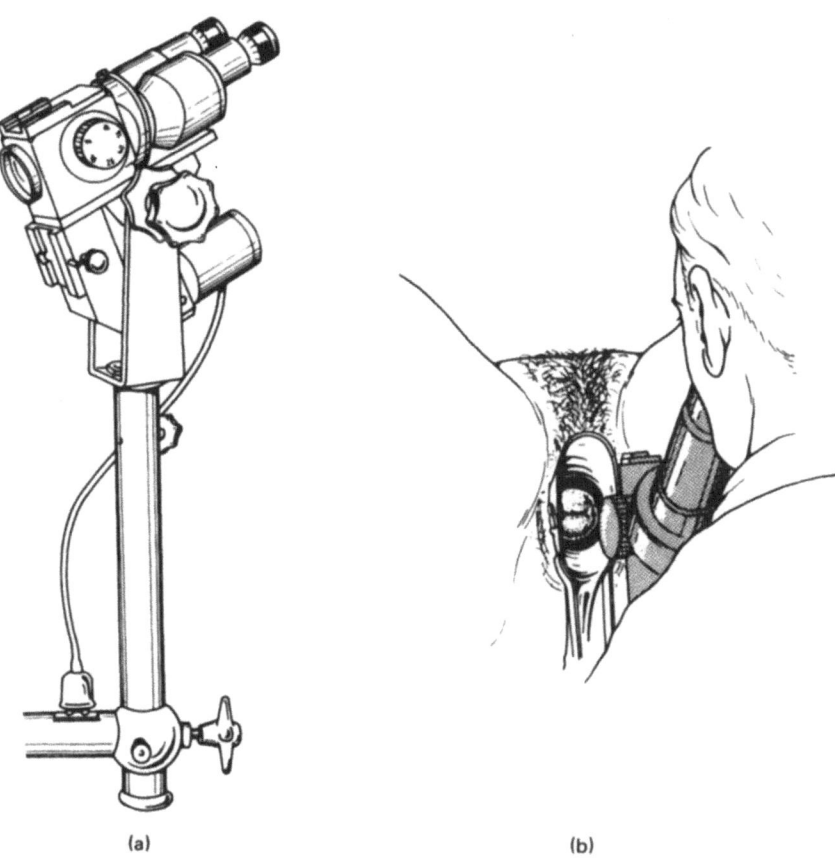

(a)   (b)

Fig. 57. Colposcopy: (a) the optical portion of the colposcope; (b) investigation of the portio with the colposcope after introduction of the self-retaining speculum.

The examination begins with cleansing of the cervix with a 3% acetic acid solution to remove cervical mucus.

Briefly, the following changes can be identified under low magnification.

Physiological variations:

(1) Squamous epithelium covers the portio, indicating that the squamocolumnar junction is located inside the cervical canal (Fig. 58).

(2) If the squamocolumnar junction is located outside of the cervix, the condition is called *ectopy* (a more appropriate term than "erosion"

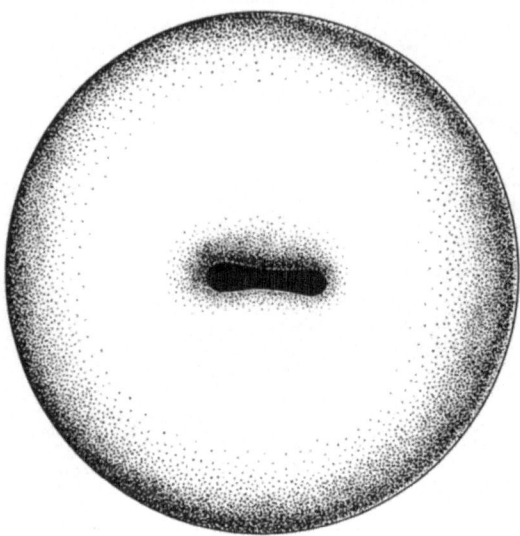

Fig. 58. Normal squamous epithelium of portio seen with the colposcope. The portio is covered by noncornified squamous epithelium; the entire surface is smooth.

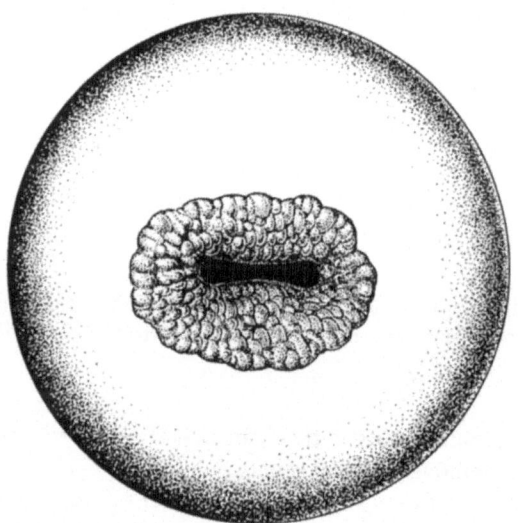

Fig. 59. Ectopy of the portio seen with the colposcope. There are delicate, regular grapelike elevations around the external os.

of the cervix). The columnar epithelium can be identified as a grapelike formation (Fig. 59).

(3) The so-called transformation zone is characterized by a thin layer of squamous epithelium covering the columnar epithelium as the result of metaplasia or direct extension of squamous epithelium overgrowing the glandular epithelium.

(4) Any lack of surface epithelium is called a true erosion. It is found colposcopically in approximately 1% of all cases examined.

Pathological changes:

(1) "Leukoplakic patches" are related to small areas of hyperkeratosis. The base of this lesion after removal of the hyperkeratosis looks like a whitish area with numerous small red dots, which are capillaries (punctuation). This lesion is frequently associated with dysplasias of all grades.

(2) A "mosaic" is a whitish lesion that is marked by thin vessels that divide the surface into small fields (Fig. 60).

(3) The irregular yellowish surface with a variety of atypical vascular patterns marks the early or grossly invasive lesion.

Fig. 60. Open and closed transformation zones seen with the colposcope. From 3 to 6 o'clock open transformation zone: the openings of the ducts of the ectopic cervical glands can be recognized. Between 6 and 10 o'clock closed transformation zone: small pearly white retention cysts shine through the squamous epithelium; the course of the capillaries is regular. Between 10 and 3 o'clock ectopic cervical glands.

Bimanual Pelvic Examination. The examination always poses problems for the beginner. It is difficult to feel the anatomical structures, but it may be even more difficult to translate the findings into three-dimensional terms. A systematic sequence of examination and recording is helpful. The findings can be described or indicated in a sketch.

Pelvic examination is performed by the physician while standing between the patient's legs. For the "vaginal" examining hand the right or left can be used, depending on personal preference. A disposable glove is worn. Since the abdominal hand has to exert more pressure, the right hand is preferred by most right-handed examiners and vice versa. Preferably two fingers are used. Only one finger is used in virgins and patients with stenosis or atrophy of the introitus.

The pelvic examination requires continuous, coordinated action of the two hands. The patient is continuously watched for signs of discomfort.

The labia are spread with the index finger and thumb of one hand. The thumb of the other (examining) hand is raised at an angle of 90° and the fourth and fifth fingers are flexed. Index and middle fingers are extended and lubricated.

The examination is begun by inserting the index and then the middle finger in the vagina and applying pressure to the perineum. It must be remembered that the most sensitive area of the vulva is around the clitoris, which should not be touched any more than absolutely necessary. Pressure exerted on the clitoris may cause the patient to become tense immediately and the pelvic examination may be ruined before it has begun.

The examiner may rest his forearm on his flexed thigh. The forearm thus creates a lengthened axis and increases the leverage, which relieves the pressure and provides relaxation for the examining fingers.

The external hand is positioned flat and relaxed on the abdomen between umbilicus and symphysis. This hand must bring the abdominal organs into the pelvis, where they are palpated by the internal fingers. The external hand can then palpate and estimate the size and consistency of the organs to be examined.

Rectal Examination

The index finger (covered by a lubricated glove) is inserted into the rectum. The patient is asked to bear down in order to overcome the resistance of the sphincter ani. The index finger is then rotated 180°.

Exerting pressure against the perineum allows deep introduction of the index finger. The external hand on the abdomen is of no value

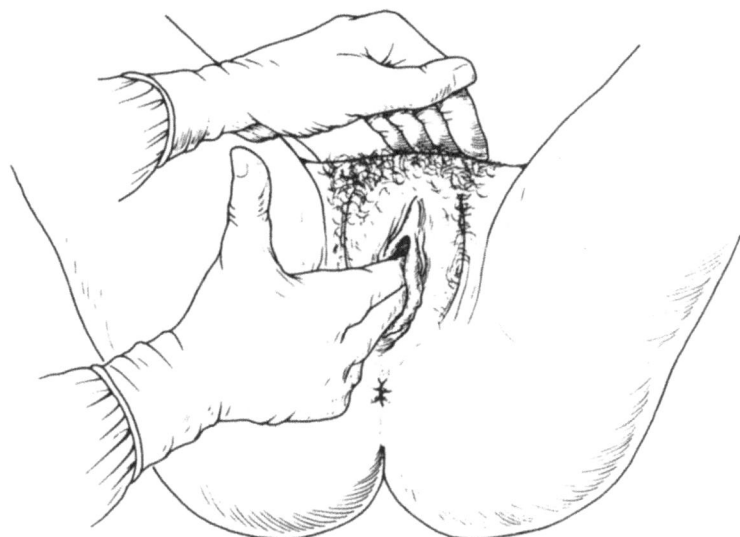

Fig. 61. Technique of vaginal examination. Index and middle finger of right hand are introduced, applying pressure against the perineum, while the left (upper) hand rests flat on the abdominal wall.

in the rectal examination. Hemorrhoids, polyps, and tumors of the rectum can be felt. Rectal examination provides also for easier palpation of the parametria and the retrouterine structures. The rectovaginal examination, in which the index finger is inserted into the vagina and the middle finger into the rectum, outlines the parametria, uterosacral ligaments, and the cul-de-sac of Douglas. *The rectal examination is part of the pelvic examination and should always be performed.*

Pelvic Examination: Details of the Examination. The pelvic examination begins with the *exploration of the vagina,* including palpation of the surface of vagina and portio vaginalis, the pelvic walls, and the posterior aspects of the symphysis. For this part of the examination the external hand is not required. Anatomical abnormalities are recorded, for instance, cysts, tumors, and vaginal septa. A cystocele and rectocele can be felt if the patient bears down, but this part of the examination is better reserved for the end because the patient may have difficulty relaxing after the uncomfortable procedure.

Palpation of the Uterus. Of interest are the size, form, consistency, and mobility of the organ. The vaginal fingers located below or behind the cervix lift the uterus against the abdominal wall. The external (abdominal) hand attempts to palpate the uterus from its posterior aspect (Fig. 61). The finger in the vagina then pushes the cervix posteriorly and therefore the corpus moves upward against the external hand. Palpation by this technique is possible only if the uterus is in an anteflexed

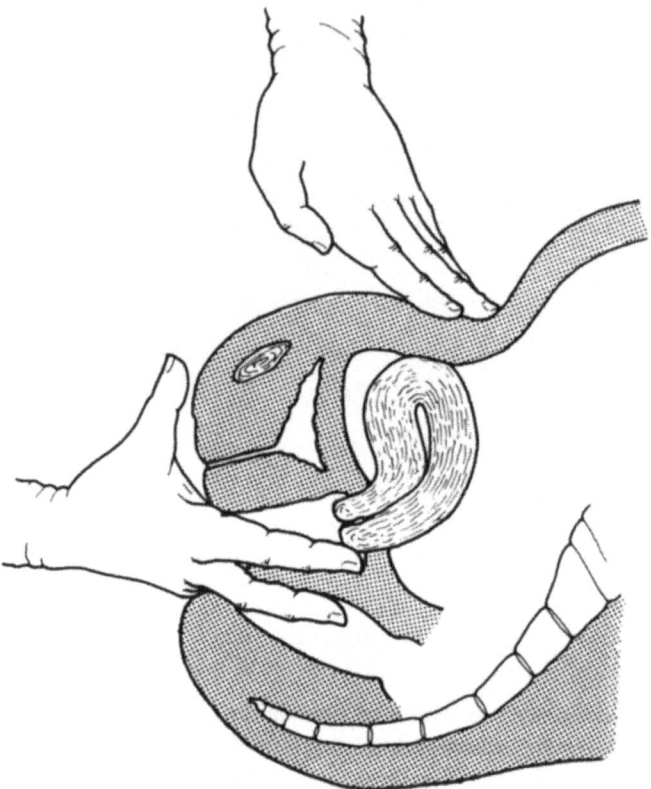

Fig. 62. Rectovaginal examination. The index finger is inserted into the vagina and the middle finger into the rectum.

and anteverted or middle position (page 245). Failure to palpate the uterus between the two hands may be caused by a retroflected uterus. This suspicion is substantiated by rectovaginal examination. (Fig. 62). One attempts to displace the corpus anteriorly by pushing with the two vaginal fingers against the posterior wall of the uterus. This maneuver often fails because pressure against the sacrouterine ligaments may be painful.

If the attempt to lift the uterus out of the pelvis causes pain or if the uterus fails to move, a fixed retroversion may be the reason. The attempt should be then terminated.

Palpation of the Adnexa. If the patient has complained about pain on one side, it is preferable to start with the examination of the other side. The fingers in the vagina are placed in the lateral fornix. They lift the adnexa toward the external hand, which in turn attempts to palpate them starting at the cornua of the fundus (Figs. 63, 64). The left adnexa are sometimes slightly elevated because of adhesions to the sigmoid. Mobile tumors may be missed if the external hand is placed

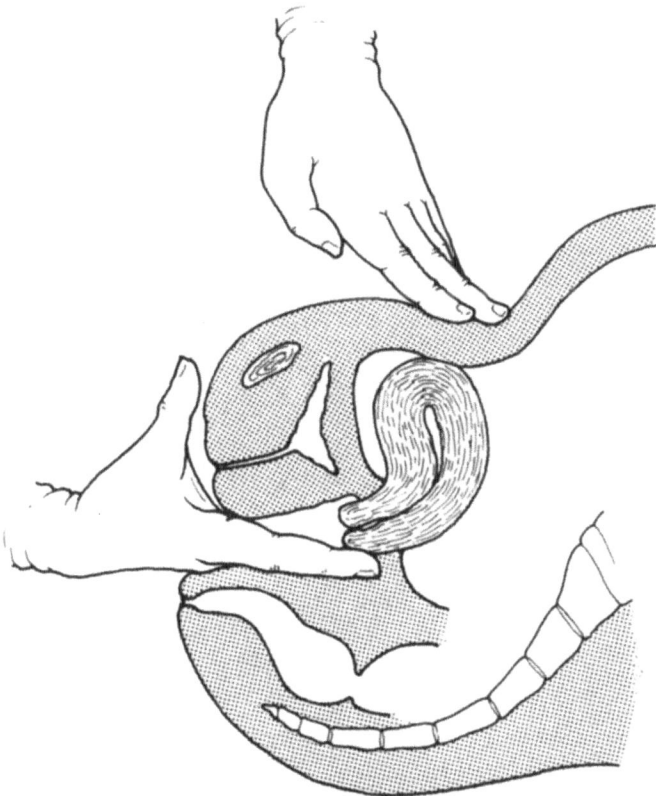

Fig. 63. Technique of vaginal examination. To palpate the anteverted uterus, one or two fingers of the internal hand lift the cervix from the posterior fornix and press the corpus against the anterior abdominal wall. The external hand attempts to grasp the posterior wall of the corpus.

too close to the symphysis. The palpation is difficult when the abdominal walls are obese. Thin abdominal walls permit easy palpation. Rough palpation of the ovary produces a characteristic pain similar to that caused by squeezing the testes (Figs. 65, 66).

Palpation of the Parametria and the Cul-de-sac of Douglas. These structures can be outlined properly only by rectovaginal examination.

The uterosacral ligaments can be felt in the pelvic congestion syndrome as pencil-thick strands (which disappear under anesthesia) and in patients with endometriosis (in which condition they persist under anesthesia). Spread of cervical cancer into the parametrium can be outlined only by rectovaginal examination. The cul-de-sac, the rectovaginal septum, and the rectal mucosa and are palpated in sequence.

Examination under Anesthesia. An anesthetic may be required if the patient fails to relax her rectus muscles or if palpation of the pelvic

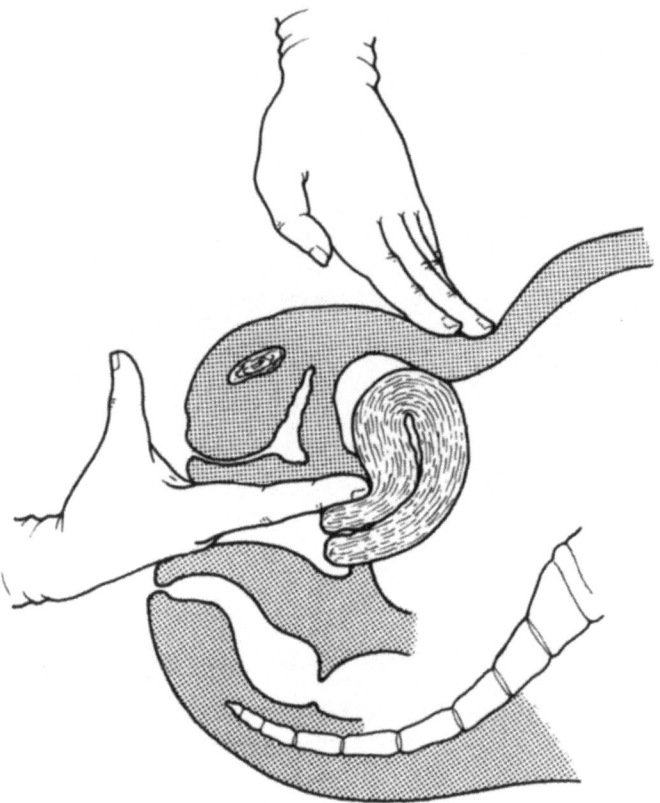

Fig. 64. Technique of vaginal examination. The examining finger or fingers are inserted into the anterior fornix; the external hand presses the fundus through the abdominal wall; thus the anteverted corpus can be defined between the fingers of the internal and external hands.

structures is inadequate because of obesity. The more skilled the examiner, the less frequent is the need for an examination under anesthesia. Anesthesia should be used liberally in small children. Examination under anesthesia is not without hazards. Since pain is abolished during anesthesia, the examination may be overly energetic. In such cases, cysts, pyosalpinges, and ectopic pregnancies may rupture. It is advisable to obtain written permission for laparatomy at the time of examination under anesthesia.

The toluidine blue test is performed when there is an abnormal lesion on the vulva or the portio vaginalis. The lesion is painted with a 1% toluidine blue solution. After 1 to 2 minutes the toluidine blue is removed with a 2% acetic acid solution. Remaining toluidine blue suggests an atypical or malignant area, which is then subjected to biopsy.

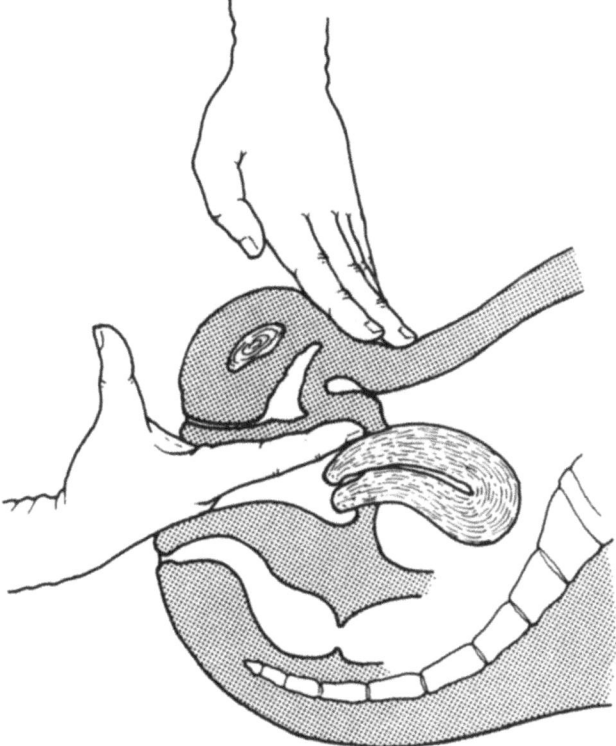

Fig. 65. Technique of vaginal examination. If the uterus is retroverted it cannot be felt between the inner and outer fingers.

The *Schiller test* is performed to outline an erosion. It is based on the glycogen content of normal squamous epithelium, which stains dark brown when painted with a 3% solution of potassium iodide. The unstained (iodine-negative or Schiller-positive) area indicates abnormal epithelium, benign or malignant, that does not contain glycogen. The Schiller test outlines the size of an "erosion" and is valuable in performing a conization.

Additional Examination. Smears and cultures are required in patients with discharge, pruritus, vulvitis, vaginitis, and cervicitis. Much information may be obtained through a *wet smear* obtained with a swab moistened with saline or saline with 1% KOH. Smears are obtained from the urethra, vagina, and rectum. The hanging drops can be examined under a microscope in the office for *Trichomonas*, fungi, and bacteria. A gram stain is required, in addition to a culture, in suspected cases of gonorrhea.

Special Considerations in the Examination of Children. Indications for gynecological examination of the child include discharge, endocrinop-

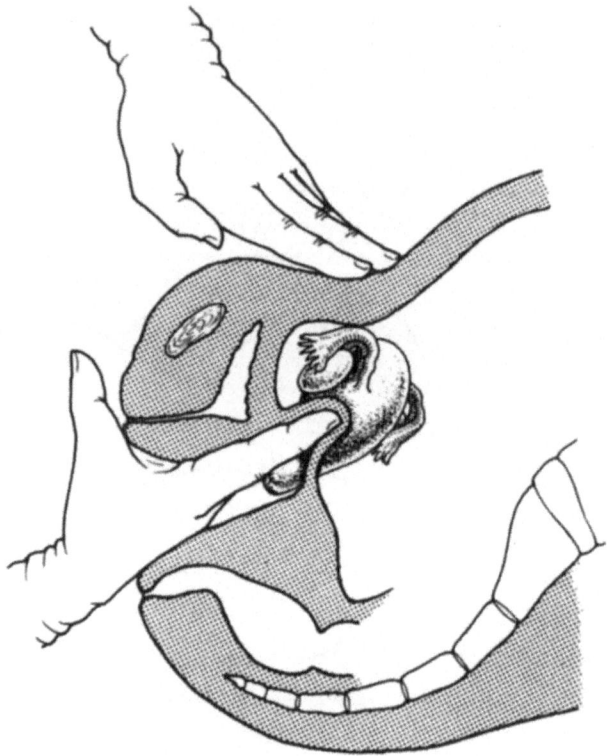

Fig. 66. Technique of vaginal examination. Palpation of the adnexa. The internal finger or fingers are inserted into the lateral fornix and lift the adnexa toward the external hand.

athies (precocious puberty and gonadal dysgenesis), suspicion of tumors, and intersexuality. The procedure depends on the age of the child.

The examination is started by observing the child and ascertaining whether behavior is consistent with age. Secondary sexual characteristics are carefully noted. The examination may be performed under anesthesia in small children to relieve fear and discomfort. The hymenal ring is not infrequently relaxed under anesthesia and it may allow easy insertion of instruments and even the small finger.

Special children's specula and a good light source are required. Specula with a direct light source such as those used in otolaryngology are very good for this purpose. If the hymenal ring is rigid, a firm rubber catheter may be used to evaluate the vagina and to obtain secretions. The smallest-caliber laparoscope provides a tool for examination of the vagina and the portio vaginalis. Special care should be taken in inspection of the vulva and the hymen (lacerations after rape and trauma, for example) and the external genitalia (clitoris, labia majora, and labia minora) in cases of suspected intersexuality. Suspicion of malformation requires special examination and karyotype analysis. It may be possible to insert a very small ear speculum or a cystoscope with a direct light source in order to remove foreign bodies. Secretions are carefully smeared and cultured. Pelvic examination may be limited to the rectal examination.

Examination of Breasts

Cancer of the breast is the most common malignant tumor in women. Examination of the breasts should therefore be a mandatory part of any gynecologic examination. Deviations from normal are indicated in a drawing in which each breast is divided into four quadrants.

Inspection. The patient sits relaxed in front of the observer, who notes the shape and differences in size of the breasts. Very careful consideration must be given to the following: retraction of one nipple, secretion from the nipple, eczematous changes, and cutaneous irregularities produced by tumors or retraction of the underlying tissues. Retraction is often better appreciated after the patient lifts and bends her arms.

Palpation. The mobility of the nipples is investigated by lifting them from the underlying tissues. The breast is then palpated with two hands; one hand supports the breast, which is palpated by the other hand. It is advisable to start laterally and proceed medially, palpating all four quadrants. The examination is performed with both of the patient's upper extremities relaxed and elevated.

Tumors and irregularities are noted, especially in regard to mobility of the tumor mass and the overlying skin and pectoralis muscle. The axillae are then palpated.

Examination of Secretions. If there is any secretion from the nipple, the material is smeared on a glass slide and stained by the Papanicolaou method. Material can be obtained for cytology by puncture and aspiration of fluid from a palpable cyst.

Mammography. Any mammary tumor requires soft-tissue radiologic examination. Early developing cancers can be identified by small densities. The area can be identified for later biopsy. In some centers mammography is now part of the routine annual check-up.

Milk-duct x-ray and thermography are recent techniques that are presently not routinely performed. In the former a milk duct is identified, contrast medium is injected, and radiologic examination is performed. The latter is a graphic representation of the infrared radiation of the skin.

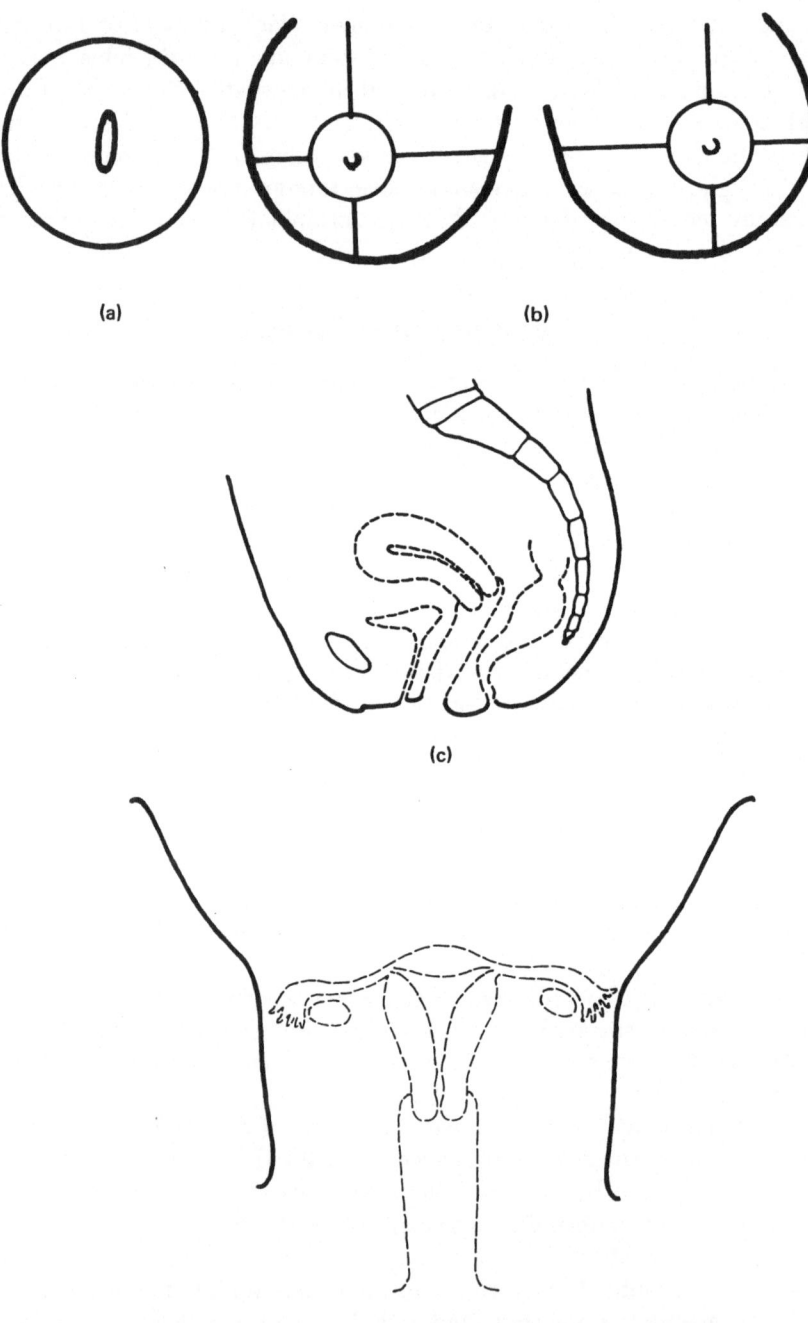

(a)

(b)

(c)

(d)

Fig. 67. History and physical examination form.

The history and physical form is shown in Fig. 67. Tables of special diagnostic examinations follow.

Table 9. Required Gynecological Examination

1. Inspection and palpation of the abdomen including the area of the kidneys
2. Inspection of the external genitalia and the introitus
3. Introduction of the speculum
   a. Inspection of the portio and vaginal walls
   b. Cytological smear (Papanicolaou smear)
   c. Colposcopy (with any cervical lesion)
4. Bimanual exam
   a. Vaginal
   b. Rectovaginal

Additional techniques

Methods	*Indication*	*Notes*
Hanging drop from vaginal fornix	Discharge Pruritus Vulvar eczema	Look for trichomonads and fungi

Stained Preparation

Secretions from:		
Cervical canal Vagina Urethra Ducts of Bartholin's glands Skene's glands Rectum	Discharge Pruritus Vulvar eczema additionally, when gonorrhea is suspected	Methylene blue or Gram-stain (trichomonads cannot be recognized) Gram-stain cultures

Menstrual bleeding

1. Culture	Suspicion of genital tuberculosis	Tassette cups (page 238)
2. Animal inoculation	(in addition to curettage)	

Further gynecological diagnostic methods:

History and pelvic examination will often lead to the diagnosis or at least narrow the spectrum of differential diagnosis. For further clarification, special diagnostic methods are presented in the following tables for rapid review.

Table 10. Special Investigation of the Urinary Tract

Method	Indication	Discussed in text on page shown
Cystoscopy	Cystitis (inflammatory, radiogenic)	262
	Preoperative, in cases of descensus and tumors	253
	Postoperative, in investigation of fistulae	261
Methylene blue or intravenous indigo carmine	Test of renal function	260
	Suspicion of ureteral occlusion or ureterovaginal fistulae	261
Intravenous pyelogram	Test of renal function (urinary obstruction)	260
	Preoperative, in cases of gynecological tumors (course of ureter)	260
	Suspicion of ureterovaginal fistula, ureteral stone	261
	Suspicion of maliormations of the urogenital system	259
Retrograde pyelogram	Diagnosis of ureteral lesions (suspicion of ureteral stenosis)	261
	Postoperative, in cases of disturbances of urinary output	261
Ureteral catheterization	Suspicion of ureteral stenosis	260
	Postoperative, with suspicion of ureteral injury	261

Table 11. Diagnosis of Carcinoma

Methods	Indication	Discussed in text on page shown
CERVICAL CARCINOMA		
Cytology	For cancer screening; repeated annually for detection and follow-up after treatment	299
Colposcopy	For any cervical lesion and for direction of punch biopsy	302
Schiller test	For aid in selecting sites of biopsy	302
Conization	For definitive diagnosis of invasion; occasionally therapeutic	303
Punch biopsy	For histologic diagnosis	304
Lymphography	For detection of lymph node metastases	
CORPUS CARCINOMA		
Fractional curettage	For histologic diagnosis and detection of extent of lesion	324
Lymphography	For detection of lymph node metastases	
OVARIAN CARCINOMA		
Exploratory laparotomy	For histologic diagnosis	345
Aspiration or lavage of cul-de-sac	For cytologic evidence of tumor cells	299
MAMMARY CARCINOMA		
Mammography	When carcinoma is suspected	113
Cytology	When abnormal secretion from breast is detected	113

Table 12. Operative Diagnostic Methods

Method	Indications	Discussed in text on page shown
Endometrial biopsy	Diagnosis of phase of cycle	141
Curettage	Diagnosis of phase of cycle	141
	Disturbances of bleeding	141
	Suspicion of endometrial tuberculosis	235
	Suspicion of corpus carcinoma (fractional)	322
	Suspicion of endometrial metastases from ovarian carcinoma	322
Culdocentesis	Suspicion of abscess of cul-de-sac	187
	Suspicion of extrauterine pregnancy	187
Posterior colpotomy	Suspicion of extrauterine pregnancy (findings not clear)	187
Laparoscopy	Unclear findings	187
	Suspicion of extrauterine pregnancy	351
	Infertility	181
	Suspected ovarian causes of endocrine disturbance	351

SPECIAL DIAGNOSTIC METHODS IN INFERTILITY

Basal body temperature	Diagnosis of phase of cycle	64
Spinnbarkeit	Detection of ovulation	63
Sims-Hühner test	Information about number and motility of sperm in cervical secretion	182
Kurzrock-Miller test	Cervical factor; early knowledge of male factor	186
Rubin test	Proof of patency of tubes	187
Hysterosalpingography	Proof of patency of tubes	187
	Detection of uterine anomalies	133
Laparoscopy	Inspection of the adnexa	229

CHAPTER 9

Developmental Anomalies of the Female Genitalia

Understanding the developmental anomalies of the female genitalia requires knowledge of the stages of normal development.

(1) Sex is determined by the complement of sex chromosomes provided by the maternal and paternal gametes.

(2) Differentiation of the gonads depends on the presence or absence of a Y chromosome. The presence of a Y chromosome initiates the development of a testis; the absence of a Y chromosome results in female development.

(3) Differentiation of the internal and external genitalia depends on the differentiation of the gonad and is controlled by hormones. Testes and their hormones lead to formation of a male genital tract and external genitalia. Absence of testes and their hormones leads to development along female lines.

"Errors" may occur at each stage of development. In the extreme case they lead to complete arrest of embryonic development. If viability is preserved, a broad spectrum of developmental anomalies of the reproductive organs may occur with varying effects on the phenotype.

Disorders of Chromosomal Sex Determination

Deviations from the normal sex chromosomal complement are accompanied by disturbances in the development of the reproductive organs. Aberrations of the sex chromosomes may occur during gametogenesis of the parental germ cells or, after fertilization, during one of the early postzygotic mitotic divisions.

Aberrations of Sex Chromosomes during Meiosis. Total or partial loss of a sex chromosome or addition of one or more sex chromosomes results in abnormal development of the gonads.

Changes in the normal number of sex chromosomes are explained by an error in division of the diploid chromosome set during the first meiotic division of the paternal or maternal gamete. These errors are caused by either a failure of homologous chromosomes to separate during the anaphase of the first maturation division (meiotic nondisjunc-

119

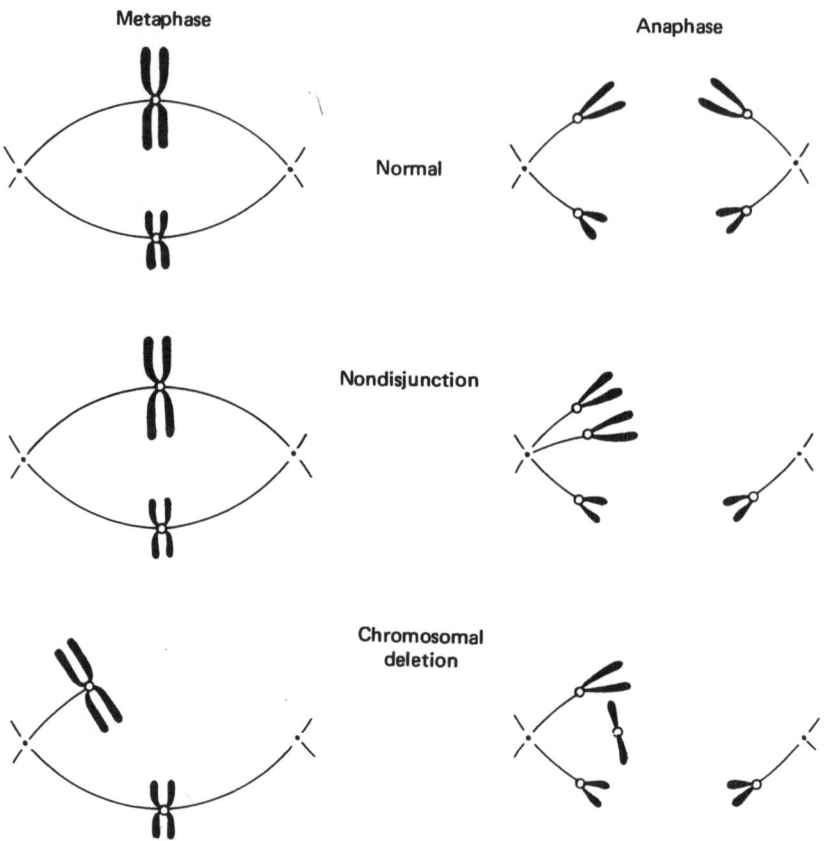

Fig. 68. Scheme of nondisjunction and of chromosome loss during anaphase.

tion) or loss of a chromosome during migration in anaphase (anaphase lag). These errors result in a germ cell with an extra chromosome or a gamete deficient in one chromosome. At fertilization the zygote then may contain three (trisomy) or only one (monosomy) of the affected chromosomes (Fig. 68).

Nondisjunction during meiosis of the oocyte results in either an extra X chromosome or no X chromosome. If an ovum with two X chromosomes is fertilized, a zygote results with either three X chromosomes (Triple X) or an XXY complement (Klinefelter's syndrome) (Figs. 69, 70).

An ovum containing no X chromosome results in an XO or YO complement, depending on the sex chromosome present in the spermatozoon at fertilization. The YO combination is considered lethal. Zygotes without an X chromosome are not viable because of the missing autosomal genes located on the X chromosome. Nondisjunction of the sex chromosomes

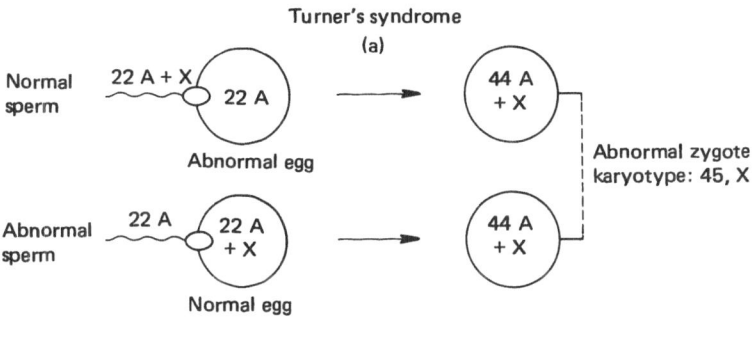

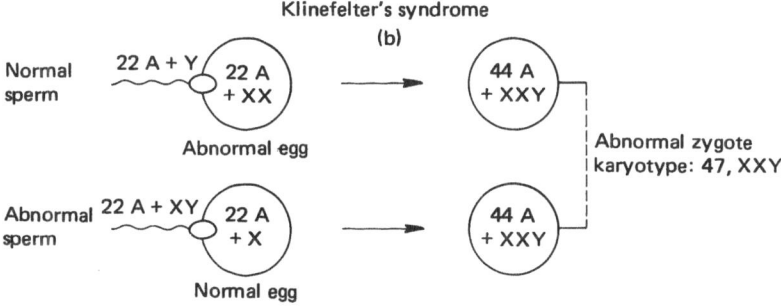

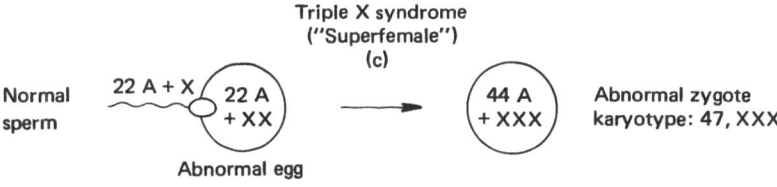

Fig. 69. Scheme of sex chromosomal anomalies of the zygote as a result of meiotic nondisjunction or sex chromosomal deletion in anaphase lagging during gametogenesis of paternal or maternal germ cells. (a) Pattern in Turner's syndrome; (b) pattern in Klinefelter's syndrome; (c) pattern in the Triple X complement.

during spermatogenesis results in one spermatozoon with an XY complement and the other with no sex chromosome. After fertilization of a normal ovum the sex chromosomal complement of the zygote will be either XXY (Klinefelter's syndrome) or XO (Turner's syndrome). Little is known about the causes of these maldistributions. The age of the mother appears to be a predisposing factor in Klinefelter's syndrome and the triple X complement. In some cases X-linked hereditary disorders such as red-green blindness, glucose-6-phosphate-dehydrogenase deficiency, and Xg blood factor indicate whether the error occurred during

Mature Ovum / Sperm	X	XX	O
X	XX normal female	XXX Triple X (Superfemale)	XO Turner's Syndrome
Y	XY normal male	XXY Klinefelter's Syndrome	YO (lethal)
XY	XXY Klinefelter's Syndrome	XXXY Variants of Klinefelter's Syndrome	XY normal male
O	XO Turner's Syndrome	XX normal female	OO (lethal)

Sex chromosomal complement of the zygote

Fig. 70. Sex chromosomal complement of the zygote as a result in separation during gametogenesis of one or both parental cells. Turner's syndrome and Klinefelter's syndrome can thus arise as a result of nondisjunction during oogenesis or spermiogenesis. The Triple X complement arises only through faulty separation during oogenesis (after A. K. Sohval, 1963).

paternal or maternal gametogenesis. X monosomy is compatible with further development only under certain conditions. It is usually associated with abortion rather than delivery of a viable infant with Turner's syndrome. The XXY complement is compatible with life but is associated with infertility because of the maldevelopment of the gonads. Women

Table 13. Sex Chromosomal Complement of the Zygote

Sperma-tozoon	Ovum					
	X		XX		O	
X	XX	normal female	XXX	triple X syndrome	XO	Turner's syndrome
Y	XY	normal male	XXY	Klinefelter's syndrome	YO	(nonviable)
XY	XXY	Klinefelter's syndrome	XXXY	male triple X	XY	normal male
O	XO	Turner's syndrome	XX	normal female	OO	(nonviable)

with an XXX complement may occasionally be fertile. Sex chromosomal abnormalities may be associated with various degrees of mental retardation. Sex chromosomal complements resulting from erroneous distribution during gametogenesis of one or both parental germ cells are associated with several clinical syndromes. Turner's and Klinefelter's syndromes may result from nondisjunction during either oogenesis or spermatogenesis. The triple X complement is always caused by erroneous distribution during oogenesis.

Rare forms of polysomy of sex chromosomes such as XXXX or XXXY are caused by errors of segregation of chromosomes during both maturation divisions of one parental germ cell or by union of paternal and maternal germ cells both of which show a sex chromosomal aberration.

Structural anomalies may result from loss of a piece of a chromosome (deletion) or translocation of a fragment. In either case there may be serious loss of genetic material. Exogenous factors such as irradiation and viral infections may be the cause of such errors in gametogenesis. Among the structural anomalies, deletions of the X chromosome and the Y chromsome are of particular interest.

An iso-X chromosome is normally associated with viability; developmental retardation depends on the loss of the short or long arms. In most instances the short arm is lost (Fig. 71).

Structural abnormalities of the sex chromosomes are infrequent, but they must be kept in mind when dealing with incomplete forms of gonadal dysgenesis.

Sex Chromosomal Aberrations Occurring after Fertilization. The same numerical and structural disorders described for meiotic divisions may occur during early mitotic divisions after fertilization.

At this stage of development numerical aberrations of sex chromosomes are caused also by mitotic nondisjunction or chromosome loss

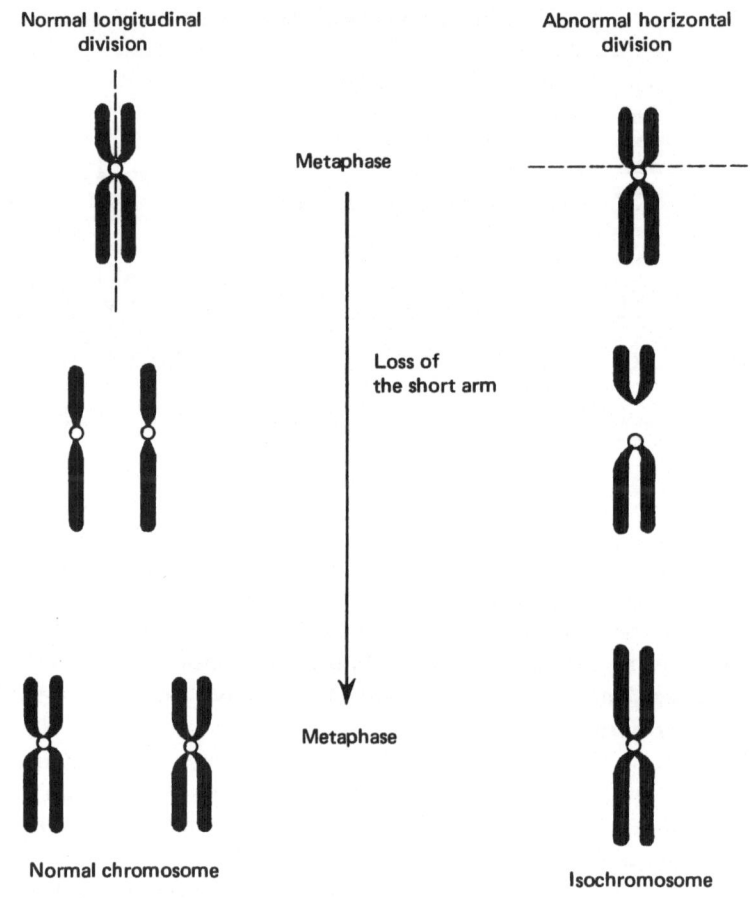

Fig. 71. Formation of an iso-X chromosome. This structural anomaly results if horizontal instead of normal longitudinal division of an X chromosome takes place. If the short arm is lost during anaphase of this division, there results an abnormally long X chromosome because of chromosomal reduplication of the remaining long arms.

in anaphase. Maldistribution may occur during the first mitotic division of the zygote or one of the following early postzygotic divisions. Mosaics may thus be formed.

Two cell populations result from a female-determined zygote with loss of one X chromosome in the anaphase of the first postzygotic division. One has a normal sex chromosomal complement and the other lacks one X chromosome, providing a mosaic 46, XX/45, XO. Identical mitotic maldistribution of the gonosomes may occur later in the postzygotic phase. The proportion of the abnormal cell line in tissues depends on the stage of early development at which the disorder occurred. Multiple mosaics result from repeated errors in division in the postzygotic stage. Mosaics with an XO component usually present signs of Turner's syndrome. Mosaicism of a male embryo with one Y chromosome results in resemblance to Klinefelter's syndrome.

Among the rare structural mosaic patterns with abnormal sex chromosomes, the iso-X is of clinical interest. Loss of the short arm during anaphase of this division results in an abnormally long X chromosome when replication occurs before the next cell division by duplication of the long arm. The X chromosomal complement of this cell line has a long-arm trisomy and a short-arm monosomy. An excess of genes on the long arms is, therefore, accompanied by a deficit in genetic material of the short arms. Turner's syndrome with growth retardation may result from this chromosomal anomaly.

Analysis of sex chromatin provides a clue to diagnosis of numerical and structural anomalies of X chromosomes, including mosaicism. Since each extra and each abnormal X chromosome is selectively inactivated, the anomaly may be suspected from the number and size of the Barr bodies.

Absence of Barr bodies associated with female phenotype suggests Turner's syndrome.

Positive sex chromatin (Barr body) associated with a male habitus suggests Klinefelter's syndrome.

A reduced proportion of cells with sex chromatin in phenotypic females suggests mosaicism with an XO cell line. Increased numbers of sex chromatin bodies indicate polysomy (for example, triple X pattern with two Barr bodies).

A large sex chromatin body suggests an iso-X chromosome with a long arm.

Unusually small sex chromatin bodies suggest a deletion in an X chromosome.

A varying number of sex chromatin bodies per nucleus with a male phenotype may be associated with a Klinefelter mosaic (Fig. 72).

To prove a mosaic pattern, chromosomal analysis should be performed on two different tissues, for example blood and skin cultures.

Disorders of Gonadal Development and Differentiation

The following are major causes of disorders of gonadal development and differentiation:

(1) abnormal complement of sex chromosomes
(2) genetic mutation
(3) exogenous factors (for instance, infection and irradiation).

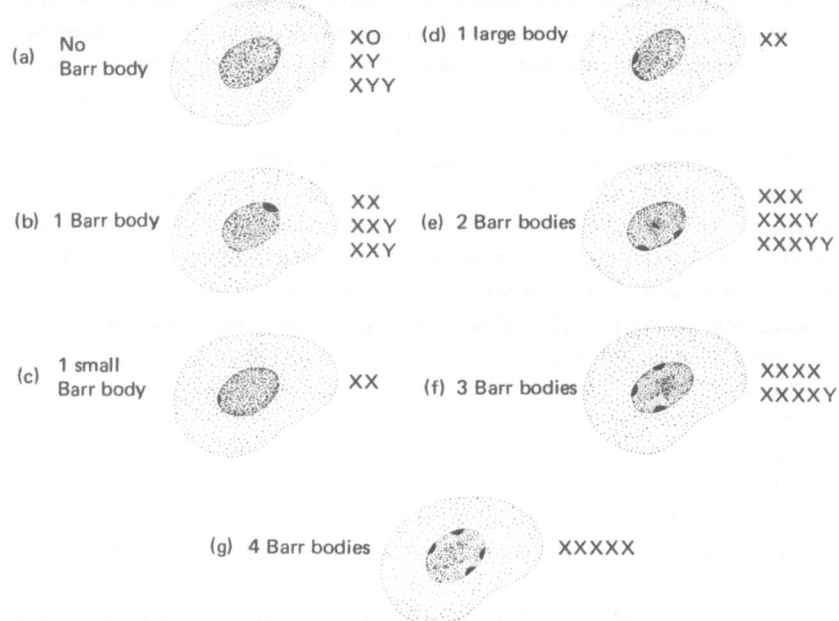

Fig. 72. Barr bodies in sex chromosomal anomalies. The number of Barr bodies is usually one less than the number of normal X chromosomes. Structural anomalies of the X chromosome can be recognized by abnormally large or small Barr bodies.

Gonadal Agenesis. Gonadal agenesis is a rare phenomenon in which the gonadal anlage is missing or damaged by a local endogenous or exogenous agent. The genital tract and external genitalia differentiate in the female direction. If the damage is so limited that it does not interfere with further development of the embryo, an infantile habitus becomes evident in adolescence or adulthood, as well as primary amenorrhea, severe genital hypoplasia, and lack of secondary sexual characteristics. The psychic orientation is female. The chromosomal complement in most cases is that of a normal male (46, XY) or female (46, XX). Sex chromatin pattern is consistent with chromosomal complement. Diagnosis is obtained by laparoscopy or laparotomy.

These patients must be informed about their infertility and the possibility of the need for permanent estrogen replacement.

Gonadal Dysgenesis. Gonadal dysgenesis is defined as an arrest in gonadal development before or immediately after gonadal differentiation. Neither ovarian nor testicular tissue is found.

If an arrest in gonadal development occurs after differentiation of the embryonal gonad into an ovary or testis, the syndrome may be called either ovarian or testicular dysgenesis. Appearance and structure of these ovaries or testes is highly variable, depending largely on the time of developmental arrest.

Turner's syndrome (45, XO) represents the classical form of gonadal dysgenesis. The abnormal sex chromosomal complement causes early arrest of gonadal development. Observations of early abortions reveal the beginning of ovarian differentiation, with germ cells and early development of follicular cords. Degeneration of the follicular cells is seen at the time of development of primordial follicles, however, leading to a disproportion between germ cells and follicular cells and a failure of formation of normal primordial follicles. The degenerating structures are gradually replaced by connective tissue. Finally, all that remains is a rudimentary organ on either side consisting mainly of connective tissue. These gonadal streaks are located in the posterior layer of the broad ligament.

Structures resembling primordial follicles may be seen occasionally in the gonadal streaks in Turner's syndrome, perhaps as a result of locally limited and occult mosaic patterns (cryptomosaic) with an XX cell line in addition to the predominant XO cell line. The characteristic features of Turner's syndrome (gonadal streaks and growth retardation) are also present in patients with an iso-X pattern of the long arms (with loss of the short arms of the X chromosome), and with deletion of the short arms of the X chromosome. It can be assumed, therefore, that the genes determining gonadal differentiation and normal growth are located on the short arms of the X chromosome. Failure of gonadal differentiation tends to result in development and differentiation of female secondary sexual characteristics. In Turner's syndrome and its variants, therefore, only female sexual characteristics are encountered. Since gonadal streaks are nonfunctioning organs, sexual characteristics remain underdeveloped because of lack of hormonal stimulation. The many additional phenotypic anomalies are related to the loss of autosomal genes on the X chromosome.

On the basis of examination of abortion material, it appears that the majority of embryos with the chromosomal complement characteristic of Turner's syndrome die at an early stage of development. It is estimated that only one child out of 40 with this anomaly is born alive.

"Pure" Gonadal Dysgenesis. "Pure" gonadal dysgenesis must be differentiated from gonadal dysgenesis with characteristics of Turner's syndrome.

The cause is unknown. The few subjects whose chromosomes have been examined showed a varying sex chromosomal complement, indicating that the abnormal sex determination is not the result of anomalies of the sex chromosomes but possibly of losses of genes that cannot be detected by available cytogenetic techniques.

It is also possible that unknown environmental factors may have a suppressive effect on gonadal development and differentiation. "Pure" gonadal dysgenesis should be considered a malformation resulting from a factor that interferes exclusively with gonadal development. Regardless of the primary cause, the development of the genital organs is always along female lines. Subjects with this anomaly have the gonadal streaks typical of gonadal dysgenesis but lack the other features of Turner's syndrome. These patients are phenotypic females of normal stature. The internal and external genitalia are infantile, and secondary sexual characteristics are undeveloped. The first sign of this syndrome is primary amenorrhea. Definitive diagnosis can be made only by exploratory laparotomy or laparoscopy.

Ovarian Dysgenesis and Testicular Dysgenesis

Ovarian dysgenesis and testicular dysgenesis are seen when development of the gonads is arrested after differentiation of specific female or male structures has begun. The extent of ovarian or testicular dysgenesis depends on the time and degree of interference with gonadal development. Ovarian dysgenesis is divided into afollicular and follicular forms. The afollicular type is characterized by the presence of germinal epithelium, tunica albuginea, cortical and medullary layers, and typical connective tissue of the ovary. Primordial follicles, however, are lacking. In the follicular form the gonad has an atrophic germinal epithelium and a thin tunica albuginea. Cortical and medullary structures are present. In the deep cortical layers, primordial follicles are found, but mature follicles are absent from the adult ovary. Development is arrested at a later stage than in the afollicular type, but early and severely enough to render the organ anatomically and functionally immature.

Intermediate forms may lead to ovarian hypoplasia. Microscopically these ovaries contain several small cystic follicles surrounded by granulosa cells and occasionally a luteinizing theca interna. Tunica albuginea and cortex are thick, fibrotic, and partially hyalinized. Corpora albicantia are lacking, indicating absence of ovulation.

In all these forms the sex chromosomal complement is that of a normal female (46, XX) and sex chromatin is present.

Ovarian dysgenesis is a cause of primary amenorrhea. The secondary sexual characteristics are only moderately developed but the phenotype and gender role are female. Biopsy of the ovaries is required for definitive diagnosis. The amenorrhea may be corrected, for the endometrium is responsive to hormonal stimulation. Infertility, however, cannot be reversed because it is impossible to induce ovulation. Estrogen substitution therapy is advisable.

Triple X Syndrome (previously called superfemale). This chromosomal anomaly (47, XXX) may be found with various degrees of ovarian dysgenesis and, on occasion, with normal ovarian function and fertility. In most instances, primary amenorrhea or late menarche, with secondary amenorrhea and infertility, is found. Patients with XXX patterns have an unremarkable phenotype but are usually mentally retarded. The additional X chromosome has little influence on the development of the phenotype, the gonads, or the internal and external genitalia. The prevalence in an unselected group of women is 0.15%. Among immates of institutions for the mentally retarded, the prevalence is 0.45% (three times the normal rate). X polysomies of higher degree show the same varying signs and symptoms.

Maternal age at the time of conception seems to be an etiologic factor insofar as nondisjunction during maternal gametogenesis may be increased.

Testicular Dysgenesis

Development of the testes may be arrested at any time after the initiation of differentiation, resulting in a variety of forms of testicular dysgenesis. Initiation of androgen production by the testes, at a specific time in development, is required for the normal formation of the male internal and external genitalia. If endocrine function of the testes is incompetent at an early stage of development, the embryo will develop along female lines and various forms of intersexuality may result.

Isolated stigmata of testicular dysgenesis develop if the testes cease to produce sufficient androgen at a later stage of development. The prototype of this form of testicular dysgenesis is Klinefelter's syndrome. In most instances it is caused by a chromosomal pattern of 47, XXY.

It is assumed that nondisjunction occurs more frequently in the aging ovary, possible because of overripeness of the ovum. In this connection, it is notable that Klinefelter's syndrome is related to advanced maternal age at conception. In addition, it is known that the extra X chromosome is of maternal origin. The differentiation of testes and secondary sexual characteristics at first proceeds normally. Later, however, regression and degeneration of the testes occur. The presence of an additional Y does not appear to reverse the process. For example, XXYY variants are similarly affected. The testes are small with atrophic hyalinized seminiferous tubules and Sertoli cells. The number of hypoplastic Leydig cells may be greater than normal. Signs of spermiogenesis are rarely found; azoospermia or severe oligospermia is the rule.

The syndrome does not usually become evident before puberty. Infantile genitalia (small penis and testes), eunuchoidal gigantism, and gynecomastia along with signs of feminization (scant growth of beard and high voice) are common features. Mental retardation is frequently associated.

The incidence is one in four hundred newborns. The syndrome is, therefore, the most common of the sex chromosomal anomalies. The diagnosis is based on the analysis of Barr bodies and karyotypes (Fig. 70). The gynecologist occasionally encounters these patients in the infertility clinic.

Testicular dysgenesis with intersexuality. In this case there is interference with formation of the gonads at the time of development and differentiation of internal and external genitalia. If the male gonad forms androgens but not factor X during development, wolffian and müllerian ducts may develop simultaneously. If the fetal testes produce factor X but no androgens, wolffian ducts do not develop and müllerian ducts are suppressed.

The final appearance and relation between male and female structures is determined by the degree of gonadal insufficiency and the time at which the gonads begin to fail. The relations between male and female rudimentary and well-developed structures explain the variety of forms of intersexuality and so-called pseudohermaphroditism.

The term "hermaphroditism," which stems from Greek mythology, originated in an era of purely descriptive morphology. More recently, the morphogenetic interrelations have been clarified. The various forms of intersexuality can now be defined and classified more precisely according to etiology, localization, and time of action of interfering factors. The term hermaphroditism is still used clinically to refer to the ambisexuality of the affected patients.

Testicular Dysgenesis with Partial Feminization (male pseudohermaphroditism). These patients present a male chromosomal pattern (sex chromatin negative and 46, XY). They have defective testes, which do not descend. The genital tracts display a variety of combinations of masculine and feminine structures. The external genitalia may be either predominantly male, ambivalent, or predominantly female. There is usually a problem at birth in classifying these newborns as boys or girls. Chromosomal analysis and sex chromatin studies are not sufficient by themselves. Since the psychosexual orientation is usually along female lines, it is preferable to register the child as a female in doubtful situations. A decision regarding further development along male or female lines should be made as early as possible. Operative procedures are then attempted to correct deformities of apparently male or female structures. Psychological guidance by family, pediatrician, gynecologist, urologist, and psychiatrist is most important to establish the proper gender role. The gynecologist is consulted by intersexuals at puberty when they have been raised as girls. They feel like females but the development of male stigmata makes them aware that they are different from normal girls (deepening of voice, male pattern of hair distribution, and hypertrophied clitoris). The corrective operative procedure should be tailored to psychic orientation and even a change of sex on the birth certificate may be required. Neither the chromosomal nor the gonadal sex is a determining factor in the choice of corrective procedures. Gender orientation (psychological sex) is most important.

Because dysgenetic gonads tend to undergo malignant change, they are often surgically removed. Depending on the preponderance of feminization or virilization, permanent substitution with estrogens or androgens, respectively, will be required.

Testicular Dysgenesis with Total Feminization (testicular feminization syndrome). This syndrome differs from the other forms of testicular dysgenesis because it is characterized by a complete discrepancy between chromosomal and gonadal sex, on the one hand, and phenotypic sex on the other. Chromosomal sex is, with the exception of a few mosaics, that of a normal male (46, XY; sex chromatin negative). Testes may be found in the inguinal canal or the labia majora, but the external genitalia appear female. The vagina forms a blind pouch, the uterus is missing or rudimentary, and the fallopian tubes are absent. This form of testicular feminization is explained by a genetic defect that prevents normal steroid synthesis or androgen insensitivity of the target organs. Since transmission occurs via the mother to the male offspring, it has been described as a sex-linked recessive or dominant hereditary disorder.

Intersexuality Induced by the Effect of Endogenous or Exogenous Androgens on the Female Fetus

This anomaly is characterized by masculinization of the external genitalia. It results from the influence of endogenous or exogenous androgens during differentiation of the external genital structures in a chromosomally female fetus.

The most common cause is fetal adrenocortical hyperplasia, which results in an overproduction of adrenal androgens in the fifth month of gestation. The newborn presents the stigmata of the adrenogenital syndrome (AGS). Occasionally, the overproduction of androgens is only temporary. In both cases there is a variable degree of hypertrophy of the clitoris. The anomaly is caused by a recessive enzymatic defect that interferes with normal synthesis of adrenocortical steroids. In rare cases the virilization of the female external genitalia is caused by administration of progestational agents with androgenic side effects. An exceptionally rare cause of virilization of the female fetus is a masculinizing tumor in the mother, such as an arrhenoblastoma.

True Hermaphroditism

These patients have testes as well as ovaries. An ovary may be present on one side and a testis on the other, or female and male gonadal tissue may be found together in one or both gonads. These are known as ovotestes. Depending on the predominant tissue in the gonad, the phenotype is male or female.

When a testis is present on one side and an ovary on the other, the development of the genital ducts usually follows the sex of the ipsilateral gonad. In cases with ovotestes, the level of androgen production determines the development of the genital tract along a male or female direction. In rare cases, maturation of germ cells may occur in these gonads. The appearance of the external genitalia varies. Frequently, a small phallus with hypospadias and a normal or rudimentary vagina are seen. The uterus is normally present and menstruation may occur. Secondary sexual characteristics develop at the proper time and gynecomastia may sometimes be seen. Testes as well as ovotestes may descend and reach the inguinal canal.

The cause of true hermaphroditism is not clearly understood. There is a suggestion that it represents an early gonadal defect probably caused by a genetic mutation that results in equal development of both male and female tendencies. The sex chromosomal complement varies. In the majority of cases there is a 46, XX pattern with positive sex chromatin. True hermaphroditism is rare, only about 200 having been described in the world's literature.

Selected Reading

Baramki, T. A.: *Sex Chromosomal Disorder*. In: *Obstetrics and Gynecology Annual*. Vol. I, R. M. Wynn, ed. New York: Appleton-Century-Crofts (1972).

Carr, D.: Chromosomal errors and development. *Amer. J. Obst. Gynec.*, **104**:327 (1969).

Jones, H. W.: *Chromosomal Considerations*. In: *Intra-uterine Development*. A. C. Barnes, ed. Philadelphia: Lea & Febiger (1968).

Sohval, A. S.: Chromosomes and sex chromatin in normal and anomalous sex development. *Physiol. Rev.* **43**:306 (1963).

CHAPTER 10

Malformations of the Genital Tract

Malformations of the Uterus

Anomalies of the uterus may be interpreted as arrested development. The various forms may be related to:

(1) agenesis or inadequate differentiation or regression of the uterine part of one or both müllerian ducts

(2) abnormalities in fusion of the müllerian ducts

(3) abnormalities in involution of the septum after fusion of the müllerian ducts

(4) retarded growth and development of the normally differentiated organ

Anomalies of various degrees result, depending on the time of arrest of normal development.

Aplasia of the Uterus. A cord of connective tissue replaces the uterus when differentiation in the uterine portions of both müllerian ducts fails. Absence of the uterus is usually associated with greater or lesser degrees of hypoplasia of the vagina.

Uterus Unicornis. This malformation results from an isolated aplasia of the uterine portion of one müllerian duct. One uterine horn remains deviated to one side or the other. If the tubal portion is absent also, only a cord of connective tissue remains. If differentiation of one of the müllerian ducts is incomplete, a rudimentary uterine horn results. The cavity may be lined by endometrium and in some cases a narrow tract communicating with the well-developed contralateral horn is found. In other instances the cavity of the incompletely developed horn does not communicate with the other side or with the vagina and may fill with blood (Figs. 73, 74).

Uterus Didelphys. Two completely separate uteri are formed when there is total lack of fusion of the müllerian ducts. This malformation is associated with a double cervix and a double vagina (Fig. 75).

In these uterine anomalies, the ovaries are morphologically and functionally normal, but there are often associated malformations of the kidney and the urinary tract.

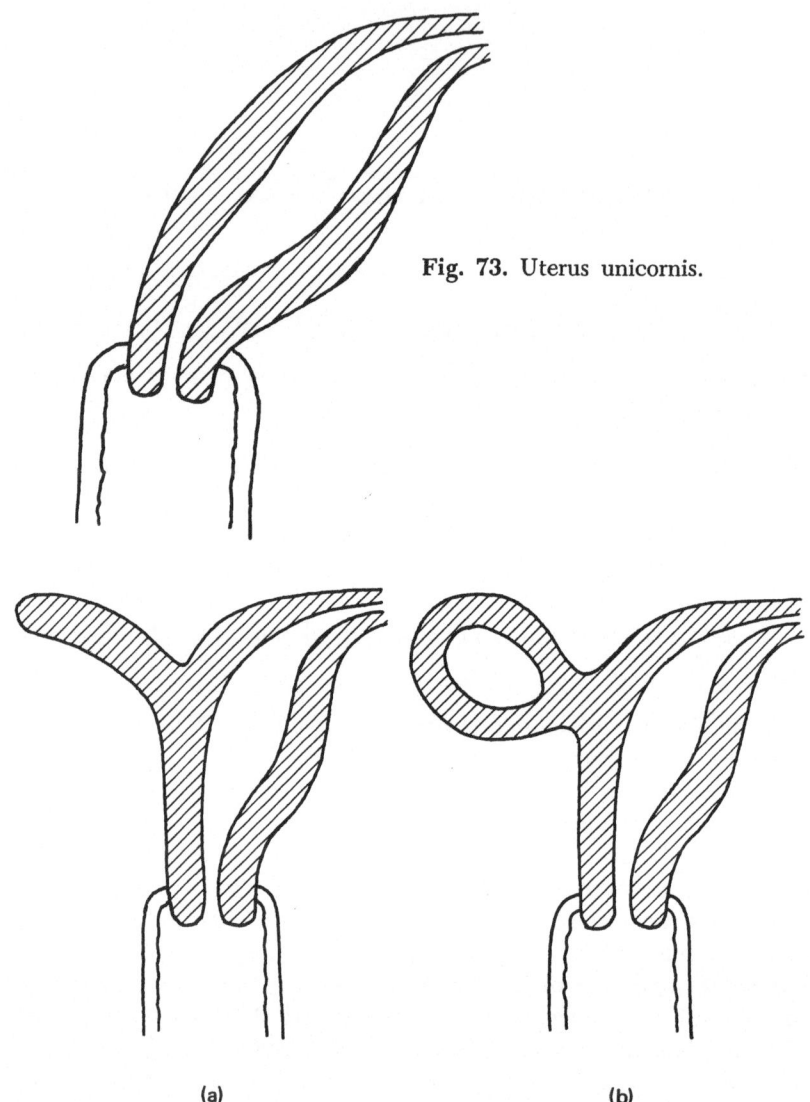

Fig. 73. Uterus unicornis.

(a)   (b)

Fig. 74. Uterus unicornis with rudimentary adjacent horn. (a) Atretic adjacent horn; (b) horn lined by mucosa with no possibility of drainage.

Bicornuate Uterus. This anomaly results when fusion of the müllerian ducts occurs only in their lower portions. The bicornuate uterus has two horns but only one cervix and one vagina. The tubes and ovaries are normally developed (Figs. 76, 77).

Septate Uterus. This anomaly results when fusion of the müllerian ducts is complete but the midline septum persists. The uterus has a

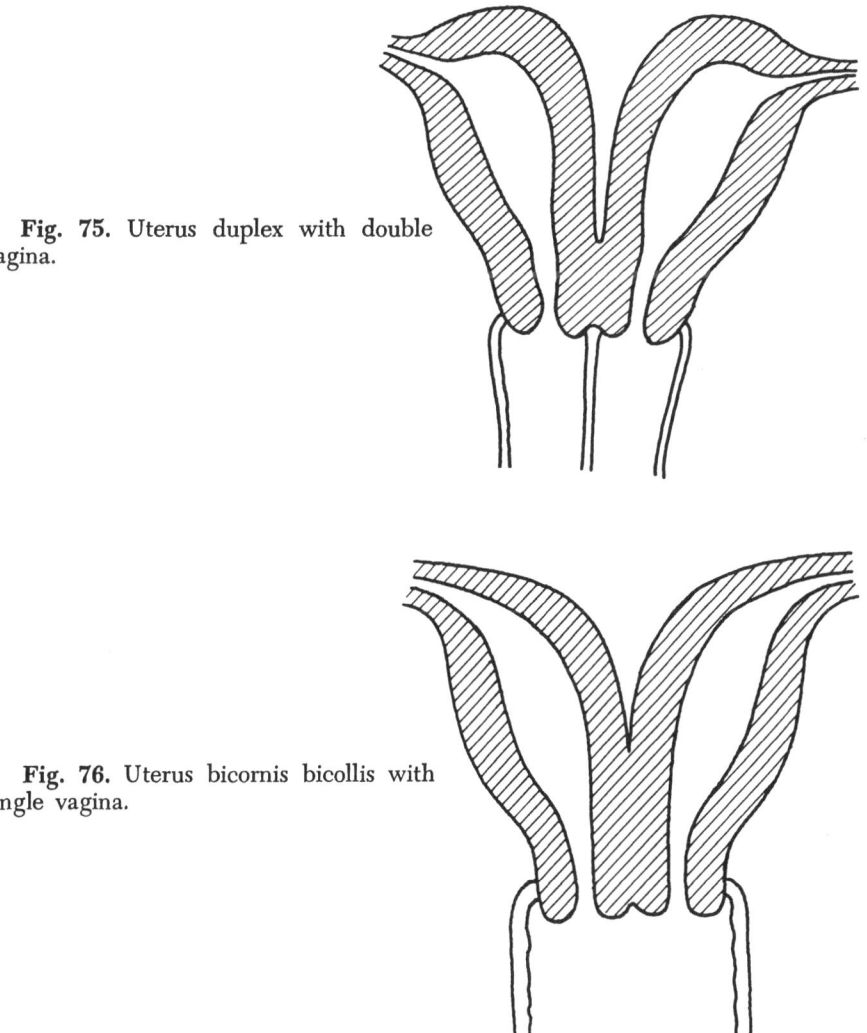

Fig. 75. Uterus duplex with double vagina.

Fig. 76. Uterus bicornis bicollis with single vagina.

normal configuration and size but the endometrial cavity and cervical canal are subdivided by the septum. The septum may also divide part or all of the vagina longitudinally (Fig. 78).

Subseptate Uterus. In this anomaly the septum is present only in the upper portion of the endometrial cavity (Fig. 79).

Rudimentary Uterus. This anomaly is a solid, small rudiment, about 1 cm in diameter. The vagina, tubes, and ovaries are normal.

Arcuate Uterus. In this condition a depression in the fundus is found but the uterus may be of normal size. It is a minor defect in an otherwise normally developed organ (Fig. 80).

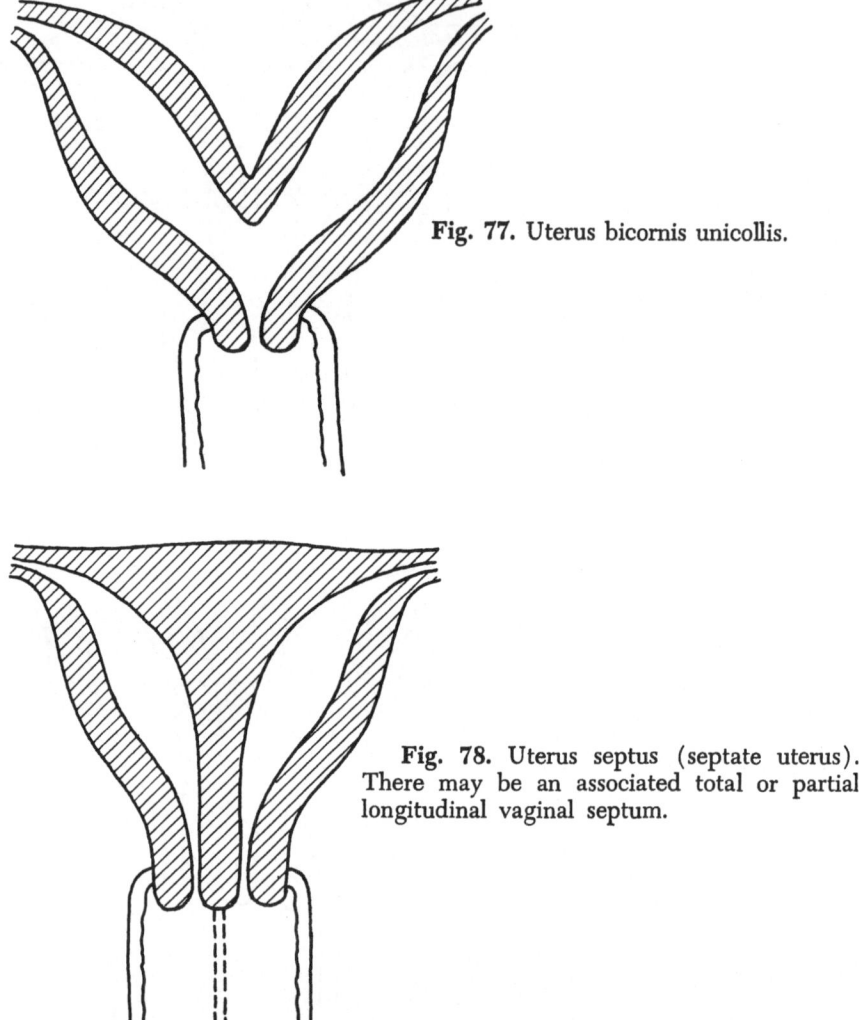

Fig. 77. Uterus bicornis unicollis.

Fig. 78. Uterus septus (septate uterus). There may be an associated total or partial longitudinal vaginal septum.

Infantile Uterus (hypoplastic uterus). The uterus is hypoplastic and of prepuberal proportions, retaining a ratio of 1:2 in size of corpus to that of the cervix. Insensitivity of the uterus to hormonal stimulation, or inadequate stimulation by the ovaries, may account for the prepuberal arrest of uterine development.

Isolated disturbances in differentiation of the fallopian tubes are rare. So-called hypoplastic tubes, which are seen on hysterosalpingography as long, thin structures, are of some clinical interest because they may be related to infertility. If conception occurs, there is an increased risk of tubal pregnancy because of a disturbance in transport of the ovum.

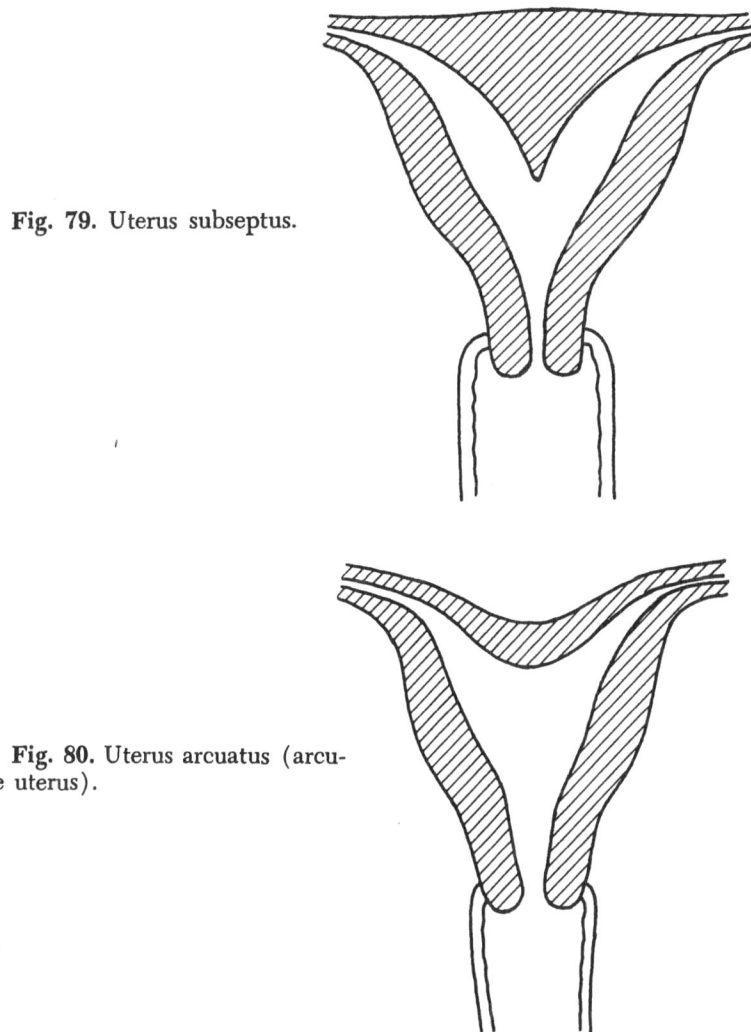

Fig. 79. Uterus subseptus.

Fig. 80. Uterus arcuatus (arcuate uterus).

Symptomatology. The symptoms are easily explained on the basis of the anatomic defects. Aplasia or severe hypoplasia of the uterus results in primary amenorrhea in an otherwise physically normal female. Malformations of the uterus such as uterus didelphys and bicornuate, unicornuate, or septate uteri are usually associated with normal menstrual function. A bicornuate uterus with a noncommunicating horn may cause serious problems at menarche. Primary dysmenorrhea is the first chief complaint. Since there is apparently normal menstrual flow from the patent horn, the pain is initially either ignored or misinterpreted. Pain becomes progressively severe with each menstrual period, however, and finally continues during the intermenstrual phase, because of the increasing hematometra of the obstructed horn and possibly ipsilateral hemato-salpinx. The blood-filled organs are palpable as pelvic masses adjacent to a normally formed uterus.

Most other uterine anomalies remain unnoticed until the reproductive age, when they are usually detected during work-up for infertility or habitual abortion. Term pregnancies, however, are not unusual. Transverse and oblique lies and breech presentations are more frequent in patients with uterine anomalies. The nonpregnant uterine horn may occasionally interfere with normal vaginal delivery and difficulties in dilatation of the cervix may be encountered (cervical dystocia). Postpartum hemorrhage is another complication.

Diagnosis. Suspicion of a malformation of the uterus should be raised by:

(1) the finding of a partial or complete vaginal septum

(2) the presence of a double cervix

(3) palpation on bimanual examination of an unusual shape of the uterus or two separate horns. The diagnosis may be confirmed by hysterography.

During pregnancy the diagnosis may be very difficult. The differential diagnosis includes myomas of the uterus and ovarian tumors.

Since almost 50% of genital malformations are associated with anomalies of the urinary tract, a urologic work-up is required.

Therapeutic Considerations. Corrective surgery is required in only a few of these conditions. It is indicated when the uterine anomaly is thought to cause repeated abortions. In septate uteri, the septum may be excised and the uterus reconstructed (metroplasty). Plastic procedures on the bicornuate uterus aim to unify the horns in a similar way. Surgical procedures are not indicated for a double uterus. An obstructed horn and its associated tubes should be removed immediately upon diagnosis.

Malformations of the Vagina

Malformations of the vagina are caused by arrests in development, either isolated or combined with anomalies of the uterus and frequently also with malformations of the urinary tract.

Aplasia, or congenital absence, of the vagina is caused by agenesis of the vaginal cord (page 16). The uterus, if present, is rudimentary. Ovaries are usually anatomically and functionally normal and secondary sexual characteristics are normal.

Atresia of the vagina results from failure of canalization or insufficient epithelialization of the initially solid vaginal cord. It is usually incomplete, involving only the upper part of the vagina.

Septate vagina. This anomaly results from persistence of the central part of the vaginal cord (page 15). It is frequently associated with a double uterus (page 138).

Symptomatology. Primary amenorrhea is the main sign of aplasia of the vagina. Occasionally the malformation is discovered when intercourse is first attempted.

Partial atresia may result in hematocolpos and hematometra.

A midline longitudinal septum may cause dyspareunia. Not infrequently it is first discovered by the obstetrician during labor. A lateral septum usually causes no difficulty.

Diagnosis. A band of tissue is found at the level of the hymen in congenital absence of the vagina. The atresia is usually found at a location corresponding to the upper part of the vagina, resembling the defect in testicular feminization.

A median vaginal septum is usually obvious on speculum examination. A lateral septum or a partial septum in the upper vagina, however, may easily be missed on inspection.

Therapeutic Considerations. Hematometra caused by vaginal atresia requires prompt surgical treatment. Reconstruction of an absent or atretic vagina is indicated if conservative measures such as dilatation are unsuccessful. Many plastic procedures have been employed for reconstruction of the vagina. The simplest procedure is the creation of a space between rectum and bladder. If the space is not kept open, there is a great tendency for restenosis. Better results are obtained by lining the newly formed space with skin grafts, using a mold of proper size. The mold is left in place for several weeks. Loops of bowel have also been used to create an artificial vagina, but they entail difficult procedures and are associated with a variety of complications.

Plastic operations are indicated when the anomaly creates psychological problems or when intercourse is desired. An artificial vagina may permit entirely satisfactory intercourse with orgasm. If conception should occur, delivery should be performed by cesarean section.

Atresia of the Hymen

The normal hymen is of variable thickness, elasticity, shape, and diameter. The only anomaly of significance is atresia caused by a failure of breakthrough of the caudal end of the vagina into the urogenital sinus. The original epithelium is replaced by connective tissue.

Symptomatology. Atresia of the hymen results in retention of menstrual blood and development of a hematocolpos and sometimes hematometra or even hematosalpinx.

The signs and symptoms are the same as those in atresia of the

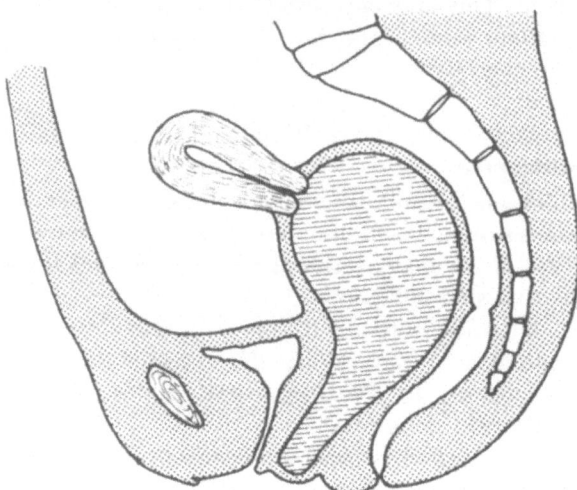

Fig. 81. Hematocolpos with atresia of the hymen. The uterus is found on top of the bulging mass as a solid resistance if there is no hematometra.

lower third of the vagina. The obstruction causes typical symptoms, which begin at menarche: lower abdominal crampy pain at monthly intervals (menstrual molimina) without external bleeding (Fig. 81).

The severity of pain depends on the quantity of blood retained and the stretching of the organs involved. Displacement of the bladder may cause urinary retention.

Diagnosis. Atresia of the hymen is usually discovered at menarche. The diagnosis is made by inspection of the vaginal introitus. On spreading the labia minora the introitus is seen to be obstructed by a bulging membrane, which appears bluish because of the retained blood. On rectal examination a cystic mass, occasionally of sufficient size to fill the true pelvis and reach to the umbilicus, is palpable. The uterus may sometimes be felt as a solid structure on top of the cystic mass.

Therapy. Therapy is essentially cruciate incision of the hymen. The dark viscous blood should be allowed to drain; thereafter, the uterus can be felt rectally. Hematometra usually empties spontaneously at the same time. Treatment should be as conservative as possible. The prognosis is generally good and fertility depends largely on the degree of damage to the fallopian tubes.

Selected Reading

Tuchmann-Duplessis, H., and Haegel, P.: *Organogénèse* 2nd. ed. Paris. Masson (1970).

CHAPTER 11

Disorders of Uterine Bleeding

Dysfunctional Bleeding

Dysfunctional bleeding is defined as irregular bleeding (compatible with altered ovarian function) from the endometrium. The hormonal dysfunction is related to disturbances of follicular maturation, ovulation, and corpus luteum formation. The basic defect in most instances is in the pituitary-hypothalamic axis. It is frequently observed just after menarche (page 67), in the postpartum period, and in the premenopausal years.

Dysfunctional bleeding may appear as a disturbance in duration or amount of flow. The various types of abnormal uterine bleeding are shown in Fig. 82. Uterine bleeding as a result of organic lesions such as endometrial or cervical polyps, adenomyosis, and endometrial cancer by definition are not causes of dysfunctional bleeding.

Polymenorrhea is shortening of the cyclic intervals to less than 25 days, if the duration and amount of flow are normal. The reason for the polymenorrhea is a shortening of either the corpus luteum phase or the proliferative phase. Another possibility is a short monophasic cycle with early regression of the follicle. Diagnostic curettage in the premenstrual period reveals an incomplete secretory effect in 60% of cases and a proliferative endometrium in 20%. In the remaining 20%, a secretory endometrium develops before day 12 of the cycle.

Diagnosis is provided by the basal body temperature chart, which indicates the time of ovulation and suggests which of the phases of the cycle are normal and which are shortened (Fig. 83). The physical properties of the cervical mucus, pregnanediol excretion, vaginal cytology, and endometrial biopsy provide additional information.

Treatment is required in patients who are infertile or who are anemic because of the extent of blood loss. A patient with a 21-day cycle has four or five menstrual periods more per year than does a woman with a 28-day cycle.

A shortened follicular phase can be extended by the administration of estrogens during the first 5 days of the cycle or by clomiphene. A corpus luteum insufficiency can be corrected by premenstrual administration of progesterone, or sometimes by the injection of LH.

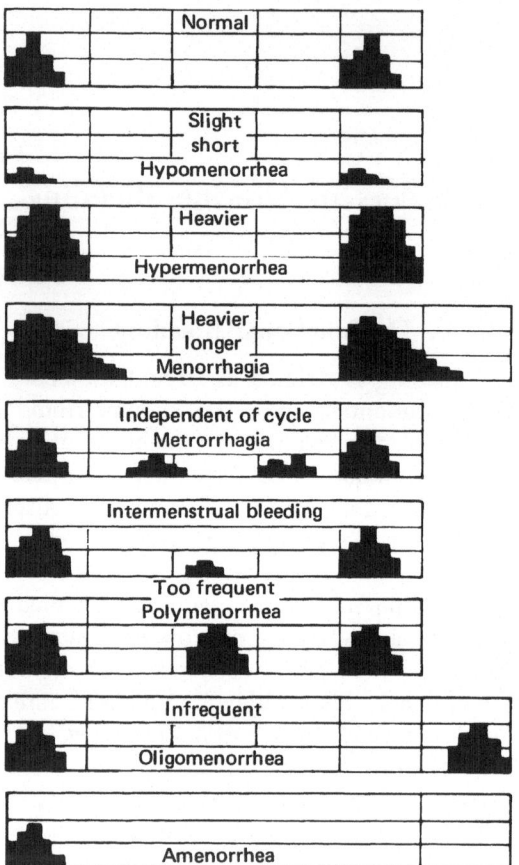

Fig. 82. Dysfunctional uterine bleeding.

Ovulation in a monophasic cycle can often be induced by clomiphene. Delay in the menses without treating the cause can be achieved with estrogen-gestagen combinations contained in oral contraceptives.

Oligomenorrhea. This is prolongation of the menstrual cycle beyond 38 days. This irregularity is an expression of ovarian insufficiency and may be a precursor of amenorrhea. The abnormality may be primary or secondary. Oligomenorrhea is diagnosed only after at least six normal menstrual periods have occurred. Secondary oligomenorrhea is a common sign of the Stein-Leventhal syndrome. In approximately 70% of other cases, the follicular phase is delayed. Maturation of the follicle may be inadequate. The follicles reach the size of only 5 mm and then undergo atresia. Ovulation is delayed but the corpus luteum phase is of normal length in 50 to 70% of cases. It is shortened in 10 to 20% and absent in 10 to 30%.

Treatment is required in patients with infertility problems. Ovulation can be induced at day 14 by the administration of clomiphene on days 3 to 7 or by the injection of LH and FSH (Pergonal and HCG). The cycle can be shortened also by an estrogen-gestagen combination; but this treatment does not correct the underlying cause and may occasionally cause progression to a pill-induced amenorrhea.

Menometrorrhagia. Prolonged menstrual bleeding is called *menorrhagia,* and excessively heavy bleeding for a normal length of time is termed *hypermenorrhea.* Uterine bleeding unrelated to the menstrual cycle is *metrorrhagia.* In many instances *metrorrhagia* extends into the normal menses to produce *menometrorrhagia.*

Prolonged dysfunctional bleeding in most instances is caused by a persistent follicle without ovulation. The prolonged production of estrogens results in simple hyperplasia of the endometrium or even cystic and glandular hyperplasia (page 320). After atresia of the follicle, estrogen withdrawal bleeding is the result, with delayed shedding of the endometrium.

Treatment is effected by so-called medical curettage. The hyperplastic endometrium is transformed by progestational agents into a secretory type of endometrium. Since the endogenous estrogen concentration is unknown, it is advisable to use an estrogen-progestin combination in large doses. Bleeding terminates within the next 24 to 48 hours and is independent of the transformation of the endometrium. The estrogen-progestin combination is given for 5 to 10 days depending on the dose. Two to 4 days after termination of the hormone administration, uterine bleeding occurs. This bleeding is similar to a menstrual period in that the endometrium has undergone some secretory changes. Failure to obtain hemostasis suggests another diagnosis and a proper curettage must be performed to rule out endometrial cancer, submucous myomas, and endometrial polyps.

To prevent recurrence after successful hemostasis and medical curettage, ovulation should be monitored by measuring the basal body temperature. If there is no ovulation by day 20, an estrogen-progestin combination is administered for 5 days, after which uterine bleeding will occur.

Ovulation Bleeding. Some women experience slight spotting for 1 to 2 days during ovulation because of a slight decrease in estrogen after the ovulation peak. This spotting can be prevented by small doses of estrogen between days 12 and 16.

Premenstrual Staining. Slight staining for several days before the onset of menstruation results from an early decrease of estrogen production in the corpus luteum, although progesterone is still produced. The normal withdrawal bleeding is delayed and the shedding of the endometrium is superficial. It is not yet clear whether the cause is primarily

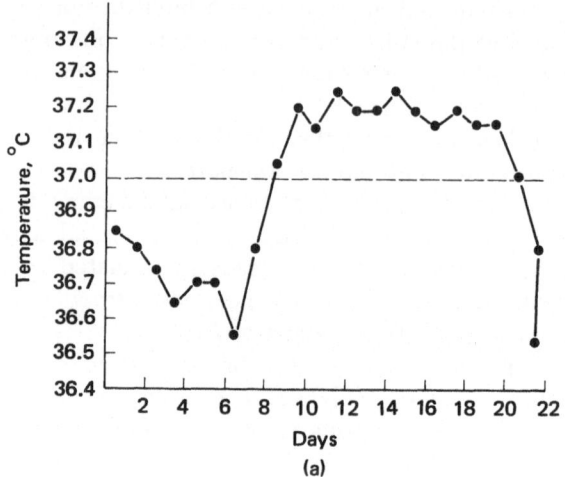

(a)

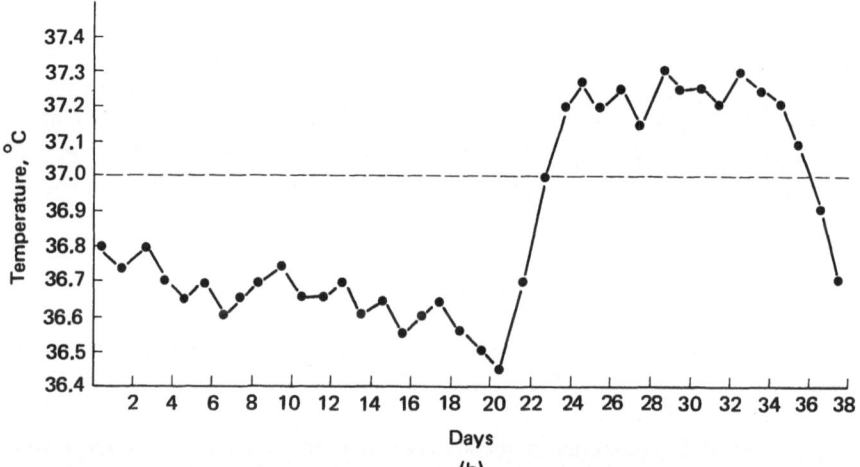

(b)

Fig. 83. (a) Basal body temperature curve in polymenorrhea with shortened follicular phase; (b) oligomenorrhea with prolonged follicular phase; (c) oligomenorrhea with anovulatory cycle (monophasic); (d) corpus luteum insufficiency (staircase phenomenon) with a short plateau phase.

ovarian or hypothalamic. Treatment is provided by administration of estrogen-progesterone from days 20 to 25 of the cycle.

Postmenstrual Staining. This problem is explained by an incomplete shedding of the endometrium resulting from an incomplete secretory transformation. Another possible mechanism is the delayed regeneration and proliferation of the desquamating endometrium as a result of previous curettage, endometritis, or delayed growth of a subsequent follicle.

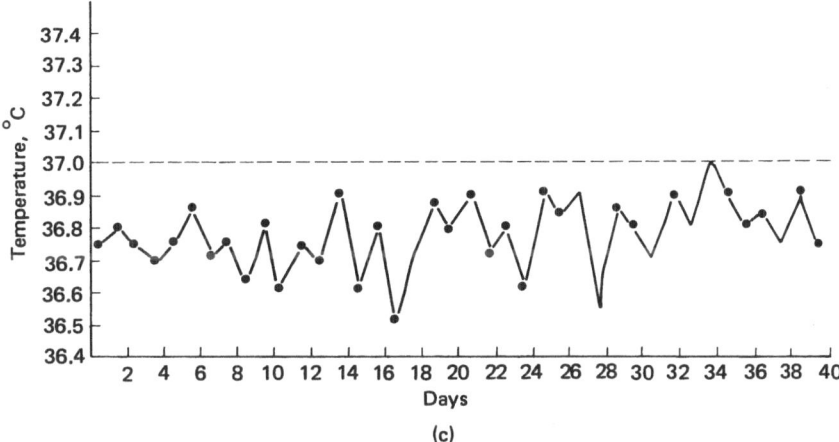

(c)

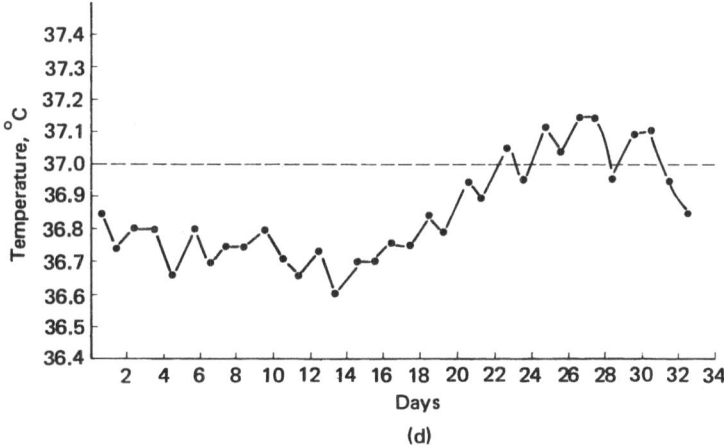

(d)

Treatment is administration of small doses of estrogens from days 1 to 7 of the new cycle or an estrogen-progestin combination from days 20 to 25 premenstrually.

Irregularities of Flow. An excessive menstrual flow is called *hypermenorrhea,* and blood flow below normal is called *hypomenorrhea.*

Hypermenorrhea, in most instances (80%), results from organic disorders and is usually not a result of endocrine imbalance. Its cause must therefore always be ascertained before treatment. Hypermenorrhea is associated with submucous and intramural myomas, adenomyosis, endometrial and cervical polyps, and bleeding disorders such as thrombocytopenia.

Hypomenorrhea is frequently associated with the administration of birth control pills. In general, hypomenorrhea does not require treatment if the menstrual cycle is regular.

Amenorrhea

The lack of any menstrual period for more than 3 months is called amenorrhea. It is normal in childhood, the menopause, pregnancy, and lactation. On the basis of history, amenorrhea is divided into primary and secondary forms. A diagnosis of primary amenorrhea is made if the girl has not experienced menstrual bleeding by age 16. The gonadotropin level provides a clue to the outcome: the higher the level, the worse the prognosis.

Etiology of Amenorrhea. Primary amenorrhea is usually related to chromosomal abnormalities and genital malformations, whereas secondary amenorrhea is most frequently a sign of pituitary-hypothalamic failure. A classification of amenorrhea is given in Table 14 (see page 147).

Central Nervous System. The psychogenic amenorrheas are the most frequent in this group. They originate in the cerebral cortex and via the hypothalamus interfere with releasing factors. Etiologic factors include separation from home and academic or professional stresses. The so-called war amenorrhea is well known, for many patients in concentration camps were amenorrheic. Amenorrhea is frequent also in houses of detention. Other reasons include disturbances of interpersonal relationships such as death of a close relative. The extreme form is anorexia nervosa, a neurasthenic disorder.

Another interesting type of a central nervous system amenorrhea is *pseudocyesis*, in which the patient, convinced that she is pregnant, exhibits many signs and symptoms of pregnancy including amenorrhea.

The hypothalamus is also involved in some forms of postpartum amenorrhea that are not infrequently associated with obesity. The Chiari-Frommel syndrome is characterized by postpartum amenorrhea, persistent galactorrhea, and genital atrophy.

Organic causes for hypothalamic amenorrhea include tumors (craniopharyngioma), which may be associated with other hypothalamic signs and symptoms such as increased appetite and obesity. Rarely, infections such as meningitis and meningoencephalitis may cause amenorrhea.

The olfactogenital syndrome includes arhinencephaly and hypoplasia of the hypothalamus, and is often associated with epilepsy and oligophrenia.

Pituitary Amenorrhea. The classic example is Sheehan's syndrome, or pituitary cachexia, which is the result of necrosis of the anterior lobe of the pituitary. Most likely, the necrosis results from fibrin deposits

Table 14. Classification of Amenorrhea

A. *PHYSIOLOGIC*
    Pregnancy
    Prepuberal state and delayed puberty*
    Menopausal state
B. *IATROGENIC*
    Surgical castration or hysterectomy
    Drug suppression
    Radiation
C. *OBSTRUCTIVE TYPE*
    Imperforate hymen*
    Cervical atresia, vaginal atresia*
    Cervical stenosis, congenital or acquired
D. *CENTRAL NERVOUS SYSTEM*
    1. Psychogenic (Hypothalamic)
        War (strain) amenorrhea
        Pseudocyesis
        Anorexia nervosa
        Psychoses, nonspecific
    2. Space-occupying congenital, inflammatory, or neoplastic lesions
        Craniopharyngioma
        Gliomas, teratomas, tuberculous and luetic granulomas, hemangiomas
    3. Familial disorders
        Laurence-Moon-Biedl Syndrome*
        Hand-Schuller-Christian Disease*
    4. Hypothalamus
        Chiari-Frommel Syndrome
        Adiposogenital dystrophy
E. *PITUITARY*
    Pituitary cachexia (Simmond's disease, Sheehan's syndrome)
    Pituitary adenomas (chromophobe, basophilic, acidophilic)
F. *OVARIAN*
    Ovarian dysgenesis*
    Stein-Leventhal syndrome (polycystic ovarian disease)
    Masculinizing tumors
    Testicular feminization
    Hilus cell tumors
G. *UTERINE*
    Acquired uterine atrophy
    Asherman's syndrome
    Uterine aplasia
H. *ADRENAL*
    Adrenogenital syndrome
    Acquired adrenal hyperplasia or tumor
    Cushing's syndrome
    Addison's disease
I. *CONSTITUTIONAL*
    Wasting disease (starvation, tuberculosis, diabetes)
    Obesity
J. *THYROID*
    Hyperthyroidism
    Hypothyroidism (usually menorrhagia)

---

* Primary amenorrhea only.

in the stalk. Sheehan's syndrome could be considered a form of disseminated intravascular coagulation. The original explanation that shock and hypovolemia result in necrosis is unlikely, for pituitary necrosis has not been observed in hemorrhagic shock in nonpregnant patients.

The complete clinical syndrome is found in approximately one in 10,000 deliveries and requires the destruction of 90% of the anterior lobe. Incomplete varieties, however, are much more common. There may be secondary endocrine insufficiency involving thyroid, adrenal, and ovaries. The most important early sign is agalactia. Postpartum amenorrhea, hypotonia, hypothermia, reduction of axillary and pubic hair, loss of pigmentation, decrease in libido, and personality changes are signs and symptoms of later stages.

The most frequent pituitary tumor causing amenorrhea is the chromophobe adenoma. It is hormonally inactive but it depresses the anterior lobe of the pituitary by its slow growth and thereby decreases gonadotropic function. Amenorrhea is an early sign frequently associated with galactorrhea not related to pregnancy. The syndrome then goes by the name of Forbes-Albright or Argonz-Del Castillo.

Amenorrhea may result also from an eosinophilic adenoma as well as an aneurysm of the internal carotid artery and brain tumors of various types.

The basophilic adenoma classically produces Cushing's disease. The adenomas are frequently small and are not easily shown radiologically. The disease involves women predominantly. Signs and symptoms include characteristic fat distribution, striae, hypertonus, hyperglycemia, osteoporosis, and amenorrhea. Excretion of cortisol and 17-hydroxycorticosteroids is increased.

Ovarian Hypoplasia. The ovarian tissue is decreased as a result of developmental failure. The phenotype is female; infantilism and primary or early secondary amenorrhea are additional signs. Breasts are well developed but the genitalia are hypoplastic. Gonadotropins are increased as a result of deficient feedback. Precocious menopause may be associated with ovarian atrophy. Secondary amenorrhea between the ages of 30 and 40 is associated with the typical symptoms of the menopause (page 159). Gonadotropin production is increased in these patients.

Special syndromes with ovarian amenorrhea include gonadal dysgenesis (page 128), superfemale syndrome (page 128), and Stein-Leventhal syndrome (page 161).

Hormone-producing ovarian tumors are discussed on page 336. Tumors that produce testosterone and androstenedione or dehydroepiandrosterone are clinically associated with virilism, including clitoral hypertrophy, acne, hirsutism, and a deep voice; 17-ketosteroids are either increased or normal. High testosterone levels increase after stimulation with gonadotropins.

Uterine Amenorrhea. The Asherman syndrome is discussed on page 228. Tuberculous endometritis also can cause amenorrhea.

So-called silent menstruation is rare. In this disorder the cycle is ovulatory and biphasic, but the patient does not menstruate. This rather interesting phenomenon seems to indicate a high threshold of the endometrial target organ in the presence of normal ovarian function. These patients can become pregnant.

Iatrogenic Amenorrhea. Amenorrhea can be produced iatrogenically by giving estrogen-progestin combinations. This is a desired effect in treating endometriosis, galactorrhea, or a hemorrhagic diathesis. Amenorrhea can also be produced by oral contraceptives as an undesirable side effect. Ovulation can be induced by clomiphene, restoring menses in most instances.

Diagnostic work-up for amenorrhea: Important information is obtained by a personal and family history. The general medical examination may reveal virilism, infantilism, and other constitutional stigmata. Careful examination of the breasts is mandatory (pigmentation of the areola and development of nipples and glandular tissue). Short stature, obesity, distribution of axillary and genital hair, and skin changes often lead to the diagnosis (myxedema, Turner's syndrome, or Albright's syndrome). On pelvic examination, the hymen, perineum, labia minora, and clitoris are carefully examined and malformations of the vagina are searched for. Bimanual examination provides information about the uterus and the ovaries. Vaginal cytology is a semiquantitative technique for measurement of estrogenic effects (page 29).

A variety of tests can be performed in the office to pinpoint the site of the disturbance (Fig. 84).

The progesterone test: The administration of progesterone (25 mg intramuscularly for 3 days) or an orally active progestin for 7 days results in withdrawal bleeding if the endometrium has been stimulated by estrogens. Failure of withdrawal bleeding suggests an estrogen insufficiency. A positive test excludes uterine amenorrhea.

Estrogen test: Failure of withdrawal bleeding after estrogen administration (3 × 1 tablet ethinyl estradiol: 0.06 mg per day for 10 days) indicates uterine amenorrhea.

The time needed to perform these tests can be used to complete hormone assays in plasma or urine. High gonadotropin values in association with clinical abnormalities require chromosome analysis and, eventually, laparoscopy.

If the progesterone test is positive, clomiphene is administered in increasing dosages, with monitoring of basal body temperature charts, to test the pituitary-hypothalamic response. If the patient ovulates, the clomiphene test can be used therapeutically to achieve a desired pregnancy (page 72). In addition, curettage is required to obtain information

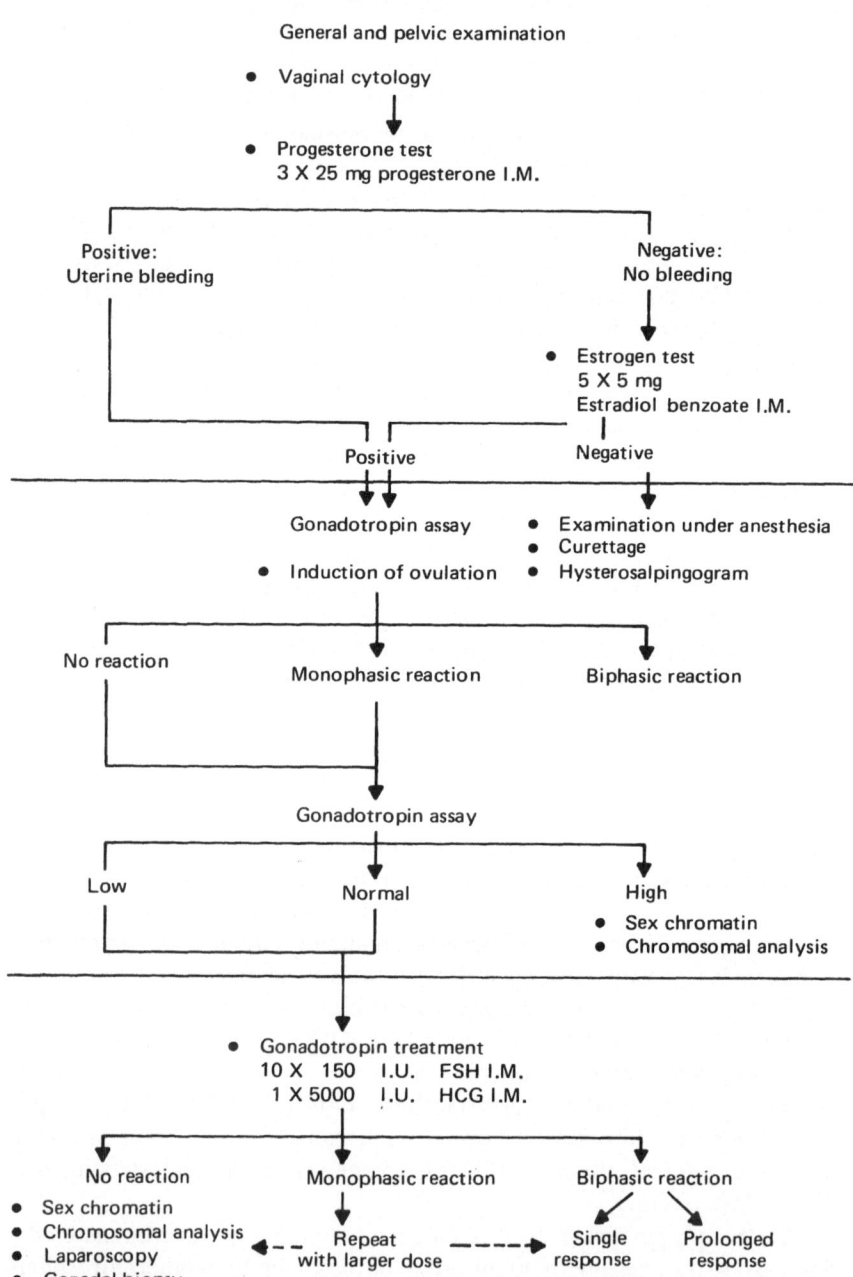

Fig. 84. Scheme of the diagnostic investigation of amenorrhea.

about the endometrium and the length of the uterine cavity. A hystero-salpingogram is of value in detecting uterine anomalies.

The results of the gonadotropin assay aid in the diagnosis of subnormal or excessive production. Treatment with clomiphene is attempted only when the values of gonadotropins are subnormal or normal.

Functional tests: The corticosteroid inhibition test is most frequently used. In the presence of a functioning adrenal tumor, corticosteroids remain high despite the administration of dexamethasone. If the ovary is suspected of producing androgens, the gonadotropin test can be added. Androgen concentrations increase after administration of gonadotropin if the ovary is the site of synthesis.

The ACTH test is used when the adrenal is thought to produce an excess of androgens, as in Cushing's disease. 17-Ketosteroids and 17-hydroxycorticosteroids are measured after administration of 50 micrograms (μg) of ACTH.

In the presence of thyroid abnormalities, an appropriate endocrine work-up for that organ is indicated.

Prognosis: A high gonadotropin value is a bad prognostic sign, since it indicates that the ovaries are insufficient in their steroid production, inducing negative feedback. Normal or slightly decreased estrogen values are compatible with a good prognosis. In most instances, the response to clomiphene is good. Very low estrogen values are associated with a poor prognosis.

Principles of treatment: Treatment is based on the progesterone test and the estrogen and gonadotropin levels Estrogen deficiency and high gonadotropin levels in the presence of a negative progesterone test require, with a few exceptions, replacement of estrogens or combination of estrogens with progestins, without treating the underlying abnormality.

In the presence of normal or slightly decreased gonadotropin and minimal pregnanediol levels, clomiphene can be used to induce ovulation. If only menses but not pregnancy are desired, estrogen-gestagen preparations may be administered cyclically. Treatment with gonadotropins may be performed, using Pergonal (menopausal urinary gonadotropin, mainly FSH) and chorionic gonadotropin (mostly LH).

If the genitalia are hypoplastic it may be advisable to prepare the patient with continuously administered high doses of estrogen and progesterone before induction of ovulation.

Treatment with hormones should not be instituted too soon after puberty if inhibition of growth is to avoided. Estrogens are used occasionally in young girls of excessive height to terminate further growth.

Psychogenic (hypothalamic) amenorrhea often responds to a change in environment or to release from a stressful situation. Observations of patients with "concentration camp amenorrhea" revealed that a normal cycle was restored in almost 95% after liberation. Neurotic amenorrhea requires psychiatric help.

Dysmenorrhea

Pain associated with normal menstruation is called dysmenorrhea, which may be primary or secondary.

Symptoms: The pain is cramping or radiating, frequently associated with back pain. It begins a day before menstruation or with the onset of bleeding. Often nausea, vomiting, dizziness, headache, and occasionally depression or restlessness are associated.

Etiology: Organic causes are malformations of the uterus (primary) and, in older women, adenomyosis (secondary). Myomas occasionally cause dysmenorrhea, as do endometriosis and pelvic inflammatory disease.

More frequently dysmenorrhea has no organic cause (dysfunctional). It has recently been observed that the administration of prostaglandins produced a nearly identical form of dysmenorrhea, and it is now believed that these substances may be important etiologically.

Treatment: Organic dysmenorrhea requires treatment of the underlying disease. The dysfunctional form, especially in the presence of ovulatory cycles, responds favorably to oral estrogen-progestin combinations in the form of the pill.

If all these attempts fail and psychotherapy is unsuccessful, drugs for pain are unavoidable. Codeine and more powerful opiates should be withheld as long as possible.

In *membranous dysmenorrhea* severe pain is caused by the expulsion of the entire functional layer of endometrium. Long-term successful treatment for this condition is unusual.

Menstrual molimina may occur in the absence of bleeding or even as "phantom pain" after hysterectomy.

The Premenstrual Syndrome

More than 50% of women experience, 5 to 8 days before menstruation, tension associated with depression and nervousness. If these complaints are exaggerated, the diagnosis of premenstrual syndrome is made. The main symptoms are irritability, restlessness, fear, and depression. Somatic complaints are swelling and pain in the breasts, distension of the abdomen, constipation, and flatulence. Swelling of the lower extremities may occur, as well as cardiovascular symptoms, generalized itching, and headache. There is an increased tendency to asthmatic attacks, migraine, vasomotor rhinitis, and epilepsy. The syndrome is present exclusively in patients with biphasic ovulatory cycles and is more frequent in women past their late thirties. The symptoms cease with the onset of menstruation.

Many theories have been invoked to explain the premenstrual syndrome. One of the etiologic factors may be related to retention of water, presumably a prolonged estrogen effect as a result of insufficient inactivation of steroids by the liver. Other factors may be retention of sodium or a secretion of gonadotropins or aldosterone. None of these concepts, however, has been scientifically verified.

Treatment. Since the cause is basically unknown, many regimens are employed. Diuretics, tranquilizers, aldosterone antagonists, small doses of progesterone, and limited intake of sodium have all been tried, but sufficiently controlled studies are unavailable.

Selected Reading

Dalton, K.: *The Premenstrual Syndrome.* London: Heineman (1964).

Davis, M. E., and Frederick, H.: "Diagnosing the Amenorrhea," in *Gynecologic Endocrinology.* J. J. Gold, ed. New York: Hoeber (1968).

Israel, S. L.: *Diagnosis and Treatment of Mentrual Disorders and Sterility,* 5th ed. New York: Hoeber (1967).

Mayer, M.: "Premenstrual Tension," in *Progress in Gynecology.* Vol. III. J. V. Meigs and S. H. Sturgis, eds. New York and London: Grune & Stratton (1957).

Scommegna, A.: "Therapy of Mentrual Disorders," in *Gynecologic Endocrinology.* J. J. Gold, ed. New York: Hoeber (1968).

CHAPTER 12

Pathology of Puberty

Precocious Puberty

Signs of puberty such as breast development, axillary and pubic hair, and menarche before the age of 8 indicate precocious puberty or sexual precocity. The variation in time of onset of puberty makes it difficult in some cases to differentiate normal from abnormal. The syndrome is observed in 1 to 20,000 girls. It is four times more frequent in girls than in boys.

Etiology. True precocious puberty should be differentiated from the pseudosyndrome.

The genuine syndrome is caused by a central nervous system disturbance. Puberty is apparently normal, although it simply occurs prematurely. Gonadotropins can be demonstrated; production of estrogen by the ovaries is normal; and occasionally ovulation, formation of a corpus luteum, and even pregnancies have been observed.

The true syndrome is organic, resulting, for example, from brain tumors, hydrocephalus, neurofibromatosis, and inflammation of the hypothalamic region.

A rare form of true precocious puberty is Albright's disease. Precocity is associated with fibrotic bone dysplasia and skin pigmentation. The cause is unknown but the long-range prognosis is good. Bone changes end in adolescence.

The pseudopuberal precocity is a result of hormone-producing tumors of the ovary or adrenal. Maturation of the sexual organs occurs, although there is no ovulation or corpus luteum formation. Whether the tumors produce androgens or estrogens, determines phenotypic development.

Symptoms. Precocious puberty, regardless of the cause, begins with breast development and appearance of pubic hair. Onset of menstrual bleeding occurs somewhat later. The children mature psychologically and intellectually at a normal rate. As a result of the precocious hormonal production they are initially taller, but they are ultimately shorter than average because of earlier closure of the epiphyses (Fig. 85).

Diagnosis. After an extensive history and careful examination, x-ray studies are performed to obtain information about the bone age. Skull

155

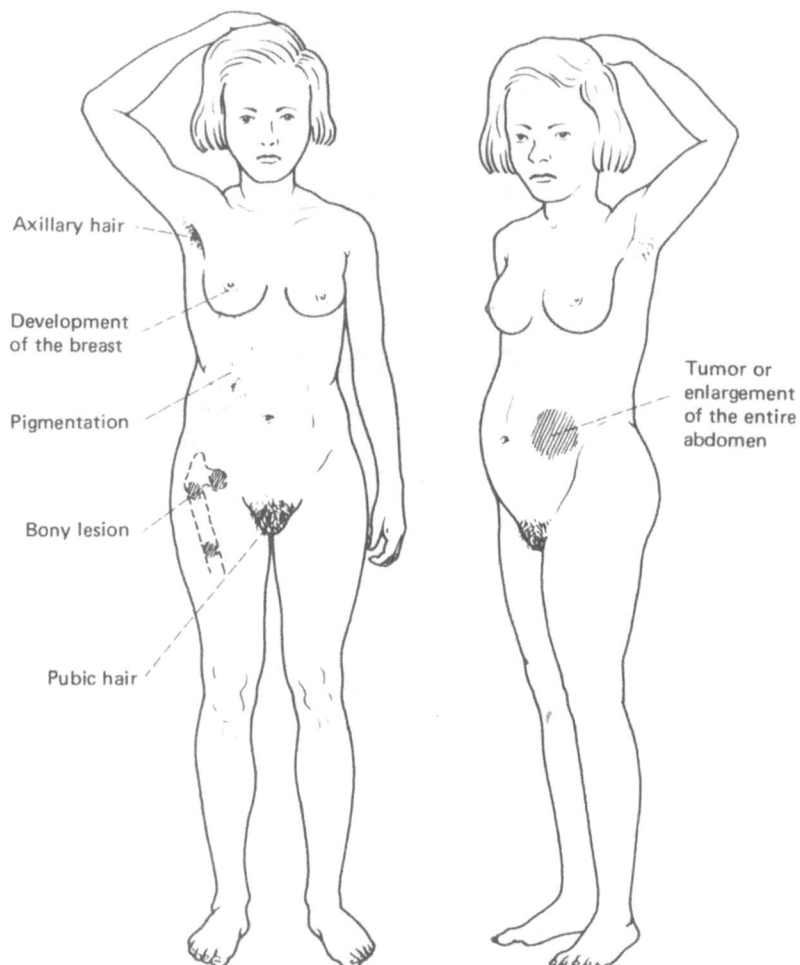

Fig. 85. Clinical aspects of precocious puberty. *Left:* True precocious puberty=hypothalamic-hypophyseal. Pigmentation and bony lesions=Albright's syndrome. *Right:* Nonconstitutional form produced by an estrogen-secreting ovarian tumor.

roentgenograms to judge the sella turcica and intravenous pyelography are performed. Vaginal cytology and assays of gonadotropins, 17-ketosteroids, testosterone, and pregnanediol are included in the hormonal work-up. In the true syndrome, these values should be normal. In the pseudosyndrome, gonadotropins are lacking but high levels of either estrogen or testosterone are found.

Laparoscopy may be indicated to rule out ovarian tumors, and pneumoperitoneum to rule out adrenal tumors. Electroencephalography, pneumoencephalography, ventriculography, and angiography may be helpful for the final diagnosis. If they are all negative, regular check-ups are mandatory to avoid missing slow-growing tumors.

Treatment. Treatment depends on the underlying disease. Whether precocious puberty without an organic lesion will require treatment depends on the age of the patient. There is no treatment for Albright's disease. Removal of tumors causing the pseudosyndrome results in involution of the secondary sexual characteristics.

Progestins may be tried in the treatment of constitutional precocious puberty. The girls become amenorrheic, the estrogen-dominated cytologic findings regress, and breast development ceases or regresses. Bone growth is not affected. The help of a child psychologist is advisable.

Retarded Puberty and Delayed Menarche

Delayed development of secondary sexual characteristics, especially breast development, between the ages of 14 and 16 is nearly always associated with a delayed menarche. Growth in height, obesity, or excessive thinness may be associated.

Etiology. Retarded puberty may be of central nervous system or ovarian origin. In general, no lesion is found, but genetic and prenatal factors and parenchymal defects of the ovary may be involved.

Diagnostic evaluation must be performed by age 15 at the latest to prevent psychological difficulties. Gonadal dysgenesis must be excluded as well as testicular feminization and lesions of the diencephalon.

Treatment. If there is no underlying lesion, assurance is essential and loss or gain in weight as indicated is advised. It is important to explain to the patient that she will develop normally. Clomiphene or estrogen-progestin preparations may be tried. It may be helpful to show the patient that uterine bleeding can be induced at any time. Delayed puberty, however, is frequently associated with ovarian insufficiency at a later age.

Selected Reading

Jolly, H.: *Sexual Precocity.* Springfield, Ill.: Thomas (1955).
Wilkins, L.: *The Diagnosis and Treatment of Endocrine Disorders in Childhood and Adolescence,* 3rd ed. Springfield, Ill.: Thomas (1965).

CHAPTER 13

Pathology of the Menopause

Bleeding Irregularities

Menometrorrhagia is explained by the failure of ovulation and formation of the corpus luteum. Persistence of the follicle with estrogen withdrawal bleeding is the etiologic factor in metrorrhagia. The endometrium is proliferative or shows glandular and cystic hyperplasia (page 322).

Vasomotor Symptoms. There are three main signs and symptoms frequently associated with premenopausal and, less frequently, menopausal women: hot flashes, sweats, and dizziness. In addition, the following complaints are frequently elicited: numbness in fingers, functional cardiovascular complaints, headache and migraine, depression and fear, and insomnia and nervousness. Most of these symptoms are nonspecific. They may be annoying but they do not form a disease entity. Many patients, however, experience these symptoms to a degree that they produce severe psychosomatic reactions. Only in a few instances do these functional disturbances progress to organic disease. Ovariectomy and ovarian irradiation usually produce exaggerated menopausal symptoms.

Somatic manifestations of the postmenopause include involution of breasts and atrophy of the introitus and vagina, which can cause pruritus and dyspareunia. Loss of turgor of the skin, relaxation of the vagina with stress incontinence, and cystourethritis may be additional features.

Hypertension, osteoporosis, and other medical diseases may be related to the relative lack of estrogens.

Treatment. Many of the menopausal symptoms respond to the administration of estrogens. A few years ago a group of gynecologists proposed "estrogens forever," a concept acclaimed by the lay press. The majority of gynecologists, however, were less enthusiastic, especially with regard to the "rejuvenation" that the proponents claimed. Many patients do not want to take the hormones for a prolonged period of time. It is generally agreed that hormones are indicated if there are problems that require treatment, especially vasomotor symptoms and atrophic changes resulting in loss of sexual function. Contraindications must be considered (Table 23). The selection of the proper estrogen is important and there is still need for developing estrogenic compounds that have little effect on the endometrium but are effective in treating vasomotor symptoms. Many women in the premenopause have slight vasomotor symptoms. As long as there is cyclic bleeding, these symptoms may be treated by pills containing combinations of estrogen and progesterone. This regimen should not be continued after the menopause, however, since the majority of women do not want to bleed regularly for an indefinite length of time.

After cessation of the menstrual cycle, menopausal symptoms are treated by estrogens, preferably orally. Intramuscular injection, when the uterus is present, may result in unpredictable and sometimes uncontrollable bleeding requiring curettage. Estradiol valerate, which is administered by injection after hysterectomy, and conjugated equine estrogens (Premarin, which is given orally in doses of 0.3, 0.6, or 1.25 mg per day) are the main drugs used for treatment of climacteric symptoms. A variety of regimens are in common use; for example, estrogen may be given for 21 or 25 days after interruption of medication for a week or more. Neither interruption of the drug nor progestational agents are of proven value in treatment of the menopause.

Tranquilizers. In some instances the estrogen dose can be reduced by the addition of tranquilizers such as Valium or Librium.

Estrogen-androgen Combinations. Formerly, androgens and estrogen-androgen combinations were used to treat menopausal symptoms, but they may produce side-effects such as acne, deepening of the voice, hirsutism, and hypertrophy of the clitoris. Androgens may be indicated in exceptional circumstances in an attempt to increase libido and to treat mastodynia.

Any treatment of the menopausal syndrome requires a careful medical examination including breast and pelvic examinations and vaginal cytology. Uterine bleeding is seen in less than 5% of patients if the dosage is properly regulated. Recurrent uterine bleeding precludes further treatment with estrogens. Repeated curettages would be required since in this age group the incidence of endometrial carcinoma is highest. Medical curettage (page 175) is contraindicated in this age group. In a few instances, especially in the presence of myomas or adenomyosis, hysterectomy may be indicated in order to treat vasomotor symptoms hormonally without ill effects.

Estrogen-induced carcinogenesis has not been proved (page 210). Nausea, abdominal bloating, and pain in the breast are usually related to overdosage.

CHAPTER 14

Clinical Aspects of Endocrine Disorders

Stein–Leventhal Syndrome

In 1935, Stein and Leventhal described a syndrome characterized by menstrual irregularities, infertility, hirsutism, moderate obesity, and polycystic ovaries. Wedge resection of the ovaries resulted in ovulatory cycles and regular menses in the majority of cases.

The syndrome is said to occur in 4 to 30% of women with hirsutism and in 0.6 to 3.4% of patients with infertility problems. The pathological finding of polycystic ovaries is probably more frequent than the full-blown clinical syndrome, for polycystic ovaries are found in 3.5% of all women at autopsy. The syndrome usually manifests itself first between the ages of 20 and 30.

Symptomatology. Secondary (only occasionally primary) oligomenorrhea or amenorrhea is characteristic. Anovulatory infertility is a primary manifestation. Basal body temperature is monophasic in 60 to 90% of cases. Hirsutism involves the chin, the chest, the area between symphysis and umbilicus, and the upper and lower extremities. Acne is common. Obesity is inconstant (Table 15, Fig. 86).

The uterus is normal or slightly hypoplastic. The ovaries are enlarged 2- to 5-fold. Enlargement of the ovaries is, however, not necessarily

Table 15. Symptomatology of the Stein–Leventhal Syndrome

Symptom	Frequency in %	
	Mean value	Variation
Obesity	41	16–49
Hirsutism	69	17–83
Virilization	21	0–28
Sterility	57	15–77
Dysfunctional bleeding	29	6–65
Dysmenorrhea	23	—
Biphasic body temperature curve	15	12–40
Corpus luteum at operation	22	0–71

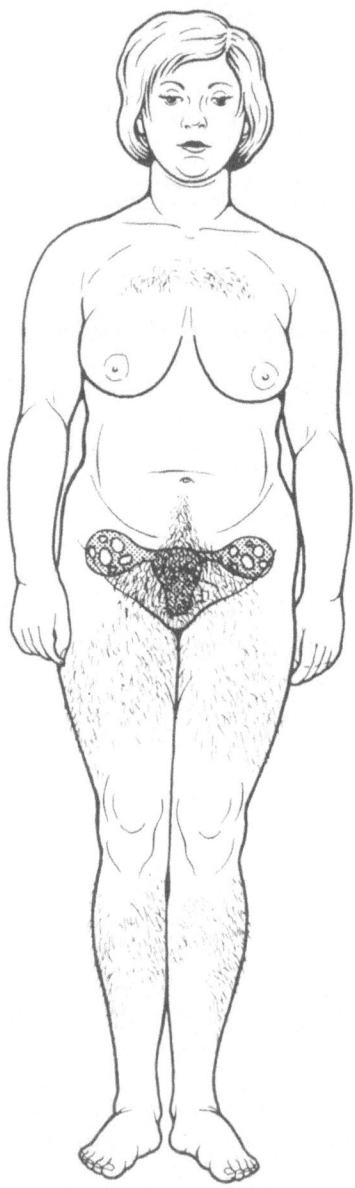

Fig. 86. Clinical aspects of the Stein–Leventhal syndrome. Polycystic ovaries, hirsutism, obesity, oligomenorrhea or amenorrhea, and infertility.

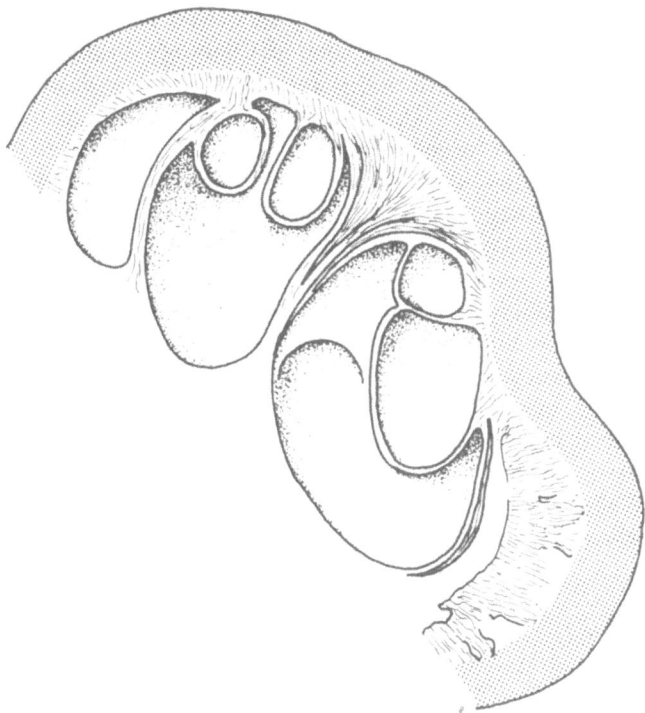

Fig. 87. Wedge resection of a polycystic ovary. Thickening of the tunica albuginea, multiple cysts in the ovarian cortex, hyperplasia of the theca, atrophic ova, and fibrosis.

required for the diagnosis because polycystic changes may occur in almost normal-sized ovaries. Macroscopically, the ovaries have a gray surface and the capsule is thickened as much as three times normal size. Below the surface there are multiple cysts of various sizes containing clear fluid (Fig. 87). The theca interna is hypertrophied (hyperthecosis).

Etiology and pathogenesis are still unclear. A popular concept is pituitary-hypothalamic dysfunction, resulting in overstimulation of the ovaries by gonadotropin.

Diagnosis. Gonadotropin excretion is at the upper limit of normal or slightly elevated. 17-Ketosteroids are increased in only one third of patients, despite the commonly observed hirsutism. Testosterone in urine and plasma may be increased up to 5-fold. The values increase further after stimulation with human chorionic gonadotropin (HCG). Dexamethasone does not cause a drop in the androgen concentration. Excretion of 17-hydroxycorticosteroids is normal. Estrogen excretion is low, increasing after stimulation with HCG. Differential diagnosis includes all clinical syndromes characterized by increased androgens with menstrual irregularities and hirsutism, such as Cushing's disease and psychogenic amenorrhea with virilization. In addition, hyperplasia, adenoma, and carcinoma of the adrenal, androgen-producing ovarian tumors, and familial hirsutism must be considered.

Treatment. Ovulation can be induced in more than 70% of cases with clomiphene. The dosage should be kept low and monitored by measurement of plasma estrogen or estrogen excretion in order to avoid overstimulation of the ovaries. Frequent pelvic examinations are mandatory. Induction of ovulation by clomiphene, and in occasional cases by HCG, should be reserved for treatment of infertility. Wedge resection is often successful in restoring ovulatory cycles and fertility (ovulatory cycles 80%; pregnancy rate 63%), but it is still uncertain how long this effect lasts. The wedge has to be large, including one-third to one-half the ovarian tissue. Even then the clinical syndrome may recur after 1 to 3 years. Wedge resection, if indicated, should therefore be performed at a time when the patient wishes to have children.

Gonadal Dysgenesis

Definition and etiology are discussed on page 127.

Symptomatology. Turner's syndrome can be diagnosed, in most instances, in the newborn. Characteristic features include the webbed neck, lax skin, edema of the upper hand and foot, deformed ears, pigmented nevi, and other anomalies. Cardiovascular malformations such as coarctation of the aorta and bony defects may be the cause of death (Fig. 88).

Secondary sexual characteristics such as pubic and axillary hair and breasts are underdeveloped because of the absence of functional ovaries. The child with Turner's syndrome is short (under 5 feet) and the genitalia are infantile. Bone age is retarded, and the vagina, uterus, and fallopian tubes are very hypoplastic. Ovaries are absent and in their place are whitish structures ("streaks") consisting of connective tissue without parenchyma (Fig. 89).

Diagnosis. FSH and LH are very high because of the negative feedback. Estrogens are very low whereas 17-ketosteroids, 17-hydroxycorticosteroids, and ACTH are normal. Final diagnosis is made by analysis of sex chromatin and karyotype. Laparoscopy is advisable.

Differential Diagnosis. In the presence of a normal chromosomal pattern the following entities must be considered: ovarian hypoplasia, delayed puberty, and pituitary dwarfism; in the last condition excretion of gonadotropins is decreased.

Treatment. Treatment is substitution with estrogen or sequential hormonal therapy. Hormonal treatment, however, should not be initiated

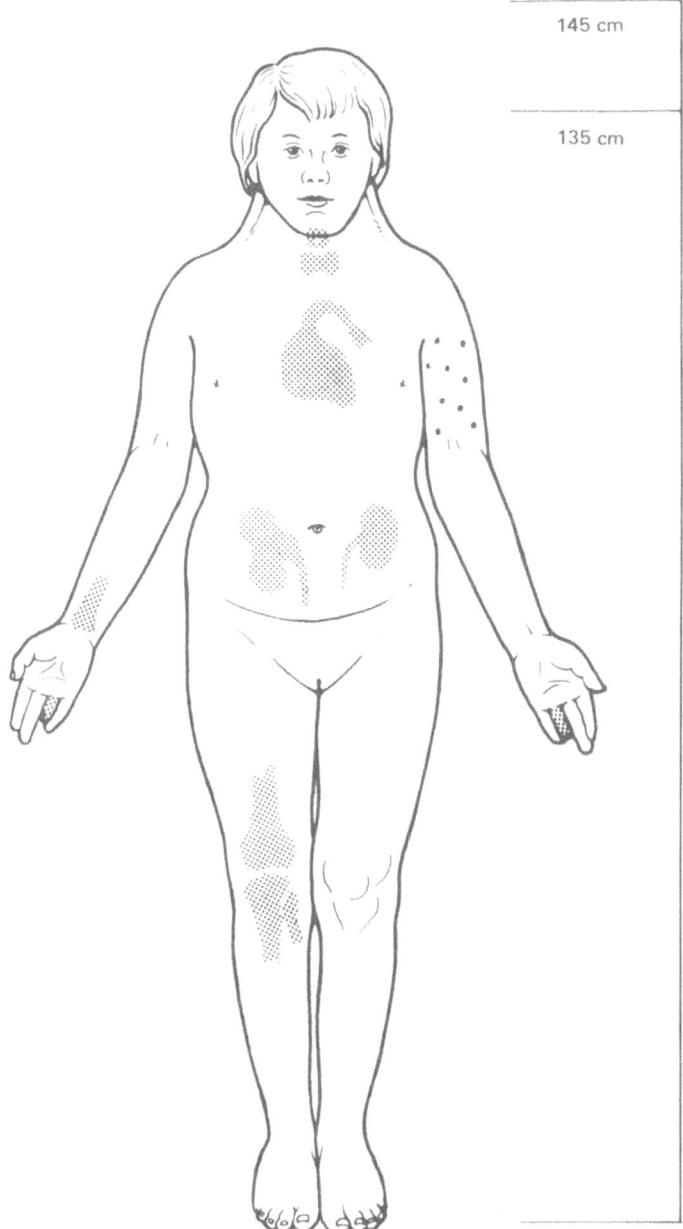

Fig. 88. Clinical aspects of Turner's syndrome. Gonadal dysgenesis with short stature, webbed neck, barrel-shaped thorax, cubitus valgus, bony defects, malformations of the heart, vessels and kidneys, and cutaneous anomalies.

before age 14 in order to avoid premature closure of the epiphyses. The administration of anabolic steroids can stimulate growth to a certain extent, but these patients remain shorter than normal girls of the same age. Growth hormone fails to increase height. The patient must be informed that she cannot become pregnant but that sex life may be normal.

Tumors in the streak gonads require hysterectomy and removal of the streaks because of the frequency of malignant change.

Adrenogenital Syndrome

The classical syndrome is caused by a congenital enzymatic defect. The postpuberal acquired syndrome is, in most instances, caused by hyperplasia of the adrenal and less frequently by an adenoma or carcinoma. Special enzymatic defects result in the adrenogenital syndrome with hypertension and sodium loss. If not recognized early, this form of the disorder may be fatal.

Etiology. The uncomplicated adrenogenital syndrome is caused by a defect or insufficiency of the enzymes involved in the synthesis of cortisol. The defects are in C-21 hydroxylase, 11-β hydroxylase, and 17-α hydroxylase. The result is a block in the metabolism of 17-α hydroxyprogesterone to compound S and cortisol. Because of the lack of cortisol (negative feedback), the ACTH-releasing factor is liberated and the production and secretion of ACTH is not sufficiently inhibited. The excess secretion of ACTH overstimulates the adrenal cortex and causes hyperplasia. The hyperplastic adrenal in turn produces steroid pre-

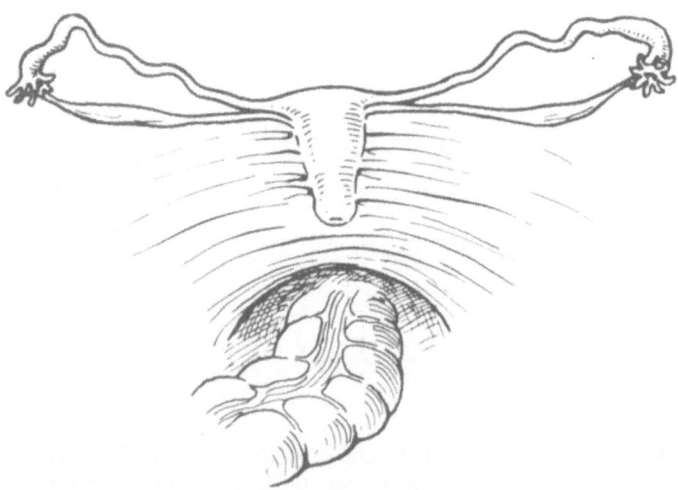

Fig. 89. Internal genitalia (seen from behind) in Turner's syndrome: gonadal dysgenesis (streak gonads), hypoplasia of the uterus and tubes.

cursors and androgens in excess. The androgens not only block the production and release of FSH and LH but also are responsible for virilization (Fig. 90).

Frequency. The congential adrenogenital syndrome is a recessive hereditary disorder. It is found in both sexes at the rate of about 1 case in 5000 newborn infants.

Symptomatology. In the congenital syndrome there is virilization of the external genitalia, ranging from slight hypertrophy of the clitoris to the formation of a phallus. The urethra may be normal or there may be hypospadias and the labia majora may be fused (scrotum-like) (Fig. 91). Extreme virilization of the female external genitalia may result in the incorrect assignment of the newborn's sex. Children with the adrenogenital syndrome initially grow faster than normal children because of the anabolic effects of the excess androgen, but in later developmental stages growth may cease prematurely because of early closure of the epiphyses. Patients with the adrenogenital syndrome are, therefore, in general, shorter and more muscular than average, with broad shoulders and a small android pelvis. The breasts are hypoplastic, the trunk is long, and the extremities are short. Hirsutism involves chin, breasts, and extremities. The uterus is small and cervical mucus is scant. The ovaries are hypoplastic and occasionally polycystic.

The magnitude of virilization is less pronounced in the acquired postpuberal adrenogenital syndrome (Fig. 92). There is only hypertrophy of the clitoris in most instances. Irregularities of the cycle are common, whereas in the congenital form of the syndrome primary amenorrhea is an important sign.

Diagnosis is based on the pronounced virilism and male phenotype in a chromosomal female. Hormonal excretion studies reveal an increase in 17-ketosteroids (20 to 100 mg per 24-hour urine), dehydroepiandrosterone, and androsterone. There is also an increase in pregnanetriol, the urinary metabolite of 17-hydroxyprogesterone. The increased hormone values return to normal after administration of corticosteroids, which provide a negative feedback to ACTH.

The hormonal values in the acquired postpuberal adrenogenital syndrome do not return to normal after administration of corticosteroids because androgens produced by the adenoma or carcinoma are not controlled by ACTH. Diagnosis may be aided by roentgenograms and angiograms.

Differential Diagnosis. Males with genital malformations such as hypospadias, bifid scrotum, and inguinal testes may have the same clinical appearance but do not exhibit the hormonal excretion patterns found in the congenital adrenogenital syndrome.

Treatment. Lifetime treatment with corticosteroids is required in the congenital adrenogenital syndrome. The resulting drop in androgens

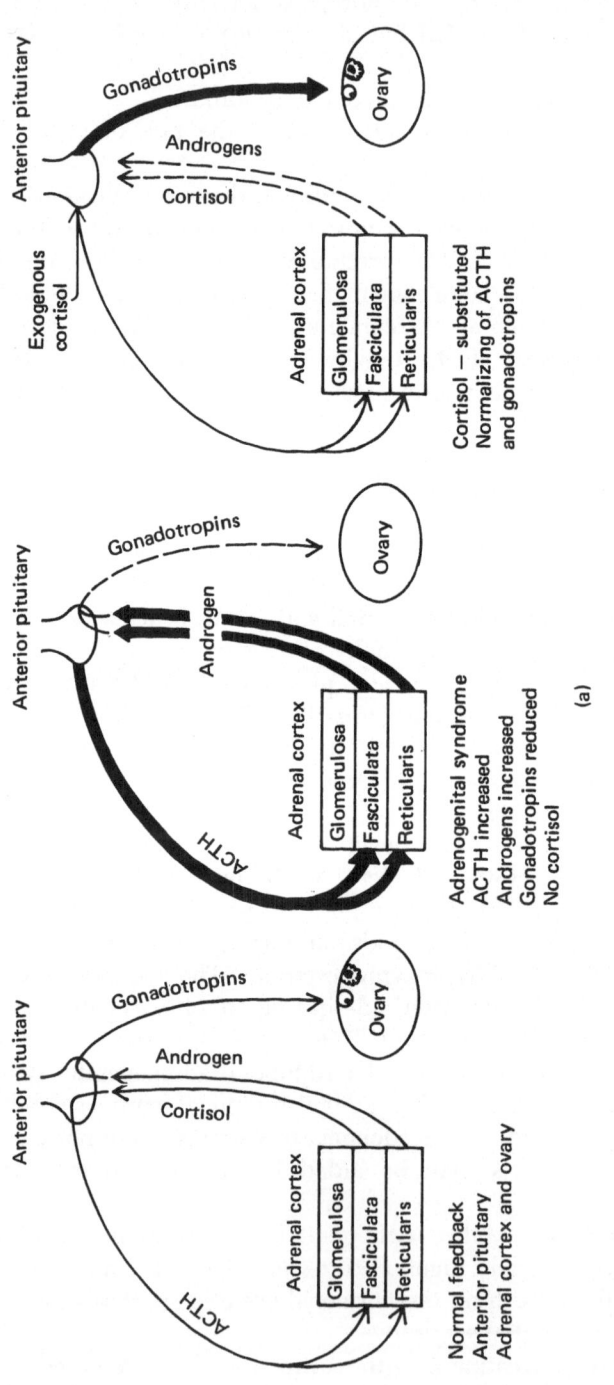

(a)

Fig. 90. (a) *Left:* Normal relations of hypophysis, adrenal cortex, and ovary. *Center:* In AGS there is an elevation of androgen production from the zona reticularis and enlarged zona fasciculata. Lack of cortisol synthesis produces increased output of ACTH, which results in adrenal hyperplasia. The high level of androgen inhibits gonadotropin secretion and causes physiologic quiescence of the ovary. *Right:* Exogenous corticosteroid administration inhibits ACTH secretion. Adrenal androgen production and hyperplasia decrease. Endogenous cortisol is low. Gonadotropin secretion and ovarian function return toward normal. (b) A frequent form of the AGS results from a C-21 hydroxylase deficiency, which partially blocks the production of glucocorticoids and mineral ocorticoids. The main pathway of hormone production is then through endogenous metabolites, which accumulate. Occasionally the syndrome is associated with C-11 or C-17 hydroxylase deficiency.

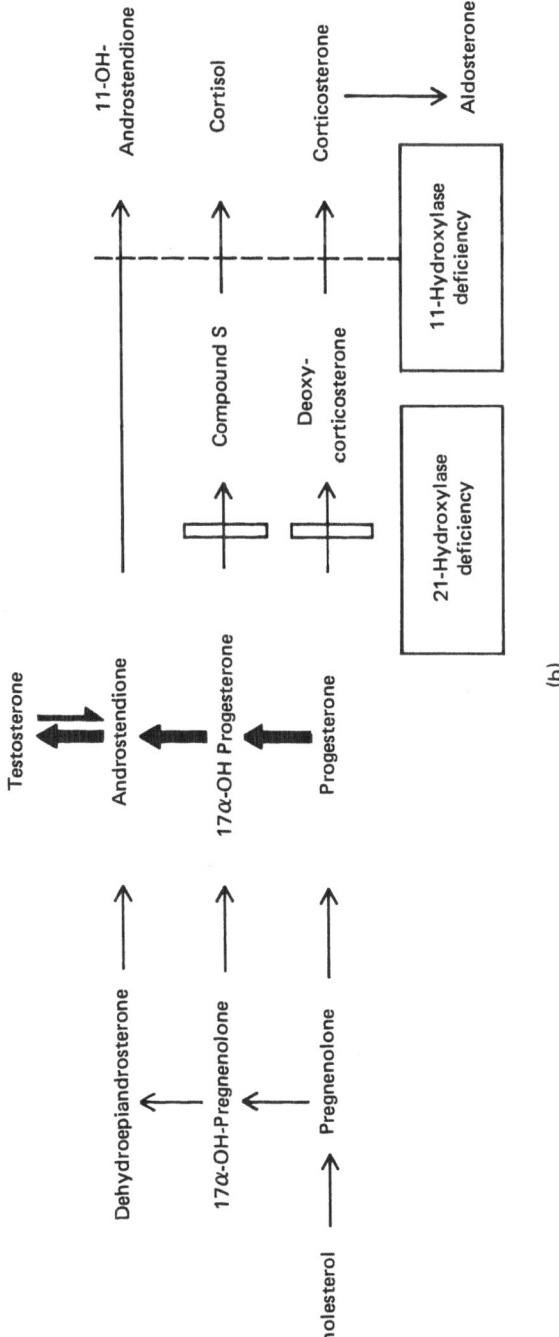

(b)

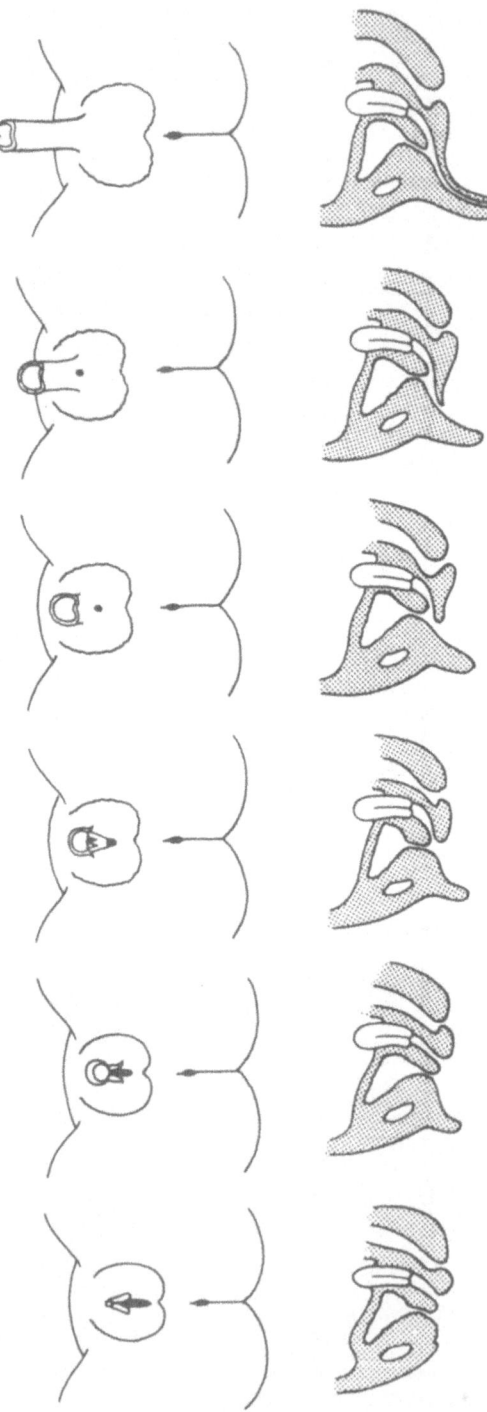

Fig. 91. Schematic representation of various manifestations of virilization of the female genitalia in the AGS. Views from above and in median cross section. Typically the urethra opens into the urogenital sinus with hypospadias of the phallus and scrotum-like labia. The degree of virilization is dependent on the severity, the onset, and the duration of androgenic influence.

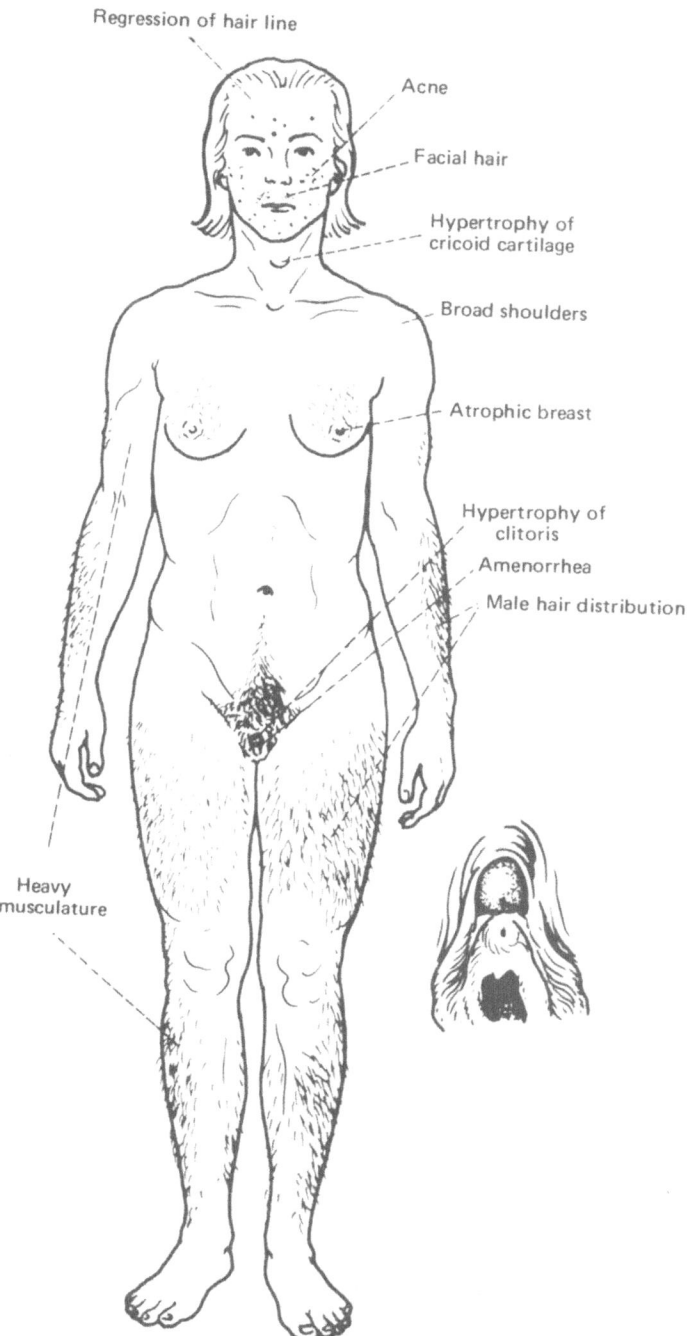

Fig. 92. Virilization in the AGS.

allows the pituitary to secrete and release normal amounts of FSH and LH. As a result, the patient may menstruate normally and even become pregnant. Prolonged treatment also restores a female phenotype, although the hirsutism is usually only slightly decreased. Plastic surgery may be necessary to reduce the hypertrophic clitoris or provide a normal introitus.

An adenoma or carcinoma requires surgical removal. Postoperative systemic administration of glucocorticosteroids may be necessary because the contralateral adrenal is sometimes atrophic.

Prognosis is good with adenoma but poor with adrenal carcinoma.

Testicular Feminization

The etiology and pathogenesis are discussed on page 131.

Symptoms. In the typical "hairless woman" the phenotype is female. Fat distribution is similar to that of normal women and the breasts are sometimes large although they have juvenile nipples. The extremities are long. Pubic and axillary hair are absent or greatly diminished. The external genitalia are female with hypoplasia of the labia minora (Fig. 93). The clitoris is small. The uterus is missing or at least rudimentary, resulting in a primary amenorrhea. Testes may be abdominal or inguinal. Histologically, the seminiferous tubules are narrow and frequently lacking a lumen, as in embryonic gonads. Leydig cells are normally developed. Spermiogenesis is usually absent. By the age of 30 approximately 30% of these patients will have developed adenomas, seminomas, or dysgerminomas of the gonads.

The patient feels like a female but sexual disorders may complicate the clinical picture.

Diagnosis is made by the lack of pubic and axillary hair and of the uterus. The chromosomal pattern is 46/XY and, therefore, Barr bodies are absent. Steroid values are within normal limits but gonadotropins are incrased.

Treatment. Testes are removed because of the frequent malignant tumors that develop within them and the patient is given estrogens. The vagina may be sufficient for intercourse or it may require mechanical or surgical enlargement. It is sufficient to advise the patient that she cannot have children but it is pointless to tell her that she is a genetic male.

Transsexuality and Transvestitism

The transsexual patient appears to be a phenotypically normal male or female with appropriate internal and external genitalia but feels and behaves as though he or she were of the opposite sex. The patient tries by all means to achieve the opposite sex. Males undergo surgery

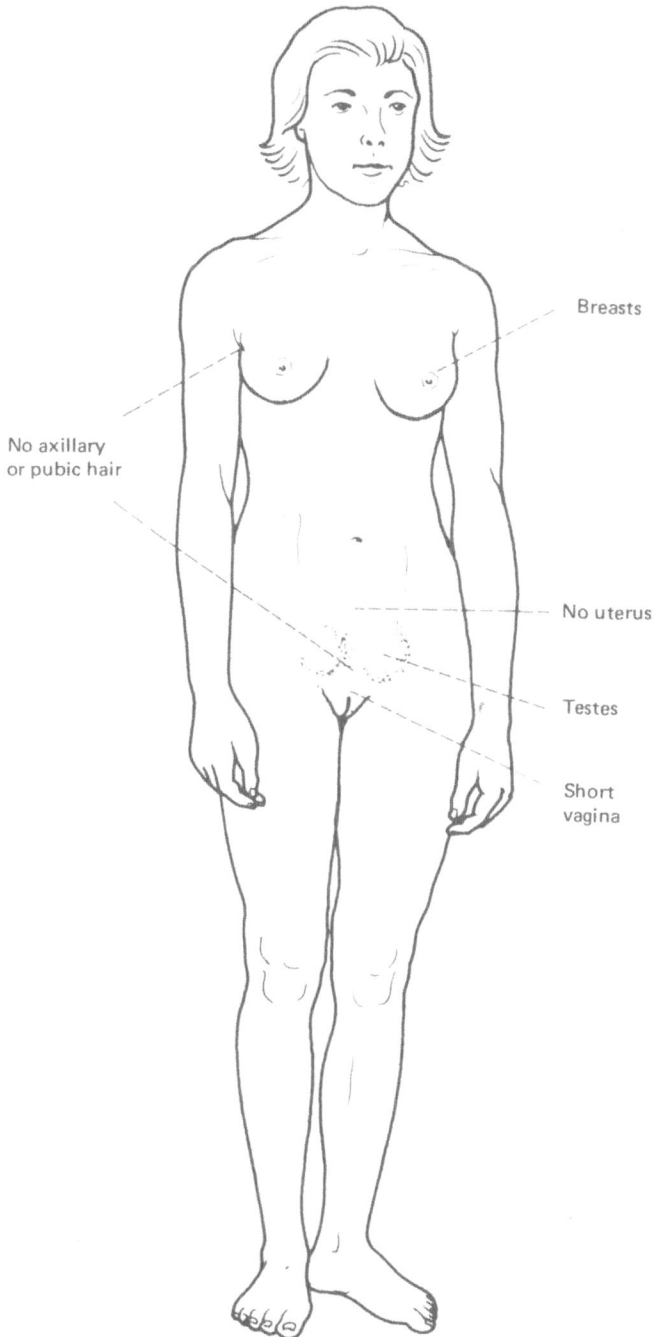

Breasts

No axillary
or pubic hair

No uterus

Testes

Short
vagina

Fig. 93. Clinical appearance in testicular feminization, female phenotype. Vagina without uterus, inguinal testes bilaterally, absence of axillary and pubic hair.

for removal of the penis and creation of female external genitalia. Surgical procedures in the female transsexual are more difficult.

The transvestite does not necessarily feel sexually like a person of the opposite sex, but merely wears clothing of the opposite sex. Such patients are not necessarily homosexual. The problem is primarily psychiatric.

Selected Reading

Bongiovanni, A. M., Root, A. W.: The adrenogenital syndrome. *New Engl. J. Med.* **268**:283, 1963.

Goldzieher, J. W.: "Polycystic Ovarian Diseases," in *Advances in Obstetrics and Gynecology*. S. L. Marcus and C. C. Marcus, eds. Baltimore: Williams and Wilkins (1967).

Morris, J. M., Mahesh, V. B.: Further observations of the syndrome testicular feminization. *Am. J. Obst. Gyn.* **87**:731, 1963.

CHAPTER 15

Principles of Hormonal Therapy

In many instances, natural hormones are replaced for therapeutic purposes by synthetic steroids or nonsteroids with hormonal activity because they exert greater or more prolonged activity and because they are cheaper.

Hormones can be used for three basic purposes.

(1) *Substitution.* Absent or insufficient endogenous production can be supplied exogenously. Substitution as a rule does not correct an imbalance and must be continued for a long period of time, occasionally for a lifetime. Examples of substitution are the cyclical administration of estrogen and progesterone in patients with Turner's syndrome and of progesterone in the presence of a corpus luteum deficiency.

(2) *Stimulation.* An inadequately functioning organ is stimulated by exogenous hormones. Stimulation is used not only for treatment but also for diagnosis (function tests). The use of gonadotropic hormones to stimulate ovarian hormone production is an example.

(3) *Inhibition.* Hormones may be used to reduce excessive function of an endocrine gland. Ovulation inhibition by estrogen-progesterone preparations or inhibition of pituitary functon by estrogenic hormones are examples. The principles of substitution, stimulation, and inhibition are sometimes applied simultaneously.

The quantity and quality of hormone-induced reactions at the target organ level depend on the dosage and duration of the exogenous hormone and the sensitivity of the target organs (Fig. 94).

Blood and tissue concentrations depend on the amount administered, the rate of absorption, metabolism, storage, blood flow, and excretion.

For a hormonal effect a minimal reaction time is necessary, which is different and characteristic for each target organ and for each hormone.

The intensity of effect of a hormone on a given target organ is determined by the dosage and duration of administration of the hormone, as well as by its distribution and biological activity.

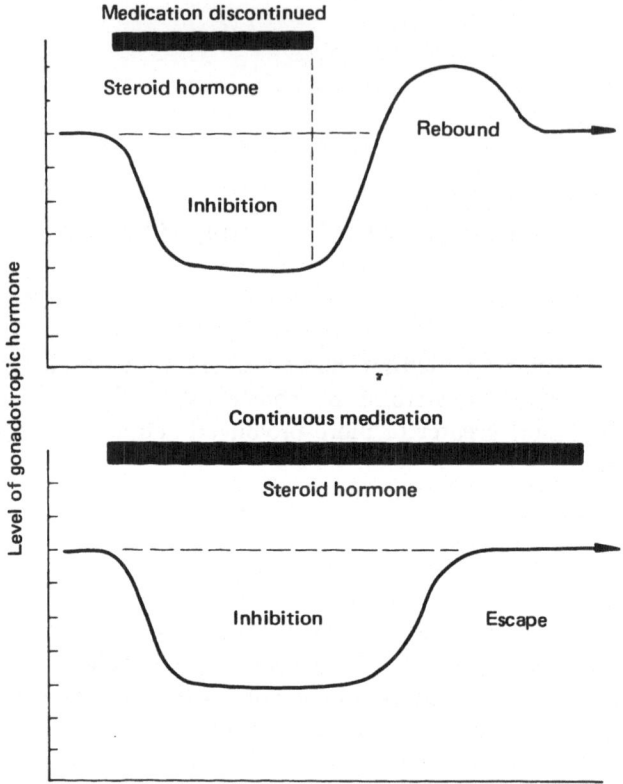

Fig. 94. *Top:* Rebound phenomenon. Excessive increase in gonadotropins after termination of estrogen-progestin medication. *Bottom:* Escape phenomenon. Adjustment of the gonadotropin level to the previous level by continuous equivalent dosage of estrogen-progestin medication.

The duration of hormonal effect depends on the half-life, metabolism, solubility, storage, protein-binding, conjugation, and excretion of the hormone and its metabolites.

Half-life is defined as the time in which the quantity or the activity of a hormone in the circulation is decreased by half.

Turnover time is the time in which the total amount circulating in the organism is replaced by new production or exogenous administration.

Clearance is defined as the amount of blood that is cleared of the hormone in a given time. The value is expressed in milliliters per minute.

Solubility and absorption are of obvious significance in determining effectiveness.

Most natural steroid hormones exert little activity by the oral route. Absorption is improved by adding hydroxyl, methyl, or halogen groups, which delay metabolism by the liver and prolong effectiveness. Oral administration is convenient in that dosage can be individualized and the administration is painless. Furthermore, oral medication does not require the presence of medical personnel. Gastrointestinal reactions, especially nausea, however, are disadvantages. Furthermore, absorption by the intestinal mucosa is somewhat unpredictable. Parenteral treatment has the advantage of well-controlled administration and is indicated in patients who have difficulty swallowing tablets, who vomit, or who are unreliable in taking oral medication. The injection may be painful, however, and require assistance of medical or paramedical personnel. The effect of a long-acting injectable medication cannot be readily reversed, and absorption from tissues may be unpredictable, especially in the presence of edema, cardiovascular disease, or obesity. If a combination of hormones is given, such as estrogen and progesterone, synchronization of the effect is difficult to predict, and delayed and heavy withdrawal bleeding is common.

Criteria for the evaluation of a hormone:

(1) The drug must be harmless, especially in pregnancy.

(2) Side-effects and adverse reactions should be minimal.

(3) Action should be reproducible.

(4) Chemical purity must be established.

(5) Dosage should be on the basis of weight or international standard biological units.

(6) The drug should be stable.

(7) The drug should be economical.

Antagonistic action of a drug means the diminution or abolition of the effect of one hormone by another. If both hormones act on the same target with opposite effects, true antagonism results. Functional antagonism means that the hormones act on different targets with the resulting opposite effects. If the antagonistic action involves the same enzyme reaction, competitive inhibition results. An example of a synergistic effect is the relation of estradiol benzoate to progesterone in a ratio of 1 to 20. An increase in either component results in antagonistic effects.

Side-effects and Adverse Reactions. Undesired side-effects or adverse reactions must be differentiated from drug (allergic) reactions. The most frequent side-effects of hormones are given in Table 16. Some of them can be abolished by changing the method of administration, the particular drug, or the dosage.

Dosage: In most instances hormones are not administered according to body weight but rather on the basis of clinically appreciable effects on the target organs. A criterion of the effectiveness of estrogen is the proliferation of the endometrium, and, for progestational agents, the

Table 16. Side-Effects of Treatment with Steroid Hormones

Estrogens	Progestins	Androgens	Corticosteroids
Water retention	Diuresis	Weight gain	Hyperglycemia
Pigmentation	Dry vagina	(N-retention)	Hypertension
Fibrocystic breast	Tendency to	Hirsutism	Edema
disease	fungous infection	Loss of hair	Osteoporosis
Cervical discharge	Norlestrin	Deepening of voice	Catabolic
Growth of myomas	derivatives:	Acne	effects
	Increase in	Increase in libido	
	appetite		
	Weight gain		
	Acne		
	Hirsutism		

High dosage			High dosage
Nausea	Fatigue	Hypercalcemia	Acne
Headache	Depression	Cholestasis	Euphoria
Tension	Migraine	Hypercholesterol-	Restlessness
Insomnia	(Gestagen	emia	Insomnia
Cholestasis	withdrawal)		Psychosis
Hypertension	Decrease of libido		Stomach and
Hyperglycemia	Hypomenorrhea		intestinal
Increase of	or amenorrhea		ulceration
phospholipids			Thrombo-
Increase of			embolic disease
triglycerides			
Thromboembolic			
disease			

transformation into a secretory endometrium. Gonadotropins must be administered according to their effects on the ovary, as indicated by excretion patterns of hormones.

Hormones can be differentiated according to their effects on various target organs. Conjugated estrogens and estradiol have little effect on the endometrium but are still effective in the treatment of vasomotor symptoms. Certain progestational steroids exert little influence on the basal body temperature or the pituitary-hypothalamic axis (for example, retroprogesterone and allyl estranol). In contrast to natural progesterone or 17_α-hydroxyprogesterone, nortestosterone derivatives may cause virilization (acne and hirsutism).

Common Uses of Hormonal Therapy

Combined treatment with estrogens and gestagens is most frequently used for the purpose of oral contraception. These combinations also may delay menstruation for convenience (for example, during a honeymoon). Withdrawal bleeding occurs two to three days after termination of the pills. This regimen may produce side-effects, especially prolonged premenstrual tension, fatigue, and depression. It is preferable to start oral contraception with compounds containing a larger dose of estrogen, giving it until two days before the expected menses.

Pseudopregnancy Regimen. This type of treatment requires large doses of estrogens and progestational agents continuously for months. Treatment is started with 5 or 10 mg of Enovid daily. The dose is increased to 20 to 40 mg per day depending on breakthrough bleeding. Treatment is continued for 6 to 9 months. The main indications for such continuous therapy are endometriosis and certain cases of hypoplastic genitalia and breasts. Some patients with Sheehan's syndrome may benefit from this treatment.

This regimen is frequently accompanied by the side-effects of nausea and weight gain.

Ovulation induction: This treatment should be performed only by a specialist in gynecology. *Clomiphene* is the main drug used for this purpose. It has weak estrogenic as well as antiestrogenic activity. Treatment is usually begun using 50 mg a day for 5 days. Ovulation usually occurs in the following 2 weeks. Effectiveness of clomiphene requires intact pituitary-hypothalamic function (Table 17).

Table 17. Clomiphene Treatment

Prerequisites

---

(1) Normal or slightly decreased gonadotropin values ⎫ As signs of intact hypo-
(2) Normal or slightly decreased estrogen values   ⎬ thalmic-hypophysial-
                                                        ⎭ ovarian system
(3) Other endrocrine glands: normal or only slightly
    elevated adrenal androgen production and normal
    thyroid function

Indications

Disturbances of the cycle ⎫ because of ⎧ Anovulation
Infertility   ⎭   ⎨ Inadequate corpus luteum

. ***Gonadotropins.*** Preparations with predominantly FSH activity are derived from urine of menopausal women (Pergonal). Human chorionic gonadotropin is used as a source of LH-type hormone. Treatment must be monitored by estrogen assays to prevent overstimulation.

Selected Reading

Bettendorf, A., Insler, V. (ed.): *Clinical Application of Human Gonadotropins.* Stuttgart: A. Thieme (1970).

Kistner, R. W.: *The Use of Progestins in Obstetrics and Gynecology.* Chicago: Year Book Med. Publ. Inc. (1969).

Shearman, R. P.: *Induction of Ovulation.* Springfield, Ill.: Thomas (1969).

CHAPTER 16

Infertility

Infertility is defined in the United States as diminished fertility, as opposed to sterility, which is the total inability to reproduce. Primary infertility is the failure to conceive at any time. Secondary infertility is the failure to conceive after one or more successful pregnancies. In both cases, the diagnosis should not be made until the patient has remained infertile for at least one year after attempting pregnancy.

Ten to 15% of all couples are involuntarily childless. The wife is the primary cause in about one-half of the cases. In about one-third of the cases the husband is the reponsible partner; the remainder are unexplained.

The wife is usually the partner who seeks medical help for infertility, but the husband must be included in a diagnostic work-up at an early stage of the investigation.

Causes of Female Infertility

The likelihood of conception decreases as the woman's age advances. By the age of 30 it will have dropped to 50% (Table 18, see page 182).

Ovarian Factors. Directly or indirectly the ovary is the cause of the majority of cases of female infertility, including failure of ovulation, infrequent ovulation, or inadequacy of the corpus luteum. The following factors are most commonly encountered:

(1) disturbances of the pituitary or hypothalamus
(2) organic or functional insufficiency of the pituitary
(3) ovarian insufficiency (primary ovarian hypoplasia)
(4) cystic change in the ovaries (microcystic degeneration or polycystic ovaries of the Stein–Leventhal syndrome)
(5) ovarian tumors, including endometriosis.

Infertility Caused by Tubal Factors. Infertility caused by tubal factors are the result of:

(1) damage of the tubal mucosa secondary to inflammation (endosalpingitis)

181

Table 18. Likelihood of Conception as Related to the Womans Age (according to Muenzer and Loer)

Age (years)	Likelihood of conception (%)
15	68
20	66
25	54
30	30
35	11
40	3
45	0.5

(2) tubal occlusion secondary to peritubal adhesions. (Common causes of postinflammatory tubal occlusion include gonorrheal and tuberculous salpingitides.)

(3) salpingitis isthmica nodosa and tubal endometriosis

(4) disturbances of tubal motility secondary to inflammatory or endocrine factors or genital hypoplasia.

Infertility Caused by Uterine Factors. The uterus is seldom the sole cause of infertility. Failure of implantation is often secondary to hormonal factors. The following organic factors are significant:

(1) myomas (submucous and intramural varieties particularly)

(2) congenital uterine anomalies

(3) inflammation or trauma to the endometrium.

Infertility Caused by Cervical Factors. The cervix, which is required to transport the semen, is a more important factor than previously believed. Major cervical abnormalities include:

(1) abnormalities of cervical secretion or composition of cervical fluid as a result of hormonal dysfunction

(2) anatomical deformities of the cervix, secondary to trauma of curettage, abortion, or delivery

(3) severe inflammatory changes

(4) immunological incompatibility between cervical secretion and sperm.

Capacitation is the change that takes place in the sperm after several hours within the cervix, allowing it to penetrate the zona pellucida and the cumulus oophorus. The acquisition of this ability and thus the potential for fertilization depend on the hormonal status of the cervical secretion. It is not known whether capacitation is required in human fertilization.

Pathogenic microorganisms in the cervix may damage the sperm directly or indirectly by altering cervical mucus production. Leukocytes, furthermore, may phagocytize the sperm or decrease their motility.

Infertility Caused by Vaginal Factors. Vaginal factors causing infertility are:

(1) congenital anomalies of the vagina (aplasia, atresia)
(2) vaginal stenosis (traumatic, inflammatory)
(3) inflammation (bacterial, chemical).

Streptococci, staphylococci, and fungi may destroy the spermatozoa. Trichomonads may exert a spermicidal effect by concomitant alteration of the vaginal flora.

(4) vaginismus.

Infertility Caused by Psychogenic Factors. Psychogenic factors, although difficult to prove, are probably underestimated as a cause of infertility. Their mechanism of action may be at various levels in the reproductive process.

Psychological investigations of such women often reveal an infantile personality, that is, a woman who fears the assumption of a maternal role and who utilizes her infertility as a defense mechanism. On the contrary, an exaggerated desire to bear children may also be a factor. A third group of patients includes women in whom emotional conflicts produce somatic signs and symptoms, such as cervical hypersecretion, tubal spasm, or anovulation. Psychogenic factors may also lead to frigidity or vaginismus. Additional sociological factors include an unstable marriage, poor adjustment to the sexual partner, or environmental factors (family, profession, poverty).

Infertility Caused by Extragenital Endocrine and Other Illnesses. The endocrine organs are so closely interrelated that virtually any disturbance may have an effect on reproduction. Pituitary insufficiency or pituitary tumors affect the ovaries and thus may cause infertility. In addition, specific causes include:

(1) adrenal disorders such as hyperplasia of the cortex (congenital or acquired adrenogenital syndrome), Addison's disease, Cushing's disease, and rare tumors of the adrenal
(2) thyroid disorders including hyperthyroidism and hypothyroidism
(3) diabetes mellitus
(4) severe chronic illnesses
(5) abuse of drugs, such as alcohol and nicotine.

Infertility Caused by Immunologic Factors. Women may develop antibodies against spermatozoa. These antibodies can be demonstrated within the serum and cervical secretion. The significance and frequency of such incompatibility remain to be investigated further.

Causes of Male Infertility

After an initial work-up, including the sperm count and semen analysis, the husband is usually referred to the urologist for further investigation, particularly if an abnormality is found.

Specific causes of infertility are:

(1) primary infertility caused by disturbances in spermatogenesis: azoospermia or oligospermia (as in the Klinefelter syndrome), decrease in motility or increase in abnormal forms of spermatozoa, an unphysiological composition of the seminal fluid, or disturbance in mechanism of ejaculation

(2) impotence caused by decrease or absence of libido, inability to produce or maintain an erection, premature ejaculation or psychogenically induced inability to effect intromission.

(3) autoimmune antibodies against spermatozoa.

Diagnosis. The suggestions listed on page 95 should be followed in taking the history.

The gynecological examination is carried out according to the steps and procedures listed in Chapter 8. Endocrine factors should be elicited by careful study of the menstrual history. It is essential to ascertain early whether the patient is ovulating and whether a functional corpus luteum is produced, as outlined on page 72.

The simplest test is the recording of basal body temperature (page 185). Success of this method depends upon regular recording of the temperature before arising each day for several months. The occurrence and time of ovulation may thus be ascertained. The optimal time for conception is the interval including 2 days before and 1 day after the temperature rise (Fig. 95).

Absence of a temperature rise or a biphasic curve suggests lack of ovulation. A prolonged hypothermic phase indicates delayed or inadequate maturation of the ovum. A stepwise or protracted temperature rise or a shortened hyperthermic phase suggests insufficiency of the corpus luteum, a more common cause of infertility in older women.

Unfortunately, the time of ovulation within the cycle can be ascertained only retrospectively, but in a patient with regular biphasic menstrual cycles, this method may permit accurate prediction of the optimal time for conception.

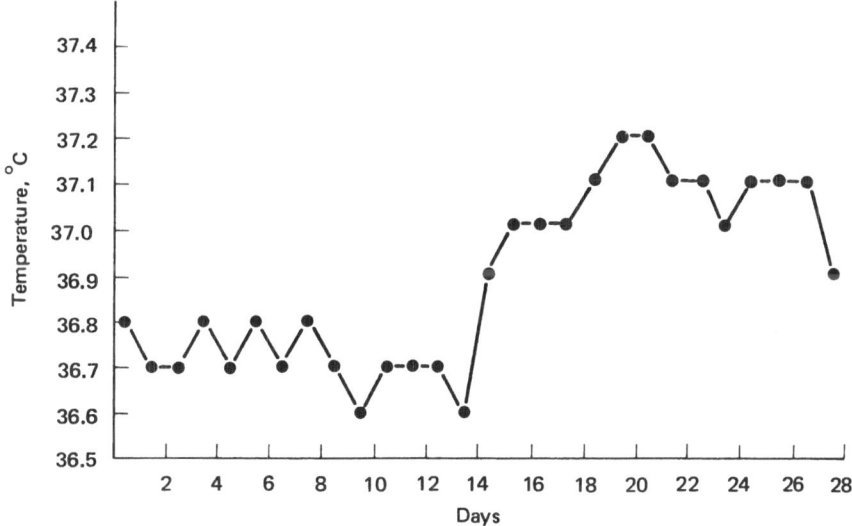

Fig. 95. Basal body temperature curve in the normal biphasic cycle.

Information obtained by basal body temperature curves are supported by repeated analyses of the cervical secretion by Spinnbarkeit and fern tests (Fig. 95). The Spinnbarkeit test consists in stretching a drop of cervical secretion to test its ability to be drawn out into long threads. In the preovulatory phase the Spinnbarkeit increases to 6 to 8 cm; immediately before ovulation, it may increase to approximately 15 to 20 cm. The cervical mucus arborization (fern) test consists in air-drying a drop of cervical secretion on a slide. The presence of crystals, or a fern pattern, as seen with the microscope suggests impending ovulation (page 63); its absence suggests anovulation.

During collection of a sample of cervical secretion, the cervical os should be inspected carefully. At the time of ovulation it is usually somewhat dilated.

Hormonal cytology is a semiquantitative means of ascertaining ovarian function. At the same time the vaginal flora are screened. Vaginal cytology is supported by endometrial biopsy, which may give an accurate prediction of the development of the endometrium during a particular phase of the cycle. Endometrial biopsy is required when there is any suspicion of an atrophic endometrium or endometrial tuberculosis. Should these office procedures be negative, quantitative hormonal analyses may be indicated (page 72). Specific laboratories should be utilized for these tests, which are often difficult and time-consuming.

The husband's fertility is then ascertained before the wife is investigated further. After a semen analysis, which is preferably performed on a specimen obtained by masturbation, the gynecologist performs

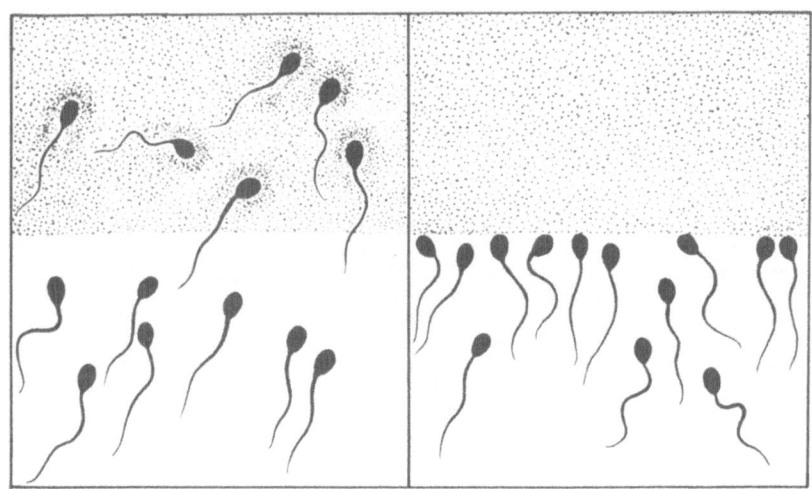

Fig. 96. Kurzrock–Miller Test. *Left:* Penetration by sperm of the cervical mucus (positive test). *Right:* Failure of penetration (negative test).

the Sims–Hühner test, which is carried out as follows. Within 4 to 6 hours after coitus, a sample of cervical secretion is obtained and analyzed under the microscope to detect motile spermatozoa. This test should be performed during the preovulatory phase, at which time the cervical mucus presents optimal conditions for sperm penetration. After 6 hours approximately 60% of the sperm should be motile. The husband's fertility should be further investigated by quantitative sperm count and estimation of morphology and motility. If the husband's fertility has been ascertained, a negative Sims–Hühner test suggests an immunological incompatibility. To demonstrate this problem, the Kurzrock–Miller test (Fig. 96) can be used. In this procedure a drop of semen from a fertile man is placed adjacent to the patient's preovulatory cervical secretion and the penetration is recorded. As a control, a drop of the husband's sperm is checked against the preovulatory cervical secretion of a fertile woman. If the husband's sperm fails to penetrate the cervical secretion of both patients, the spermatozoa are probably defective. If the donor's sperm fails to penetrate the patient's cervical secretion only, a cervical factor may be the cause of infertility.

Several procedures are utilized to establish tubal patency. They include tubal insufflation (Rubin's test), hysterosalpingography, and laparoscopy or culdoscopy.

Tubal insufflation consists in introducing carbon dioxide under controlled pressure through the fallopian tubes. A vacuum adapter is utilized to seal off the portio. A cannula is inserted with a vacuum adapter into the portio vaginalis of the cervix. A drop in pressure, as indicated by the attached manometer, suggests patency of the tubes. The graphically recorded pressure curves indicate whether there is tubal closure or patency.

Hysterosalpingography consists in injection of a contrast medium (preferably water-soluble) into the cervix. The uterine cavity and the fallopian tubes may be visualized either by roentgenogram or fluoroscopy. Compared with tubal insufflation this method has the advantage of localizing the site of tubal occlusion. At the same time, anomalies of the uterus and defects such as submucous myomas may be detected.

Laparoscopy and culdoscopy permit a direct observation of the internal genitalia. Lesions of the tubes, particularly ampullary occlusion and peritubal adhesions, may thus be directly inspected. (In addition, tubal pregnancy can be observed directly if a contrast medium is simultaneously injected into the uterus through the cervix with direct laparoscopic or culdoscopic visualization.)

Laparoscopy and culdoscopy frequently replace hysterosalpingography but each of the methods has its own merits in the diagnosis of infertility and the extent of tubal impairment.

Principles of Treatment of Infertility

The success of therapy, of course, depends on precise knowledge of etiologic factors. Many cases of infertility caused by anovulation may be successfully treated by induction of ovulation (page 179).

Insufficiency of the corpus luteum is often corrected by cyclic administration of progesterone or gonadotropins.

Occlusion of the fallopian tubes may require operative intervention. Operative procedures include opening of the fimbriated ends (salpingostomy), lysis of peritubal adhesions (salpingolysis), and excision of local foci of endometriosis with subsequent end-to-end anastomosis. If the obstruction is near the uterus, tubal reimplantation may be performed. There is, however, a high risk of postoperative adhesions and secondary occlusion. As a preventive measure, hydrotubation (instillation of antibiotics and steroids) may be employed. The value of this procedure has not yet been thoroughly evaluated. The success of all procedures on the tube depends predominantly on the location of the occlusion, the damage to the endosalpinx, the preservation of tubal motility, and the damage to the fimbriae.

Certain cases of infertility caused by congenital uterine anomalies may be corrected by plastic operative procedures. Myomectomy may occasionally be successful, particularly if submucous myomas are removed. A surgical correction of obstetrically produced cervical tears may be of value in the repair of an incompetent cervical os. Restitution of normal cervical secretions is of great practical importance. If the disturbance in secretion is hormone-dependent, estrogen supplementation may prove valuable. Bacterial inflammation of the cervix requires specific antibiotic therapy. Restoration of normal vaginal flora is a requirement for maximal success in the treatment of infertility. If the history, behavior, and negative physical findings suggest a psychosomatic cause, psychiatric therapy may be indicated.

If the husband's work-up reveals oligospermia, a homologous insemination should be considered. The couple should be informed, however, that inseminations may have to be repeated for several months at the time of ovulation and that the success rate may not be great.

Heterologous inseminations, or donor inseminations, should be performed by physicians who are experienced in the techniques and aware of the medicolegal implications of the procedure.

The rate of success for the treatment of all causes of infertility in women is approximately 35% in the best hands. Counseling of couples whose infertility cannot be corrected requires psychological skill. Adoption may be an acceptable alternative for may couples who seriously desire children.

Selected Readings

Behrman, S. J., and Kistner, R. W.: *Progress in Infertility.* Boston: Little, Brown (1968).

CHAPTER 17

Family Planning

Family planning includes the right to determine the number and spacing of children. To this end, temporary (contraception) or permanent (sterilization) types of birth control may be employed. The goal of family planning is a socially and medically sound, responsible family.

Contraception may be medically indicated when pregnancy would aggravate an existing disease. Genetic diseases require contraceptive advice within the scope of genetic counseling.

Planned Parenthood

Preventive medicine requires spacing of pregnancies, since the neonatal and maternal morbidity and mortality are significantly higher when pregnancies follow in rapid succession. Pregnancies are preferably spaced 2 to 3 years apart. In addition, maintenance of an optimal interval between births is important for the child's education and its adaptation within the family.

Limitation of the number of children is the most important motive for family planning. Socioeconomic factors and the prospective education of the children are major concerns.

The concept of planned parenthood also comprises contraception for the single woman. The changing sexual mores and the separation of sexuality from conception pose a complex problem.

Prevention of undesired pregnancies, illegitimate and legitimate, with the goal of decreasing criminal abortions and their complications, has become a fundamental part of preventive and social medicine.

Within the last decades there have emerged sociological, economic, and political justifications for birth control subsidized by governments or public organizations. The growth of the world's population has been explosive, especially in underdeveloped countries. Causative factors include control of epidemics, prolonged life expectancy, and decreased neonatal mortality. The population explosion threatens further economic world progress, since the population increases much more rapidly than the food reserves of the affected countries.

In 1850 the world's population was approximately 1 billion people. It had doubled by 1925 and by 1977 it is expected to increase to approxi-

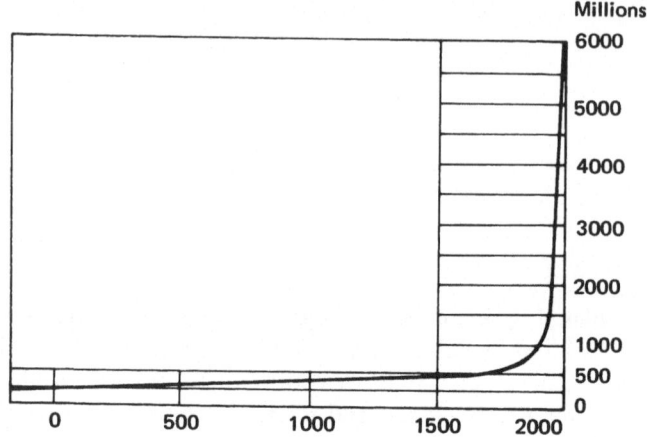

Fig. 97. The growth of world population in the last 2000 years.

mately 7 billion people (Fig. 97). Presently, 3 persons are born every second, whereas only 2 die. Thus, the world's population increases at the approximate rate of 1 person per second. Differently expressed, 80,000 additional human beings must be supplied with food each day.

Contraception is a means of influencing these events. Prior experience has clearly demonstrated that neither control by the state nor the availability of contraceptive techniques is sufficient to control the birth rate. Industrialization, socioeconomic progress, and education are much more significant factors. Technical difficulties are secondary to the solution of political, religious, and socioeconomic problems.

The Physician's Role in Family Planning

A significant portion of the modern gynecologist-obstetrician's time is occupied with counseling in family planning. The physician's function is to advise the patient or the couple about the method most suitable to them individually. If a hormonal method of birth control is desirable, the physician should select for the patient the most suitable steroids. Contraindications should be considered; instructions should be complete; and periodic pelvic examinations are required.

The prescription of oral contraceptives is a new major task for the physician. Since these drugs are rarely used for medical reasons, the possible side-effects must be weighed carefully against the benefits.

Contraception includes methods that prevent nidation of a fertilized egg. According to some definitions, such methods could be considered abortifacients. If, however, conception begins with nidation of the blastocyst, then "morning-after" pills and intrauterine devices must be considered contraceptives.

Measure of Success of Contraceptive Methods

The reliability of a contraceptive method is calculated by the formula of Pearl. It is based on a statistically significant number of cycles in which it is assumed that only one fertilization will take place. The number of undesired conceptions (failures of a given contraceptive method) is expressed in terms of 1200 months or cycles of utilization, or 100 "woman-years." It is not expressed in percent, and unprotected intercourse does not result in 100 pregnancies but in a much smaller number. The reliability of a given technique depends not only on the method (method failure) but also on errors in usage (patient failure). The failure rate derived from Pearl's formula includes both and is, therefore, a measure of the practical application of a certain technique.

Table 19 (see below) lists the failure rates of the most commonly used contraceptive methods.

Methods That Require No Preparation (so-called natural methods)

Rhythm Method (periodical abstinence). The rhythm method of the time of ovulation during a menstrual cycle is based on the observations of Ogino and Knauss. According to Ogino, ovulation occurs between the sixteenth and twelfth days before the onset of the next menstrual period. Considering the life-span of the sperm as 3 days, the

Table 19. Failure Rates of the Various Methods of Contraception (Pearl index)

Means of contraception	Failures per 100 woman-years
Oral contraceptives	
combination method	0.2 (0.0–0.8)
sequential method	1.5 (0.0–11.9)
Oral gestagens (minipill)	2.5
Intrauterine device	0.5–4.6
Diaphragm	7–10
Diaphragm in combination with spermicidal substances	5
Depot-gestagen injection	2.6
Spermicidal vaginal tablets	22.5–37
Spermicidal vaginal jelly	20
Spermicidal spray (foam)	12
Rhythm method	15–38
Condom	3–14
Coitus interruptus	35
Vaginal douching	31
Tubal sterilization	0–0.3

fertile period falls between the nineteenth and the twelfth day before the onset of the next menstrual period. In a regular cycle, the fertile period would extend from the tenth to the seventeenth days of the cycle.

According to Knauss, ovulation occurs exactly on the fifteenth day preceding onset of the next menstruation. The corpus luteum phase, or the postovulatory phase, lasts for exactly 14 days (see page 59). Knauss considered the fertile phase as a time span of 3 days before and 1 day after ovulation.

Both Ogino and Knauss cautioned that the constancy of the menstrual cycle be precisely ascertained before using this method. The greatest and smallest deviations are recorded on a calendar for 12 consecutive months. From the longest and shortest menstrual cycles the individual variation for the fertile days can be calculated by Knauss's formula:

$$\text{Longest cycle minus } 15 + 2$$
$$\text{Shortest cycle minus } 15 - 2$$

Therefore, if the longest menstrual cycle is 32 days, and the shortest 26 days, the fertile phase would extend from the ninth to the nineteenth day of the cycle. Despite careful calculations, however, unexpected variation of menstrual cycle occurs. As a result, irregular, especially delayed, ovulation can then result in an undesired pregnancy.

In summary, the rhythm method is useful in helping to ascertain the fertile days rather than the infertile period. It is therefore not surprising that the failure rate varies considerably according to the constancy of the menstrual cycle and the intelligence and reliability of the patient. Under the best conditions it is in excess of 15 conceptions per 100 woman-years.

Basal Body Temperature Measurement. The reliability of the rhythm method can be increased by the measurement of the basal body temperature (see page 185). The infertile phase can be calculated as the time starting with the third day of the hyperthermic phase and ending with the fifth day after onset of the next menstrual period. The failure rate of this method is approximately 10 per 100 woman-years. This method can be judged adequate only for intelligent couples.

Mechanical Methods

Cervical Caps. Occlusive caps of appropriate size are placed over the vaginal portion of the cervix, thereby preventing the ascent of sperm. The caps must be removed before the onset of the menstrual period and reapplied after it is over. Since only a small proportion of women are capable of applying the cervical cap themselves, a physician's help is needed, at least after each period. Endometritis or cervicitis as well as anatomical variations are relative contraindications. The failure rate is 7 to 10 per 100 woman-years.

The Vaginal Diaphragm. The vaginal diaphragm consists of a coil or spring ring covered with rubber or plastic. It is inserted into the vagina to cover the cervix mechanically (Fig. 98). It can easily change its shape because of the resilient external ring, and thus can be easily

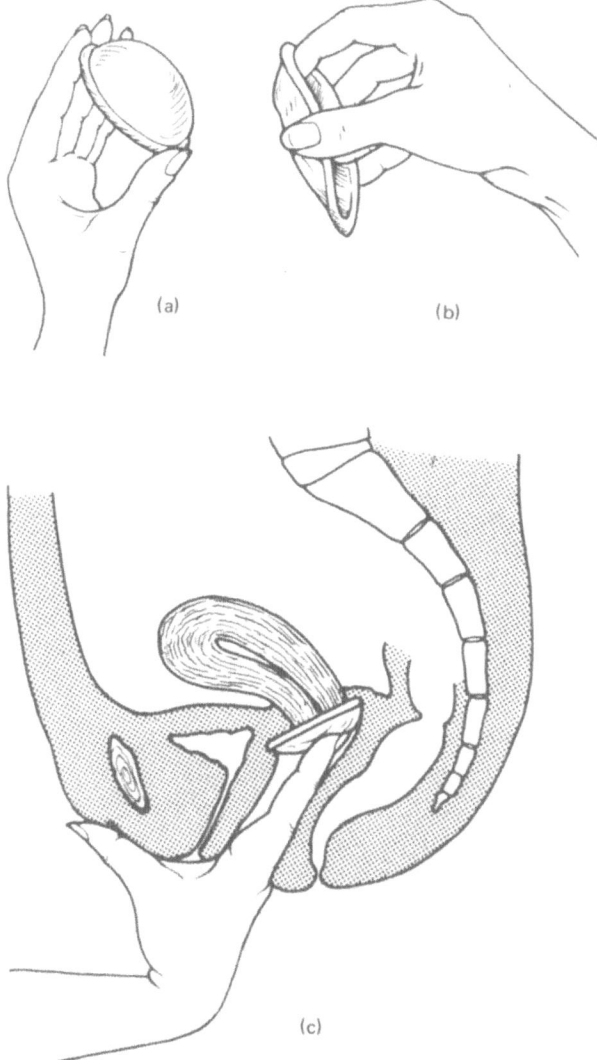

Fig. 98. The use of the vaginal diaphragm. (a) Shape of the diaphragm; (b) folding of the spring before introduction; (c) checking the position after introduction.

introduced and removed by the patient herself. Generally, it is used in combination with spermicidal jelly or cream applied to the diaphragm. The diaphragm is introduced before intercourse and is not removed until at least 8 to 10 hours later.

The diaphragm is fitted so that the ring is snugly lodged between the posterior and the anterior walls of the vagina just behind the symphysis.

The failure rate is 7 to 10 per 100 woman-years but can be lowered to 4, provided that the diaphragm is fitted by a physician and the patient is properly informed about its use. Some women reject the vaginal diaphragm on esthetic grounds. Contraindications are pronounced vaginal prolapse (see page 251) and scarring of the portio and vagina.

Spermicidal Jellies or Foams. When spermicidal jellies and foams are the only method of contraception used, they carry a failure rate up to 20 per 100, and thus are not considered reliable methods.

There is no evidence to suggest that spermicidal substances cause fetal malformation, should a conception occur despite their use.

Vaginal Douching. Douching immediately after intercourse is unreliable, since sperm are demonstrated in the cervical canal within seconds after ejaculation. The failure rate is close to that of unprotected intercourse.

Intrauterine Devices (IUDs). The modern intrauterine devices are made of inert plastic material or metals of a variety of shapes. They are well tolerated and can be retained in utero for indefinite periods of time. Because of their elastic properties these devices can easily be introduced in their stretched form into the uterine cavity, where they assume their original shape. The addition of zinc and copper to the IUDs improves the contraceptive effect. Zinc was found to dissolve too rapidly, but the modified Tatum-T device made of plastic wrapped with thin copper wire lowers the failure rate to 1 per 100. Side-effects such as expulsion and hypermenorrhea seem to be less pronounced than those of the larger devices without metal (Fig. 99).

Application: Insertion is best accomplished on the last day of the period or during the following two days. Asepsis is mandatory. The patient should have a pelvic examination and the uterus sounded to ascertain the length and direction of the uterine cavity and to determine the proper size of the IUD. The IUD is inserted through the cervical canal with the help of a tubular introducer (Fig. 100). Multiparous women generally do not require cervical dilatation. Nulliparous women may require dilatation of the cervix up to Hegar #4, which can be done painlessly under paracervical block anesthesia. Follow-up examinations are recommended after the first and second periods and every six months thereafter. The IUD is provided with strings that protrude through the cervical canal and allow easy removal. The strings permit

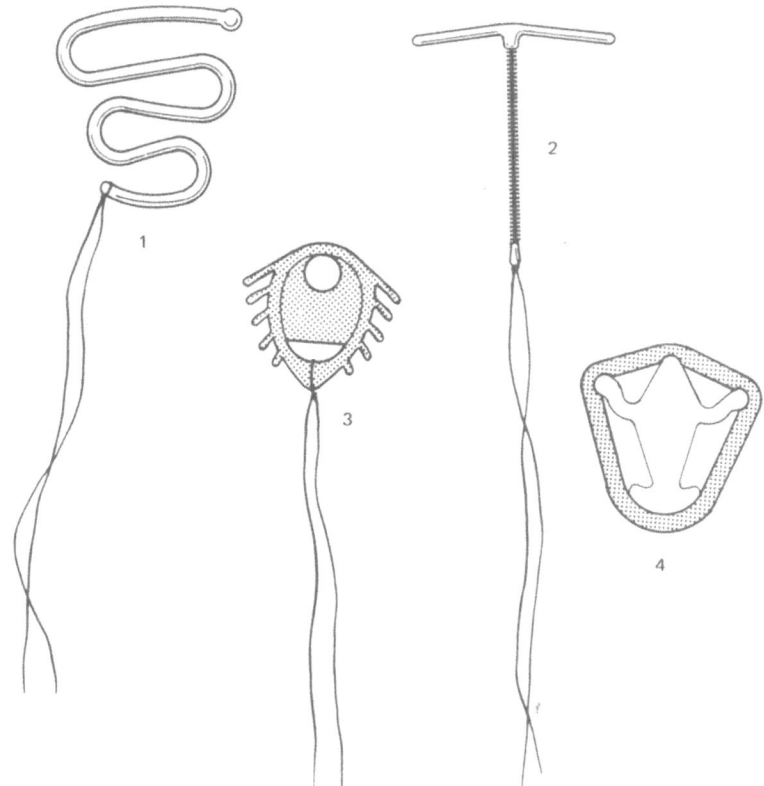

Fig. 99. Various forms of intrauterine devices: (1) Lippes loop; (2) Dalkon shield; (3) Tatum-T with copper; (4) Antikon.

the patient to feel the IUD and know that it is in place. Insertion earlier than 5 weeks after delivery is associated with a higher expulsion rate.

Complications include uterine perforation and ascending infection. The danger of perforation is greatest when the insertion of the IUD is performed by less than highly skilled personnel. The rate of this complication varies between 1:200 and 1:2000 insertions. Perforations can be recognized on a scout film of the abdomen. Pelvic inflammatory disease is caused primarily by lack of asepsis during insertion by inexperienced personnel or by exacerbation of an occult preexisting infection. It complicates about 2 to 3% of insertions.

The mechanism of action of the IUD is unknown. For some time it was assumed that the IUD influences tubal motility and thereby accelerates the transport of the egg, upsetting synchronization of development of egg and endometrium and preventing fertilization or nidation. Another hypothesis was that nidation is prevented by microthrombosis of the endometrial vessels. Neither concept has been substantiated. At present

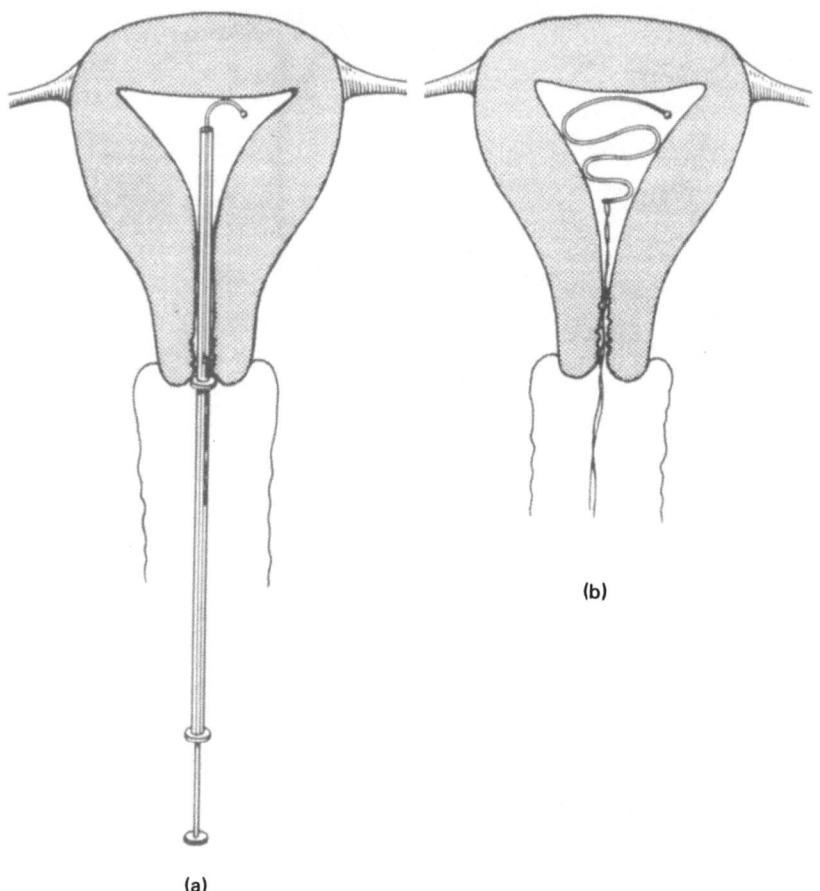

(b)

(a)

Fig. 100. Use of the Lippes Loop: (a) Introduction of the Lippes loop; (b) Lippes loop in situ.

the "leucocyte concept" is favored. Soon after insertion the IUD is coated with innumerable leucocytes. It is therefore assumed that leucocytic enzymes (lysozymes) might proteolytically destroy the blastocyst or spermatozoa.

Side effects of IUDs: One side effect may be spontaneous expulsion. The reliability of the method is compromised by the relatively high rate of spontaneous expulsion, which varies according to the shape of the IUD from 0.5 to 10%. The Lippes Loop has a spontaneous expulsion rate of 10.4% one year after insertion. Two and three years after insertion the spontaneous expulsion rates are 1.6% and 0.7%, respectively. The spontaneous expulsion rate is also higher if the IUD is inserted in the postpartum period, before complete uterine involution. Nulliparous women have a higher spontaneous expulsion rate than do multiparous women.

Another side effect may be pain and bleeding. After insertion, patients, particularly nulliparas, experience pains or vaginal spotting, which generally subside in a few hours. Prolonged or heavy vaginal bleeding requires the removal of the IUD, an event that occurs about once in 100 cases. An attempt can be made to use a different type of device.

If pregnancy occurs in patients with IUDs in place, about 60% will proceed to term. The rate of fetal malformations is not increased. Ectopic pregnancy is relatively increased but absolutely decreased in patients with IUDs.

Contraindications to the insertion of the IUD are:

(1) genital infections
(2) irregular bleeding
(3) myomas of the uterus
(4) suspicious Papanicolaou smear
(5) suspicion of early pregnancy

The failure rates of IUDs vary with the individual devices and represent, collectively, between 0.5 and 4 conceptions per 100 woman-years.

The intrauterine devices are especially appropriate for women who want to space their pregnancies. They are also indicated in women who cannot take oral contraceptives because of side-effects or other contraindications or who are unreliable (women with low intelligence or psychiatric problems). The widest use of the IUD is in developing countries.

Hormonal Contraception

A temporary inhibition of ovulation by injections of estrogen or progesterone was noted in animal experimentation as long ago as the 1930s. The progesterone preparations available at that time, however, were orally inactive. Hormonal contraception became available on a large scale with the development of orally effective synthetic steroids with strong progestational activity. Pincus and Rock were the first to undertake clinical studies of such synthetic orally active gestagens for inhibition of ovulation.

It became evident that the estrogen component reduced intermenstrual bleeding and produced withdrawal bleeding similar to that of a menstrual period. The gestagen initially employed was norethynodrel, a derivative of nortestosterone. When given in combination with estrogen, from the fifth to the twenty-fifth day of the cycle, this preparation could suppress ovulation and induce menstruation-like bleeding 2 to 3 days after withdrawal. Presently, many derivatives of nortestosterone, substituted with ethinyl or methyl groups, as well as a few derivatives of progesterone are used in "the pill" (Fig. 101). They are utilized

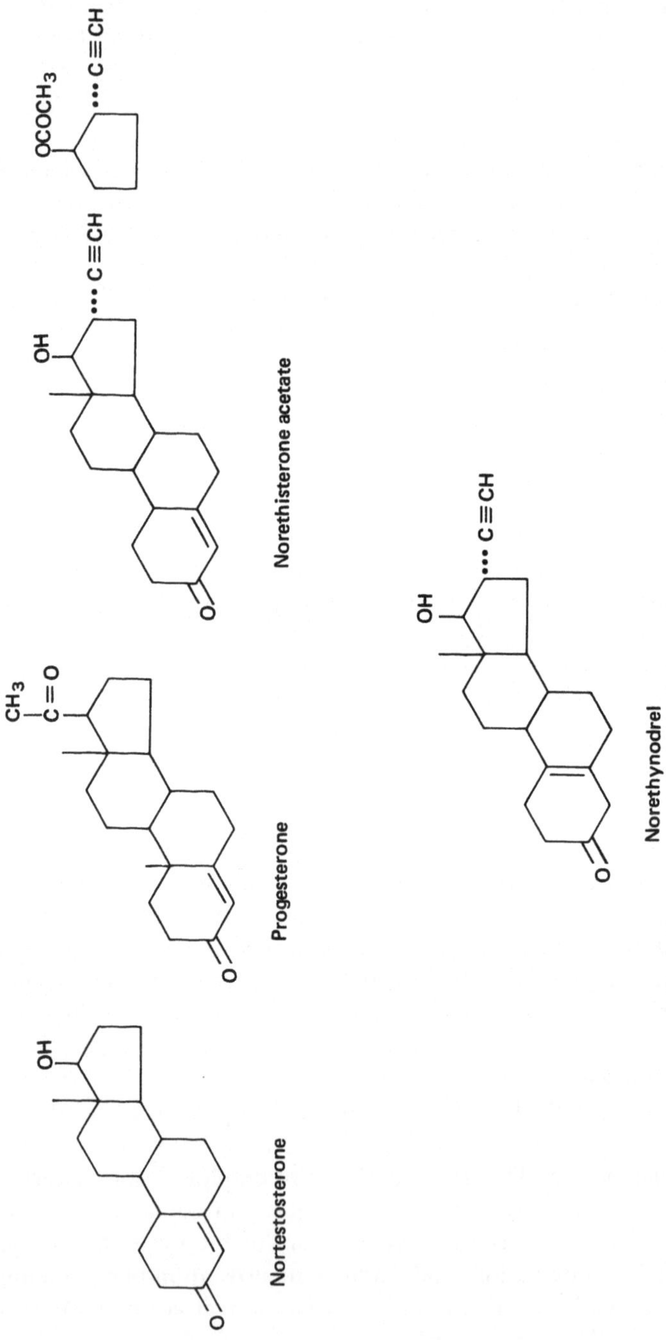

Fig. 101. The most frequently used progestins in oral contraceptives.

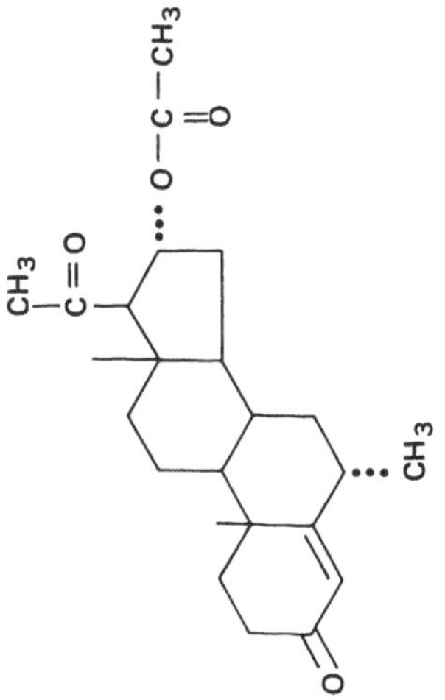

Quingestanol acetate

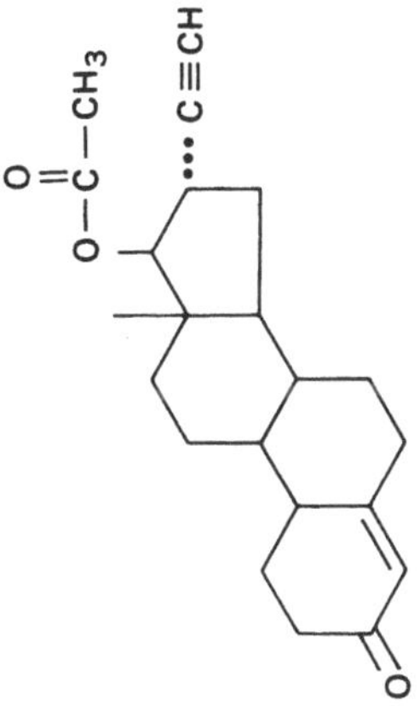

Ethynodiol diacetate

Medroxyprogesterone acetate

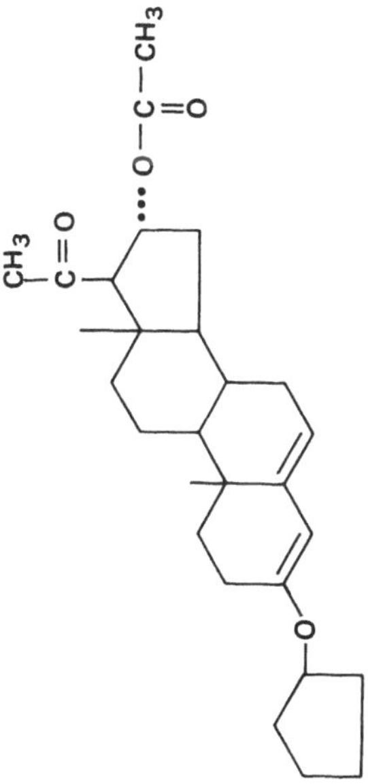

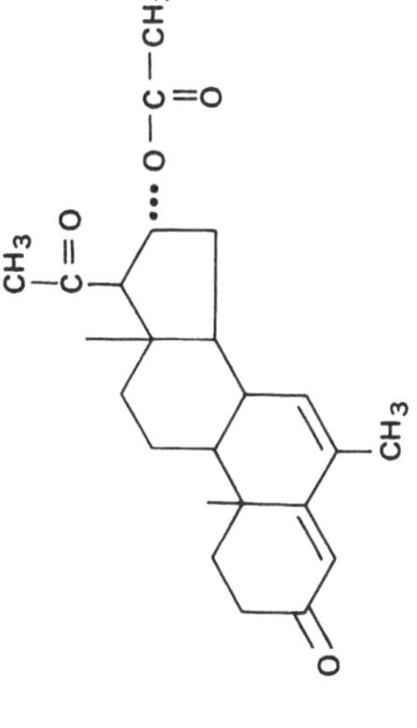

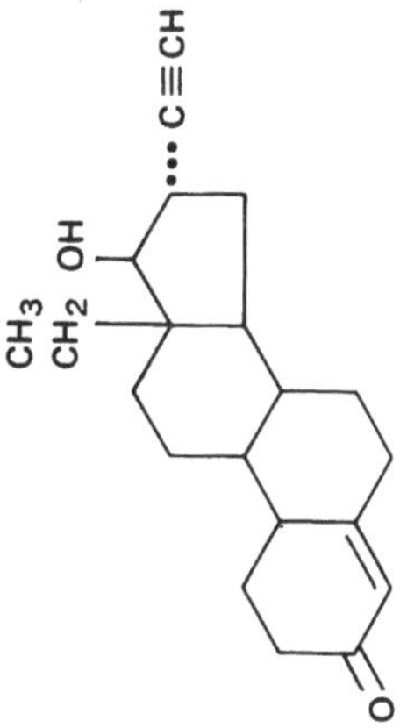

Norgestrel

Megestrol acetate

in combination with one of two main estrogens (Fig. 102). Estrogens and gestagens can be administered in two ways—in combination or sequentially.

Combination Method. Combination tablets contain a mixture of estrogens and gestagens. This form of administration corresponds to the original "pill" of Pincus and Rock. During each cycle one pill is taken daily from day 5 to 25 or 26. The proportion of estrogen to gestagen varies from preparation to preparation. The contraceptive action of the combination preparation affects the hypothalamic-pituitary system, the ovary, the endometrium, and the cervical secretion. The effect on tubal motility is not clear.

The Effect on the Hypothalamic-Pituitary System. The secretion of FSH is moderately inhibited by the estrogen component, and rupture of the follicle is suppressed (ovulation inhibition). Details of the mechanism of inhibition of the hypothalamic-anterior pituitary system are unknown. At least 10% of women who ingest estrogen-gestagen preparations will show "breakthrough ovulation." The great reliability of the combination preparation must therefore depend on other factors as well.

The Effects on the Ovary. The steroids act on the pituitary-hypothalamic centers, thereby decreasing the secretion of gonadotropins. The effect on the ovary is, therefore, indirect, but some investigators believe there is a direct action on steroid metabolism of the ovary.

The Effect on the Endometrium. The endometrium reveals rather characteristic changes. The stroma appears well developed with a pseudodecidual reaction. Nidation of the fertilized ovum requires an endometrium that corresponds morphologically and biochemically closely to the day of the cycle (see page 35). Even small disturbances of synchronization are sufficient to prevent nidation. Thus, the unphysiological action of the hormones on the endometrium must be considered an additional factor that prevents nidation.

The Effect on the Cervical Secretion. Another contraceptive effect is the hormone-induced change in the physicochemical characteristics of the cervical mucus. Estrogens render the mucus more easily penetrated by spermatozoa. Gestagens have the opposite effect. The gestagen concentration in the combination preparation is sufficient to inhibit penetration by sperm. The mucus effect, the ovulation inhibition, and the endometrial changes are probably the most significant contraceptive factors explaining the reliability of the combination preparations.

The Sequential Method (two-phase method). With this method of oral contraception estrogens are administered for the first one or two weeks of the cycle and followed by an estrogen-gestagen combination. Ovulation inhibiton is achieved with the estrogens alone, and the estrogen-gestagen combination guarantees a withdrawal bleeding similar to that of a menstrual period. The original method of administration

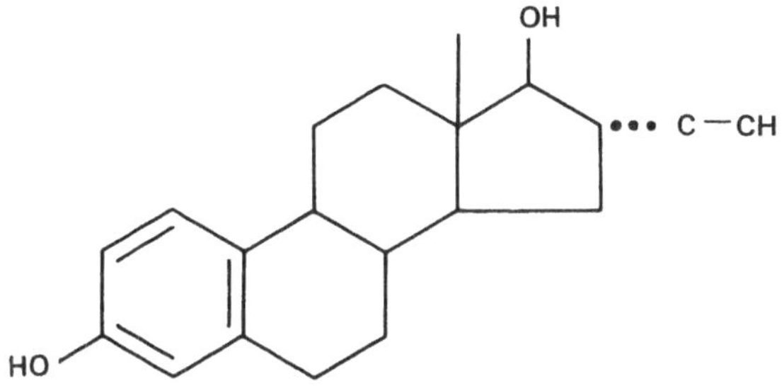

Ethinylestradiol

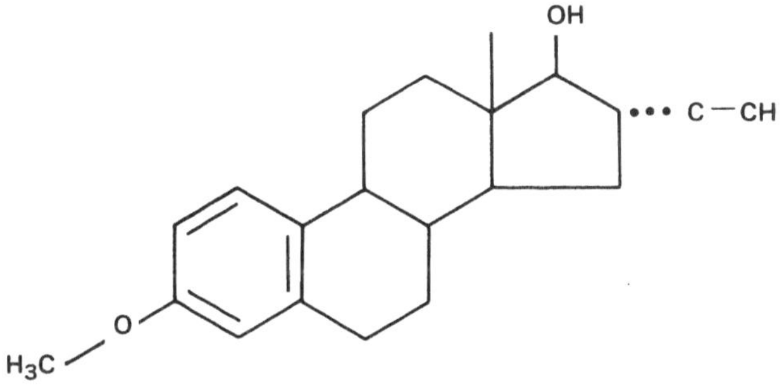

Mestranol
(3-methylether of ethinylestradiol)

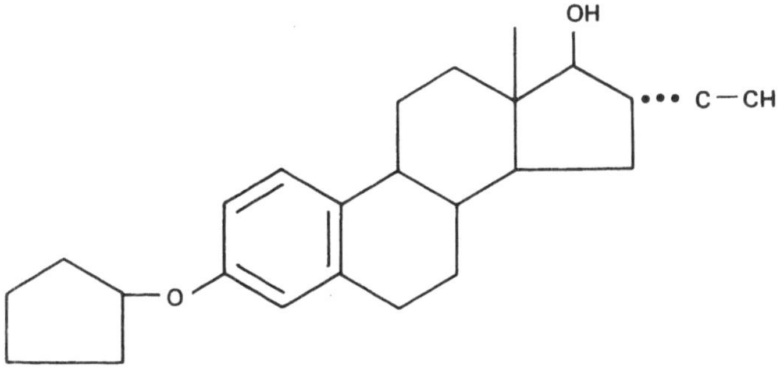

Quinestrol
(3-cyclopentylether of ethinylestradiol)

Fig. 102. The most frequently used estrogens in oral contraceptives.

was a 14 + 7 combination. Estrogens are ingested from day 5 to 18 and an estrogen-gestagen combination from day 19 to 25. This method was subsequently varied in many ways (by the 16 + 5 or 11 + 10 combination).

The endometrium is proliferative as a result of unopposed estrogens, thereby preventing nidation; but the cervical mucus is optimally penetrable by sperm. "Breakthrough ovulations" may lead, on occasion, to pregnancies and may explain the failure rate of 0.5 to 1.5 per 100 woman-years.

The "step-up" method was developed in order to reintroduce the protection of the cervical factor. Minimal doses of gestagen are added to the estrogen tablets and only during the last 7 days is the estrogen-gestagen dose given in the usual concentration.

Single Administration of Either Estrogens or Gestagens

Estrogen. Estrogen alone inhibits ovulation and is an effective contraceptive. Discontinuation of the estrogen, however, leads to prolonged estrogen withdrawal bleeding and is therefore not suitable in practice.

High doses of estrogen, when given after conception, can interfere with implantation of the blastocyst. This is the basis for the "morning-after pill." Within 72 hours after intercourse high doses of estrogen are given for 5 days (50 mg stilbestrol or 2.6 mg ethinyl estradiol per day). Approximately 4 days after the last dose, prolonged withdrawal bleeding with expulsion of the uterine contents occurs. Severe nausea, which occurs in most instances, explains why the morning-after pill is reserved for exceptional situations, for example after rape.

Gestational Agents. Most gestational agents derived from progesterone or nortestosterone inhibit ovulation only in very high doses.

The principle of the "three-month injection" is the parenteral administration of a high dose of a long-acting gestagen. The effect of an injection of 150 mg of a progesterone derivative (medroxyprogesterone acetate) lasts for at least 90 days. The daily release of the gestagen exceeds 1 mg and inhibits production of gonadotropin, especially LH, and thereby prevents ovulation. The endometrium becomes progressively atrophic and the cervical mucus impenetrable to sperm. The failure rate is 2.6 per 100 woman-years. Headache, depression, and decrease of libido are listed among its side effects. Irregular breakthrough bleeding in the form of prolonged vaginal spotting or metrorrhagia is common. It occurs after first injection in about 80% of cases, and 12 months later still affects approximately 50% of the patients. These side effects frequently require discontinuation of the method. Persisting amenorrhea after discontinuation of the gestagen is another serious complication. For these reasons, the three-month injection is not widely used.

Use of Oral Contraception

It is generally agreed that oral contraceptives can be given in a cyclic fashion for many years. The U.S. Food and Drug Administration has lifted the initially imposed time limitation. It may be desirable to interrupt the regimen for 2 to 3 months after 2 to 3 years of administration to see whether the cycles will occur spontaneously. Many patients have utilized oral contraception for much longer periods of time with no permanent, discernible ill effects.

An additional problem for the physician is the prescription of hormonal contraceptives to minors. Steroids should be prescribed to patients under the age of 16 with caution because of possible effects on somatic growth and emotional stability.

The minimal requirements for the prescription of steroid contraceptives are indicated in Table 20.

Side-Effects of the Oral Contraceptive. Oral contraception may cause a variety of side-effects. Objective consideration requires an exact evaluation of complaints that existed before the administration of the hormones. Psychological factors cannot be excluded since some women will demonstrate the identical subjective complaints when they are taking placebos.

Experience with hormonal therapy has suggested that some of the compaints are attributable to estrogen and others to the gestational agent. Estrogen seems to be responsible for nausea, vomiting, sodium and water retention, premenstrual tension (including breast complaints), headaches, cervical hypersecretion, and hyperpigmentation.

Table 20. Minimal Prerequisites for the Prescription of Oral Contraceptives

---

Before treatment
1. Exclusion of contraindications through careful history and physical examination
2. Examination of the genital organs and breasts including vaginal cytology
3. Checking the urine for sugar to exclude diabetes
4. Measuring blood pressure
5. Selection of an appropriate preparation

During treatment
1. Semiannual gynecologic examination including vaginal cytology. Breast examination. Precise inquiry into possible complaints.
2. Semiannual tests for sugar in urine
3. Semiannual measurement of blood pressure
4. Investigation of liver function if there are complaints referable to the liver or gall bladder
5. Rechecking the adequacy of the contraceptive method

Gestagens seem to be responsible for increased appetite and weight gain, fatigue, depression, and decreased libido. Nortestosterone derivatives sometimes produce acne, seborrhea, hypertension, loss of hair, headaches, and hypomenorrhea or amenorrhea.

On the basis of this outline, the most suitable preparation for a given patient can be selected. Some side effects are abolished by switching to another preparation. Table 21 lists the commercially available preparations as either predominantly estrogen, balanced, or predominantly gestagen. Since the gestagens within the various preparations have different biological activities, one cannot compare them solely on the basis of dosage in milligrams of drug. They are evaluated according to progestational activity, the antiestrogenic action, and the potential conversion of the gestagen to estrogen.

Nausea: Five to 10% of all women complain of nausea when starting oral contraceptives. With prolonged ingestion there is a decreasing prevalence of this complaint. Taking the pill after the evening meal with plenty of fluid reduces the nausea. Approximately 1% of patients experience vomiting. If vomiting occurs soon after the ingestion of the pill, contraceptive protection will be lost.

Irregular bleeding: The acyclic vaginal bleeding that occurs with oral contraception is called breakthrough bleeding, which appears mostly in the form of unpredictable vaginal spotting. In most instances, it occurs during the first months of oral contraception. The patient is advised not to stop taking the pill, but should bleeding recur, an attempt can be made to administer additional estrogens or temporarily double the daily dose. In other instances, it may be advisable to change the preparation. If the breakthrough bleeding occurs in the first half of the cycle, a predominantly estrogenic preparation may be advisable; if it occurs in the second half of the cycle, it is preferable to use a predominantly progestational preparation.

Occasionally the withdrawal bleeding fails to occur. When pregnancy is excluded, ingestion of the pill is continued after the usual interval.

Libido: Depending on the population and the preparation utilized, there is a 3 to 60% increase of libido, which can be attributed partially to lack of fear of pregnancy. A decrease of libido is noted less frequently (approximately 3 to 10%). A preparation with a higher estrogenic and a lower progestational content may eliminate this problem. Frequently, the decrease of libido cannot be reversed by changing the preparation. In that event it may be necessary to use another form of contraception.

Depression: Five to 8% of patients who are taking oral contraceptives complain of depression, especially when the preparation has a high progestational or a relatively low estrogenic content. A more appropriate preparation may effectively relieve this symptom.

Weight gain: Ten percent of women gain weight, partly because of a transient estrogen-induced retention of water. Another reason is the anabolic and appetite-stimulating effects of progestational agents. Use of pills with a lower progestational content and avoidance of derivatives of nortestosterone may be helpful.

These side-effects are considered relatively harmless. The physician should be aware, however, that continuous ingestion of unnatural hormones may alter certain physiologic functions.

Cystic disease of the breast and uterine myomas may be relative contraindications to oral contraceptives, since they may progress under the influence of the steroid hormones. If these agents are used, low-estrogen and high-gestagen compounds are preferable. Follow-up examination is required more frequently.

Amenorrhea: Prolonged amenorrhea is occasionally seen after termination of oral contraception. This type of amenorrhea is the result of interference with the pituitary-hypothalamic axis and can be treated by induction of volulation with clomiphene (see page 151).

During oral contraception, the cycles become regular, but upon discontinuation of the hormones 90% of the women experience the same cyclic irregularities as before. The steroids reduce hypermenorrhea, menorrhagia, dysmenorrhea, Mittelschmerz, and the premenstrual syndrome during treatment. Acne and hirsutism can be avoided by choosing the proper preparation.

Ovarian function: No histopathological changes in the ovaries can be detected in women taking steroids, but maturation of the follicles after discontinuation of the pill is usually delayed. The regulatory mechanism apparently requires a definite span of time to reestablish normal endocrine balance. Fertility is not adversely affected and there is no indication of any mutagenic action. The frequency of chromosomal aberrations in the lymphocytes of peripheral blood is not increased in patients taking oral contraceptives. The frequency of fetal malformations after discontinuation of oral contraceptives is not increased, but an increased frequency of numerical chromosomal anomalies was observed in abortion material from women who have conceived within 6 months after discontinuation of the contraceptives. Conception should be delayed until the cycle has again become normal.

Thyroid gland: Approximately 60% of women demonstrate a rise in PBI and ^{131}I-uptake after ingestion of estrogen-gestagen combinations. This rise is similar to that in pregnancy and does not suggest thyroid disease.

Adrenal cortex: The adrenal cortex is slightly inhibited by some steroid combinations. The capacity to synthesize corticosteroids as well as the ability to respond to ACTH remains unchanged, even after prolonged oral contraception.

Table 21. Oral Contraceptives Available in the United States, October 1970

Agent	Estrogen	Progestin	Manufacturer	Estrogenic Potency	Progestational Potency
Ortho-Novum-1	Mestranol 0.05 mg	Norethindrone 1 mg	Ortho	± Very low	+ Low
Ortho-Novum 1/80–21	Mestranol 0.08 mg	Norethindrone 1 mg	Ortho	+ Low	+ Low
Ortho-Novum-2	Mestranol 0.10 mg	Norethindrone 2 mg	Ortho	+ Low	+++ High
Norlestrin-1 Norlestrin-(21) Norlestrin-(28) (21 + 7 placebo) Norlestrin-(Fe) (21 + 7 Fe)	Ethinyl estradiol 0.05 mg	Norethindrone acetate 1.0 mg	Parke-Davis	+ Low	++ Intermediate
Norlestrin 2.5	Ethinyl estradiol 0.05 mg	Norethindrone acetate 2.5 mg	Parke-Davis	++ Intermediate	+++ High
Demulen	Ethinyl estradiol 0.05 mg	Ethynodiol diacetate 1.0 mg	Searle	± Very low	++ Intermediate
Ovulen Ovulen 21	Mestranol 0.10 mg	Ethynodiol diacetate 1.0 mg	Searle	+ Low	++ Intermediate
Enovid-E	Mestranol 0.10 mg	Norethynodrel 2.5 mg	Searle	++ Intermediate	+ Low
Enovid-5	Mestranol 0.075 mg	Norethynodrel 5.0 mg	Searle	++ Intermediate	+++ High
Norinyl-1 Noriday (21 + 7 placebo)	Mestranol 0.05 mg	Norethindrone 1 mg	Syntex	± Very low	+ Low
Norinyl-2	Mestranol 0.10 mg	Norethindrone 2 mg	Syntex	+ Low	+++ High

Product	Estrogen	Progestin	Manufacturer	Estrogen activity	Progestin activity
Provest	Ethinyl estradiol 0.05 mg	Medroxyprogesterone acetate 10 mg	Upjohn	± Very low	± Very low
Ovral	Ethinyl estradiol 0.05 mg	Norgestrel 0.5 mg	Wyeth	+ Low	++ Intermediate
C-Quens	15: Mestranol 0.08 5: Mestranol 0.08	Chlormadinone acetate 2 mg	Lilly	+++ High	± Very low
Oracon	16: Ethinyl estradiol 0.10 mg 5: Ethinyl estradiol 0.10 mg	Dimethisterone 25 mg	Mead Johnson	+++ High	± Very low
Ortho-Novum Sq	14: Mestranol 0.08 mg 6: Mestranol 0.08	Norethindrone 2 mg	Ortho	+++ High	± Very low
Norquen	14: Mestranol 0.08 6: Mestranol 0.08	Norethindrone 2 mg	Syntex	+++ High	± Very low

*Estrogen Excess
Nausea
Edema & leg cramps
Vertigo
Leukorrhea
Increase in leiomyoma size
Chloasma
Uterine cramps

*Estrogen Deficiency
Irritability, nervousness
Hot flushes
Uterine prolapse
Monilia vaginitis
Early and midcycle bleeding
Decreased amount of menstrual flow

*Progestin Excess
Increased appetite and weight gain
Tiredness and fatigue
Depression, change in libido
Oily scalp, acne
Loss of hair
Cholestatic jaundice
Decreased length of menstrual flow

*Progestin Deficiency
Late breakthrough
Heavy menstrual flow and clots
Delayed onset of menses

---

*Symptoms with Multiple Causes

Headache:
During medication cycle: estrogen excess
Between medication cycles: progestin excess
Migraine and visual disturbances: unknown
Mastalgia:
With fluid retention: same as headache
Without fluid retention: progestin excess
Spotting and breakthrough bleeding:
Early and midcycle: estrogen deficiency
Late cycle: progestin deficiency

---

* Dickey, R. P. & Dorr, C. H. II Oral Contraceptives: Selection of the Proper Pill, Obstetrics and Gynecology 33:273, 1969.

Pigmentation: Contraceptive steroids should be avoided if the patient has a tendency to facial pigmentation or if she experienced chloasma during previous pregnancies. The hyperpigmentation resulting from contraceptives is rare but does not always regress after termination of use of the pill.

Carbohydrate metabolism: Forty percent of patients show a decreased glucose tolerance, which is, however, reversible. The WHO Commission concluded that the risk of precipitating diabetes as the result of oral contraceptives is less than that of pregnancy. A diabetogenic action of contraceptives is not yet proved, but frequent glucose tolerance tests may be advisable in women who are on long-term therapy and have a metabolic or genetic predisposition to diabetes. Women with overt diabetes should use other means of contraception.

Fat metabolism: Steroids can induce changes in the levels of cholesterol, triglycerides, and unsaturated fatty acids. It is not clear whether these changes have any pathological significance.

Women who have hyperlipidemia should avoid hormonal contraceptives, since there is a correlation between hyperlipidemia and thromboembolic diseases.

Protein metabolism: Estrogen-gestagen combinations cause a significant drop in serum albumin and albumin-globulin ratio. The α_1-globulin fraction, however, especially the α_1-trypsin inhibitor, increases. These changes again are similar to those in pregnancy.

Liver function: The administration of oral contraceptives is contraindicated in patients with a history of hepatitis, jaundice, pruritus of pregnancy, and hepatic enzyme defects, as in the Dubin–Johnson and Rotor syndromes. Approximately 3% of women show a temporary rise of transaminases (SGOT, SGPT). The bromsulphalein retention is prolonged, especially at the beginning of drug administration, presumably reflecting a cholestatic effect. Failure of the transaminase values to become normal or the demonstration of elevated bilirubin values is reason for discontinuation of steroid contraceptives, but it is generally agreed that oral contraceptives do not cause liver damage.

High blood pressure has been observed in a few instances following administration of oral contraceptives. The rise is temporary and seems to be related to an estrogen-induced interference with the renin-angiotensin-aldosterone system. Persistent high blood pressure requires discontinuation of this form of contraception. In general, hormonal contraception presents a risk for women with documented hypertension. Mechanical forms of contraception are, therefore, preferable.

Migraine headaches: Approximately 8 to 12% of women on oral contraceptives complain of migraine headaches. It is assumed that oral contraceptives are initiating factors in patients with a natural predisposition to this complaint. EEG irregularities suggest changes within the hypothalamus. In many cases, there complaints can be eliminated by switching to a preparation with a low-estrogen content or to the minipill (progestin only). If no improvement is obtained, oral contraceptives should be discontinued.

Thromboembolic complications: Retrospective studies in England have focused attention on a possible correlation between oral contraceptives and the frequency of thromboembolic disorders (thrombophlebitis, phlebothrombosis, cerebral thrombosis, and pulmonary embolism).

The frequency of fatal embolic complications in women 35 to 44 years of age on oral contraceptives was estimated at 3.9:100,000, compared with 0.5:100,000 in a control group (Table 22). Studies in Scandinavia, Canada, and the United States revealed similar numbers. It is assumed that hormonal contraceptives increase the risk of embolism by a factor of at least 4.4, whereas pregnancy increases this risk by a factor of 11 or more.

There is now ample evidence suggesting that estrogen is the main causative steroid. An increase in coagulation factors appeared related to the estrogen intake, depending on the dose and time of ingestion. An increase was noted in coagulation factors II, VII, VIII, and X, as well as in fibrinogen and plasminogen. Large molecular fibrin monomers are detectable in women taking estrogen-containing birth control pills. In addition, an increase in α_1-trypsin inhibitor has been noted. The result is inhibition of approximately 90% of the antiplasmin activity of the human body. This increase is correlated with a decrease in antithrombin III activity.

All changes are similar to those in pregnancy, but it is puzzling that thromboembolic phenomena are not increased so much in pregnancy as in the puerperium. It seems, therefore, that there is a biochemical change in the direction of a hypercoagulable state, but only a few women with a predisposition may develop thrombotic complications.

It may be possible that the small risk is related not to biochemical

Table 22. Frequency of Fatal Pulmonary Emboli Calculated on the Basis of 100,000 Women on Oral Contraceptives (during and after pregnancies)

	Age group (years)	
	20–34	35–44
Controls	0.2	0.5
On oral contraceptives	1.3	3.4
During and after pregnancies	22.8	57.6

but rather to vascular factors. Nevertheless, women with a history of thromboembolic disease or venous thrombosis should avoid oral contraception. Varicose veins are not necessarily a contraindication to oral contraceptives (Table 23) unless they increase in size or number during use of the drugs.

In summary, statistical data indicate that in rare instances thromboembolic complications may occur, relative to the estrogen content of the pill. By lowering the amount of estrogen in individual preparations, the rate of complications can be decreased.

Carcinogenic effects: There is no evidence of a carcinogenic effect of these steroids.

Some data have suggested that women taking oral contraceptives have a lower incidence of suspicious Papanicolaou smears, carcinoma in situ, early cervical cancer, and cancer of the endometrium and breast, than do control groups. Other data have suggested the contrary. Results of animal experiments cannot necessarily be transposed to the human being. It may require up to 20 years for malignant transformation to become manifest in human subjects.

Preparations under Investigation

A variety of new contraceptive techniques are under investigation.

The "once-a-month injection" contains an estrogen-gestagen combination lasting 4 weeks. The side-effects are generally few, but frequent

Table 23. Contraindications to Oral Contraception

Contraindications	Relative contraindications
Thromboembolism	Severe cardiac and renal diseases
Cerebral and cardiovascular diseases	Hyperlipemia
	Epilepsy
Hypertension	Otosclerosis
Diabetes (especially with vascular involvement)	Migraine
	Tendency to hyperpigmentation (chloasma)
Hepatitis and its sequelae	Prolonged anovulatory cycles
Pregnancy	(oligomenorrhea)
Pruritus and icterus of pregnancy	
Enzymatic disturbances (Rotor syndrome, Dubin–Johnson syndrome)	Genital fungous infections
	Lactation
Porphyria	
Sickle cell anemia	

breakthrough bleeding as well as prolonged and increased withdrawal bleeding has been observed. The failure rate is 1 to 2 per 100 woman-years.

The once-a-month pill contains an estrogen that exerts depot activity for 4 weeks and a short-acting gestagen. The continuous level of estrogen suppresses ovulation, while the gestagen component causes the withdrawal bleeding. Ingestion occurs several days before the expected period.

The failure rate is low (1.9 per 100 woman-years). Severe nausea and breakthrough bleeding occur in approximately 20% of women, especially during the first few cycles.

Since the side-effects of the estrogen-gestagen combinations are attributed primarily to the estrogens, attempts are being made to develop an oral contraceptive with a low estrogen content. This experimentation has led to pills that are free of estrogens, consisting of only progestational agents. In the "minipill" the dose of gestagen is so low that it does not influence the pituitary-ovarian-uterine axis, but it is high enough to alter the cervical mucus and affect its penetration by sperm.

Additional methods involve the implantation of crystals or silicone capsules containing gestagens under the skin. Attempts have also been made to incorporate gestational agents into vaginal pessaries or intrauterine devices.

The mode of action of continuous small quantities of gestagen is still not clear.

The minipill can be ingested cyclically or on a continuous basis. Extensive observations concerning chlormadinone acetate and megestrol acetate are available. The good contraceptive effect is compromised by frequent intermenstrual breakthrough bleeding and vaginal spotting.

Silicone capsules containing gestagen, when implanted into the upper arm, facilitate a continuous release of small quantities of gestagen through the capsule into the surrounding tissue. The duration of the effect depends on the amount of gestagen and the number of capsules implanted. It can last for months or years. A common side effect (approximately 28% of women) is irregular uterine bleeding. This form of administration is probably most suitable for developing countries.

Gestagens are similarly incorporated into vaginal plastic rings and IUDs. Absorption occurs through the vaginal mucosa or endometrium. It was assumed that intrauterine devices containing gestagens would prevent uterine contractions and thereby decrease the chances of their expulsion.

Hormonal contraception initiates a continuous or repetitive interference with reproductive function. This fact and the frequency of side effects necessitates the development of additional pharmaceutical agents that could provide the same reliability with simpler and safer application and less interference with physiological functions. Prostaglandins require further study in this regard. At present these substances are under investigation as vaginal suppositories. If applied vaginally after a missed period, they induce bleeding on the same or the following day. In addition to contraction and vasoconstriction of the uterus, a luteolytic effect, which abolishes the function of the corpus luteum, may be operative.

Tubal Sterilization

Operative sterilization of the woman is performed by ligation or resection of the fallopian tubes. The operation may be performed through the vagina, the abdomen, or the laparoscope.

Methods of Male Contraception

Coitus Interruptus. Interruption of coitus before ejaculation cannot be considered a reliable method of contraception. The failure rate is 35 per 1200 cycles. Comparative data for coitus reservatus (prolonged coitus without ejaculation) are not available.

Condom. The condom is one of the most commonly used methods of contraception. The reliability, 3 to 14 per 100 years of utilization, can be improved if spermicidal creams are used. The condom provides simultaneous protection against venereal disease. This method is most applicable when the male partner wishes to take the responsibility for contraception. Failures result primarily from repetitive coitus within a brief period of time. Many men object to the condom on esthetic grounds and because of the occasional blunting of penile sensations.

Vasectomy. Male sterilization is becoming more popular in the United States, largely as a result of changing mores. The usual operation consists of dividing the ductus deferens (vas deferens) bilaterally through a scrotal incision. It has the advantage of being performed in the office under local anesthesia. There are no effects on androgen production, libido, or sexual performance. The continuity of the ductus may be restored in somewhat less than half the cases surgically (vasovasostomy). Recent reports of autoimmune disease as a result of vasectomy have not been substantiated.

Selected Readings

Calderone, M. S.: *Manual of Family Planning and Contraceptive Practice.* Baltimore: Williams and Wilkins (1970).

Ehrlich, P. R.: *The Population Bomb, 13th ed.* New York: Ballantine Books (1970).

Haller, J.: *Ovulationshemmung durch Hormone, 3rd ed.* Stuttgart: Thieme (1970).

Hardin, G.: *Population, Evolution and Birth Control.* San Francisco: Freeman (1969).

Peel, J., and Potts, M.: *Textbook of Contraceptive Practice.* Cambridge: University Press (1969).

CHAPTER 18

Gynecological Infections

Inflammation of the Vulva

Vulvitis may be caused by a variety of factors including:

(1) inflammation limited to the vulva

(2) inflammation secondary to a focus elsewhere in the genital tract

(3) a local manifestation of a dermatologic or systemic infectious disease.

Although the causes of vulvitis are variable, the symptoms in each case are similar. Most common is itching, or pruritus, followed by prolonged scratching and trauma to the skin. The inguinal lymph nodes are often enlarged in acute vulvitis and are occasionally tender.

Primary Localized Inflammation of the Vulva. Simple vulvitis is a diffuse inflammatory lesion, predominantly allergic. External factors may specifically sensitize and irritate the vulvar epithelium. The irritant must be detected by obtaining a careful history. Possible irritants are soap, detergents, and antiseptic solutions (with or without perfumes) either applied directly or used in laundering underwear. Undergarments of synthetic fibers may themselves cause allergic reactions. Drugs such as antibiotics and sulfonamides may also be responsible. Sensitization may develop through local (intravaginal) application and through excretion in the urine, as in the case of barbiturates. An isolated eczematoid vulvitis may be the result.

Diagnosis: Erythema and swelling of the vulva are features of the acute stage. Vesicles may burst either spontaneously or as a result of scratching. Oozing, formation of crusts, and secondary infection are often the results. The chronic stage of vulvar inflammation is characterized by lichenification and hyperkeratosis.

Treatment: Removal of the offending agent is essential. Identifying the agent often requires skin tests with the suspected irritants. They are usually performed by the dermatologist or allergist. After removal of the irritant, healing is usually stimulated by applying ointments containing hydrocortisone and antibiotics.

Other Types of Vulvitis. Herpes progenitalis, or herpes genitalis, is a viral disease. The causative agent, herpesvirus hominis type-2, is related to herpes labialis (type-1 herpes simplex). Herpetic vulvitis and vaginitis may be venereally transmitted. On the vulva the lesion appears as a group of vesicles surrounded by diffuse erythema and edema. The lesion has a predilection for the prepuce of the clitoris, the labia minora, and the inner aspects of the labia majora. The pruritus may lead to scratching and secondary infection. The cervical lesion, which presents fewer symptoms, is present in about 75% of all cases of herpes genitalis. The cervix may be edematous, ulcerated, or granulomatous. The disease tends to recur but to be relatively free of systemic manifestations. Local signs and symptoms include leukorrhea, abnormal bleeding, dysuria, and dyspareunia. The differential diagnosis includes herpes zoster, condylomata, erythema multiforme, and in the cervix, carcinoma. Microscopic diagnosis is often made by finding the typical viral inclusion bodies. Treatment is usually symptomatic and the disease is self-limited. Recent experience with local idoxuridine seems promising.

The significance of type-2 herpesvirus hominis has been increased by the suggestion of a possible role of the infection in the genesis of cervical dysplasia and carcinoma (see page 291). The roles of herpetic infection, promiscuity, and smegma may thus be interrelated. Women with known herpetic lesions of the cervix should be examined frequently to detect an incipient epithelial dysplasia or premalignant change.

2. *Vulvar neurodermatitis* is a rather common cause of isolated chronic vulvitis.

Diagnosis: The typical lesions are brownish red, 2 to 3 cm in diameter, and arranged in three concentric zones. The external zone is pigmented, the middle zone is papular, and the central zone shows lichenification. Vitiligo-like depigmentation may be present. The major symptom is pruritus, which is paroxysmal and often induced by the warmth of the bed.

Treatment: Hydrocortisone ointments and local infiltration with cortisone are the preferred forms of treatment.

3. *Furunculosis* confined to the vulva is usually a staphylococcal secondary infection, resulting from scratching other lesions. Diabetes must be excluded. Actinomycosis of the vulva must also be considered in the differential diagnosis.

4. *Condylomata acuminata.* These papillomatous lesions are caused by a virus. They have a typical tree-like structure with a central core of connective tissue. The moist environment of the vulva favors the growth of the virus and the spread of the lesion. Although they are primarily inflammatory, the lesions are often grouped morphologically with benign neoplasms of the vulva. Since the virus can be transmitted by intercourse, the lesions are sometimes called venereal warts. Similar lesions may be found in the anus and rectum, particularly among male homosexuals.

Fig. 103. Condylomata acuminata of the vulva, perineum, anus, and thigh.

Diagnosis: The characteristic gross appearance usually leads to rapid diagnosis (Fig 103). The identification of any associated infection, particularly gonorrhea, is required. Chemical irritants should be sought if no other cause can be found.

Treatment: In addition to treatment of the associated infection, specific therapy of the condylomata is instituted. Small lesions are often successfully treated by local application of a 20% solution of podophyllin. Only large or resistant lesions require electrocauterization or cold knife excision.

5. *Bartholinitis and Bartholin abscess.* A frequent vulvar lesion is the inflammation of the duct of the Bartholin gland. These ducts, which drain bilaterally 1 cm above the posterior commissure between the labia minora and the hymenal edge, may be easily infected by *E. coli, staphy*lococci, streptococci, and especially gonococci. The inflammatory reaction usually occludes the openings of the ducts and leads to the collection of purulent exudate. The gland itself is usually not involved in the inflammatory process.

Symptomatology: Extremely tender unilateral erythema with swelling in the area of the opening of the Bartholin duct is characteristic. As suppuration continues, the tumor may reach 8 cm in largest diameter

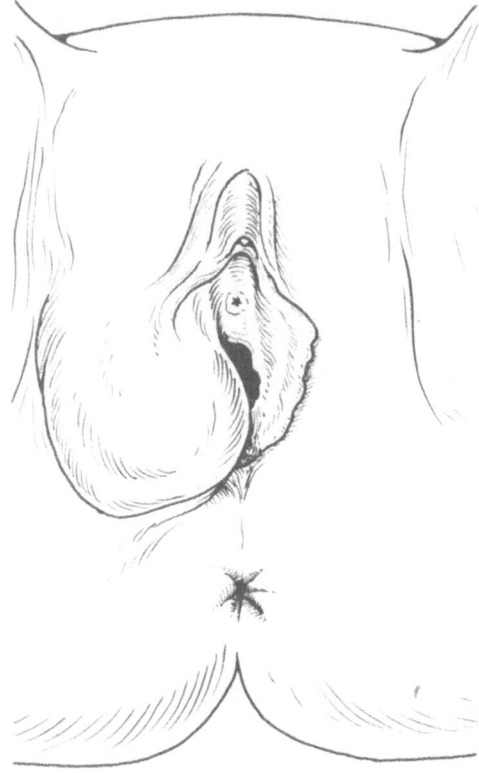

Fig. 104. Bartholin abscess. The labia majora are involved in the inflammatory swelling and the introitus is almost occluded.

and cause considerable pain during walking and sitting. It may obstruct the vaginal introitus and interfere with coitus and vaginal examination.

Diagnosis: The typical location aids the diagnosis by inspection and careful palpation (Fig. 104).

Treatment: Conservative measures such as heat may stimulate the pointing of the abscess, which is then incised near the opening of the duct. To prevent recurrence and development of a retention cyst, marsupialization is often performed (eversion of the abscess wall and suturing it to the overlying skin). Healing is usually complete in a few days, and after 3 to 4 weeks the site of the incision is no longer visible. Although marsupialization preserves the function of the gland, some gynecologists prefer excision when there is no sign of acute inflammation and the lesion is a cyst rather than an abscess. Instructions about proper hygiene of the genitalia should be given in order to prevent recurrence, and gonorrhea elsewhere must be ruled out.

Secondary Inflammation of the Vulva. In the majority of cases, vulvitis is associated with diseases of the genital tract that produce discharge. Vaginal discharge usually accompanies any chronic vulvar lesion.

Etiology: The increased prevalence in recent years of monilial vulvo-vaginitis caused by the *Candida albicans* fungus may be explained in part by the widespread use of antibiotics. As the normal bacterial inhabitants of the vagina are destroyed by antibiotics, the growth of yeast is favored. Humoral and cellular immunosuppression resulting from corticosteroids and chemotherapy are additional factors. Changes in the vaginal milieu secondary to the hormonal effects of pregnancy, Addison's disease, and hyperthyroidism may be aggravating factors. The relation of oral contraceptives to an increase in fungous infections is not yet clear. Vulvar pruritus is frequently seen in patients with diabetes mellitus. Cutaneous lacerations resulting from scratching may become infected by yeast, which causes a vulvitis that is particularly resistant to treatment.

Symptoms: Itching and burning are the principal symptoms.

Diagnosis: In monilial infection patches of grayish-white exudate are seen in the introitus, extending into the vagina and to the cervix. More frequently the monilial vulvitis appears as an acute or chronic inflammatory lesion without the characteristic patches. Without treatment, the process may extend in the form of an eczematous rash to the inguinal region and the inner surface of the thigh. Diagnosis is achieved by demonstrating the branching club-shaped hyphae in the wet preparation using 1 to 2% potassium hydroxide solution. The organism may be cultured on Nickerson's medium or a variety of other media.

Treatment: For local treatment specific antimycotic agents such as Mycostatin are available. The treatment of the underlying conditions is essential to prevent recurrence. Associated diabetes mellitus requires regulation.

Trichomoniasis of the vagina also frequently leads to a secondary vulvitis (page 223). Infestation with *Oxyuris* must be considered as a possible cause in the young girl. In addition, pediculosis pubis and scabies should be ruled out.

Vulvar Manifestations of Dermatological and Other Systemic Diseases. Dermatological conditions such as seborrheic dermatitis, psoriasis, and herpes zoster may involve the vulva. They are more frequently seen by the dermatologist than by the gynecologist, although they may appear in another form on the vulva.

Primary and secondary lesions of syphilis may also appear on the vulva.

Pruritus of the vulva with secondary infection and lichenification is usually caused by one of the previously mentioned conditions. The term "idiopathic" is frequently used to justify an insufficient diagnostic work-up. In some cases psychosomatic disease with a manifestation in the form of chronic vulvitis may respond to psychotherapy.

Recurrent inflammatory reactions of the vulva resulting from the vicious cycle of pruritus—scratching—secondary infection may actually result in dystrophic and dysplastic changes, some of which are potential factors predisposing to malignancy. Effective treatment of vulvitis may therefore be of additional value in preventing premalignant changes.

Inflammation of the Vagina (Vaginitis)

Physiological Environment of the Vagina and Its Protective Function. Exudate formed by vaginal epithelium provides a continually moist environment. The pH of the vagina is normally approximately 4.0 and it is maintained primarily by lactic acid. There are some minor variations in pH during the phases of the menstrual cycle. The acidic environment is maintained by a symbiosis between the lactic acid-producing Doederlein bacillus (*Lactobacillus*), a gram-positive nonmotile rod, and the vaginal contents. Doederlein bacilli are dependent on the desquamation of vaginal epithelial cells containing glycogen. The glycogen content and the sloughing and renewal of vaginal cells are regulated by ovarian hormones. Glycogen is broken down enzymatically to simple sugars such as maltose and dextrose, which are further broken down to lactic acid (Fig. 105).

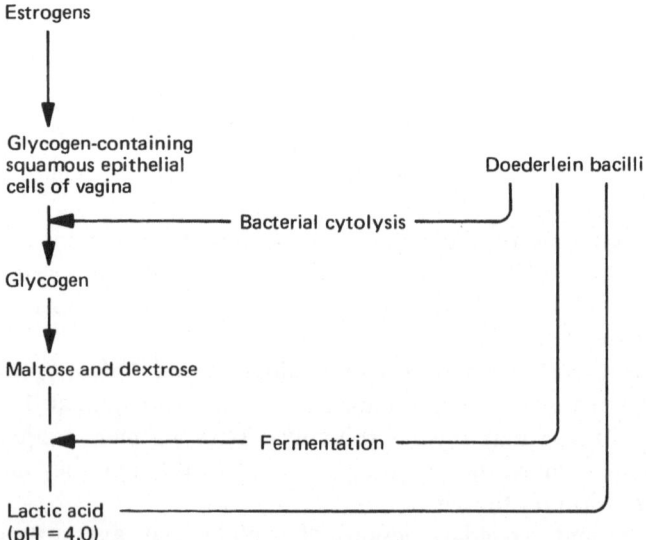

Fig. 105. Scheme of biology of the vagina: reciprocal influence of the vaginal epithelium and Doederlein bacilli on the maintenance of the acidic vaginal milieu.

Fig. 106. Hanging drop of the vagina. Doederlein bacilli in the postovulatory progestational cell picture.

The moist acidic milieu of the vagina is essential to the growth of normal vaginal bacteria. Acidity provides a selective advantage for the growth of Doederlein bacilli. At the same time it provides protection against invasion by other microorganisms and their progression upward through the genital tract (Fig. 106). The growth of pathogenic bacteria such as staphylococci and streptococci is thus prevented (Fig. 107).

Because of varying hormonal conditions, the protective mechanisms in the vagina differ in effectiveness at varying stages of life. Conditions are optimal during the reproductive age (Fig. 18).

In the newborn, the vaginal epithelium histologically resembles the adult type because of the effect of maternal estrogens (Fig. 19). Around the fifth day, Doederlein bacilli begin to appear in the vagina and the pH becomes about 4.8. After the removal of the estrogenic stimulus, the thickness of the vaginal epithelium decreases and the pH approaches neutrality, at which point it remains until puberty (Fig. 20). With reduced protection, the child is subject to a variety of vaginal infections, including gonorrhea, which is exceedingly rare in the adult vagina. Poor estrogenic support similarly predisposes the thin postmenopausal vaginal epithelium to laceration and infection (Fig. 21).

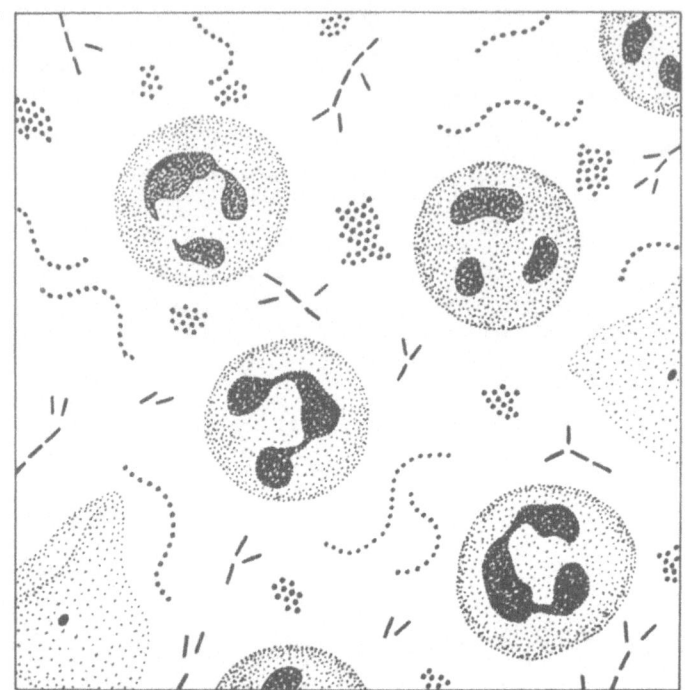

Fig. 107. Hanging drop: mixed vaginal flora (streptococci, staphylococci, and bacilli); many multisegmented polymorphonuclear leukocytes.

The following etiological factors must be considered either singly or in combination:

(1) Disturbance of the cyclic hormonal stimulation of the vaginal epithelium. For instance, persistent estrogen deficiency results in insufficient epithelial regeneration and low glycogen content of the vaginal cells.

(2) Change in acidity. Any increased secretion from the cervix results in a shift in pH toward the alkaline side, followed by the disappearance of lactobacilli and the growth of pathogenic bacteria. Acidity may be destroyed also by the injudicious use of deodorants and vaginal douches.

(3) Direct damage to the Doederlein bacilli. A side effect of antibiotics and sulfonamides is the destruction of the normal flora. Doederlein bacilli are very sensitive to antibiotics and may be distroyed before any pathogens are affected.

The most common organisms causing symptomatic vaginitis today are *Trichomonas vaginalis* and *Candida albicans*. Staphylococci, *E. coli*, and *Hemophilus vaginalis* are also occasional offenders. These changes in prevalence of the various vaginitides are explained in part by the widespread use of antibiotics and sulfonamides, which create an environment more favorable for trichomonads and yeast.

The role of the mycoplasms in genital infections is still uncertain. Some strains such as *M. hominis* can be identified frequently in the male and female urogenital tracts. They are seen in asymptomatic women as well as those with vaginitis and cervicitis. This infection appears more common in prostitutes.

Trichomonas Vaginitis. *Trichomonas vaginalis* may cause vaginitis, but the organism is frequently nonpathogenic. Among organisms causing vaginitis, the trichomonas is least affected by variation in pH. It is probably for this reason that trichomonads can often be found in cytological smears from the vagina even in the absence of symptoms or objective signs of vaginitis. Trichomonads may also be found in the ducts of Skene's and Bartholin's glands, the cervical canal, the urethra, the bladder, and the rectum.

Trichomonas is estimated to be present in the urogenital tract of the male in 15% of cases. About 40% of all nonspecific forms of urethritis in the male are caused by *Trichomonas*. More than half of the partners of women harboring trichomonads are also affected. Transfer during intercourse has been demonstrated as with some venereal diseases. Intercourse, however, is not the only mode of infection.

Symptoms: Typical signs of the infection in women are a greenish-yellow, foamy, foul-smelling discharge. The constant irritation by the discharge may result in a diffuse pruritic vulvitis.

Diagnosis: On vaginal examination, the discharge and the acute epithelial changes are typical. Severe vaginitis is often accompanied by acute vulvitis. The vaginal mucosa is either diffusely reddened, inflamed, or involved in multiple red papules. The organism can be identified in the wet smear obtained from the vagina or occasionally in urinary sediment (Fig. 108). The organism is a pear-shaped motile parasite with an undulating membrane and a long flagellum. Other organisms such as *Candida albicans* are frequently associated.

Treatment: Combined local and oral application of specific drugs such as metronidazole (Flagyl) are suggested. Vaginal medication may be inserted by the patient. During the same period tablets should be taken orally. For best results the sexual partner should be treated simultaneously. The patient should abstain from intercourse until organisms are absent from the wet smear (hanging drop).

In 90% of patients the smear becomes negative. Since resistance to metronidazole has not yet been proved, reinfection rather than persistence is the usual explanation for recurrence or the presence of the organism after treatment. Additional courses of treatment for the patient and the partner are then prescribed.

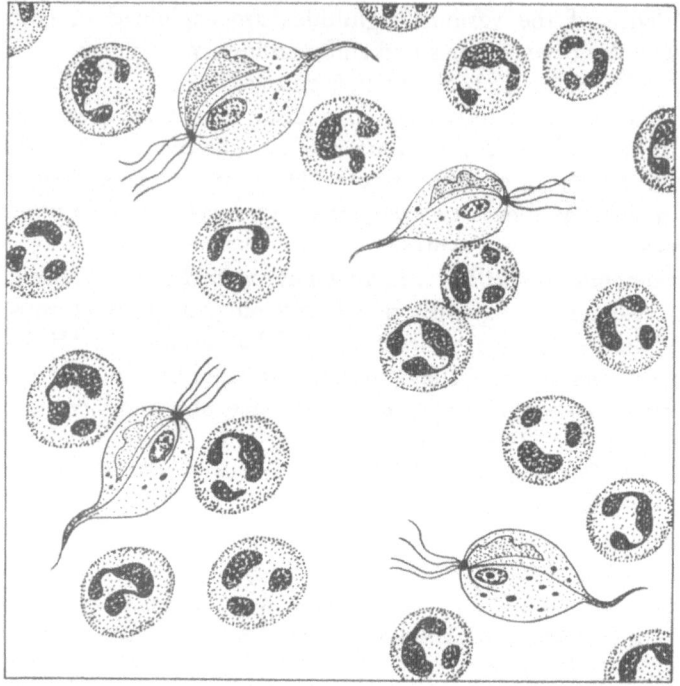

Fig. 108. Trichomonads in vaginal secretion recognized by the flagella and the undulating membranes. Larger numbers of leukocytes indicate inflammation.

Trichomonal vaginitis is frequent during pregnancy. Although metronidazole tablets may be used without apparent risk to the fetus after the twelfth week of gestation, local treatment may be preferred during pregnancy.

Moniliasis. The presence of yeastlike organisms in the vagina does not necessarily result in an inflammatory reaction. Vaginitis is frequently seen after use of antibiotics, in pregnancy, in the diabetic patient, and in the patient with severe systemic disease.

Symptoms: The major complaint apart from the whitish, cheesy discharge is an unbearable itching in the vagina, introitus, and vulva.

Diagnosis: Inspection usually reveals signs of vulvitis. On speculum examination adherent thrushlike patches may be seen. After their removal, the underlying vaginal mucosa appears reddened and edematous. Occasionally, however, inflammatory changes may be absent, the only symptom being vaginal pruritus.

The fungi may be visualized in a 1% KOH preparation of the exudate or in the stained smear (Fig. 109). A positive culture confirms the diagnosis.

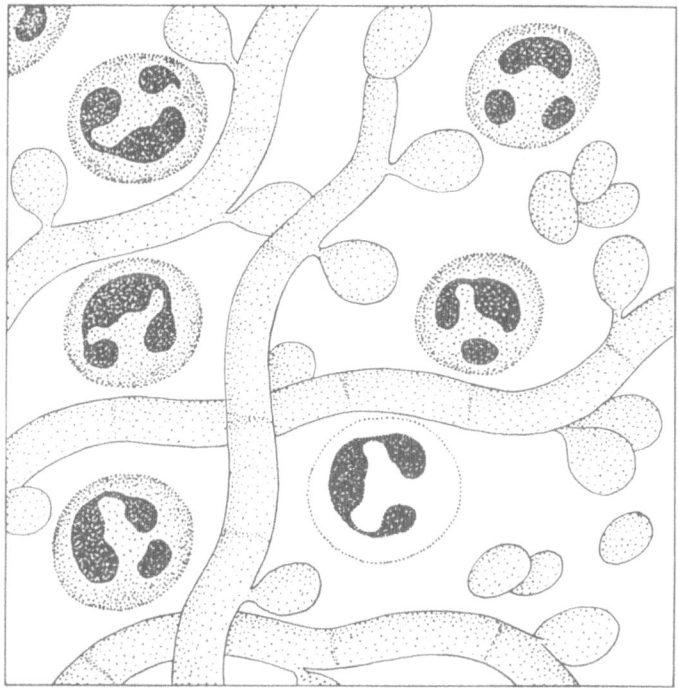

Fig. 109. *Candida albicans.* Hyphae and buds are in the vaginal secretion.

Treatment: Fungicides such as mycostatin are used locally in the form of vaginal suppositories, creams, or ointments. The partner should be treated by applying fungicidal ointment to the glans penis.

Nonspecific Vaginitis. The main symptom of nonspecific vaginitis is discharge, as in any form of vaginal infection. Vulvar pruritus and dysuria may develop secondarily. The vaginal mucosa is diffusely reddened in the acute stage. A mixed bacterial picture is observed under the microscope.

Treatment: Short-term local antibiotics usually control nonspecific vaginitis. To prevent recurrence, the cause of the vaginitis has to be identified (for example, estrogen deficiency or hypersecretion of cervical glands) and appropriate treatment initiated.

Senile Vaginitis. Postmenopausal vaginal atrophy may cause an acute "senile" vaginitis. Cocci, *Hemophilus vaginalis,* and *Escherichia coli* are commonly found. The thin vaginal mucosa may be infected by gonococci, as in childhood.

Symptoms: A blood-tinged, serous, occasionally purulent discharge is observed, sometimes associated with vulvar pruritus, dysuria, and dyspareunia.

Diagnosis: The atrophic vaginal epithelium shows numerous red spots, occasionally with superficial ulceration, that bleed on contact. Cytology reveals atrophic cells, any number of polymorphonuclear leukocytes, and bacteria. Malignant diseases of the cervix or corpus uteri must, of course, be ruled out in any case of bloody discharge. Ulcerated lesions iin particular must be subjected to biopsy.

Treatment: Apart from the elimination of pathogenic organisms, treatment of senile vaginitis requires building up a resistant vaginal epithelium, usually by application of vaginal creams containing antibiotics and estrogens. In some instances, it may be preferable to administer estrogens in small doses orally, although the possible complication of uterine bleeding must be considered.

Cervicitis

The main symptom of cervicitis is increased cervical discharge, although increased cervical secretion does not necessarily mean inflammation of the cervix. Cervical mucorrhea may be seen under physiological as well as pathological conditions. Secretion of the cervical gland is influenced by ovarian hormones (page 63). Cervical mucus, therefore, varies not only with the different phases of the cycle but also with the stages of a woman's life. Secretion of mucus is physiologically increased when estrogen levels are high, as at the time of ovulation. Progesterone reduces the quantity and changes the physical properties and composition of the mucus. In patients with anovulatory cycles, stimulation of the production of mucus by estrogens persists during the cycle unopposed by progesterone.

One of the local etiological factors in increased secretion by cervical glands is the displacement of the squamocolumnar junction onto the surface of the portio vaginalis. Tears and scars from traumatic deliveries may also distort the external os with the resulting ectropion producing mucorrhea.

Infiltration by polymorphonuclear leukocytes and lymphocytes in tissue sections leads the pathologist to make the diagnosis of chronic cervicitis. This is a pathological term, however, which does not reflect the clinical situation. There is little correlation between the histological observation of inflammation and bacterial invasion of the cervical canal.

Local treatment of cervicitis by electrocauterization or caustic solutions is indicated only when the erosion is extensive and productive of significant mucorrhea.

The acute inflammatory reaction of the cervix is characterized by purulent yellowish discharge. Dysuria is frequently present as well. The general well-being of the patient is unaffected. Gonorrhea as a cause of the acute cervicitis should be ruled out.

Diagnosis: Purulent secretion in the cervical os is seen on speculum examination. Cultures identify the organism, but special media are required for the growth of gonococci. Inflammatory and neoplastic diseases of the upper genital tract should be considered.

Treatment: Nonspecific inflammation usually responds to local medical treatment. Surgical procedures such as conization should be deferred until after the acute infection has subsided.

Endometritis

Nonspecific inflammation of the nonpregnant endometrium is an uncommon clinical entity. Bacteria invading the genital tract apparently skip the uterine cavity except in the puerperal uterus. In the nonpregnant uterus the commonly seen lymphocytic infiltration of the endometrium is a result of the cyclic sloughing and regeneration of this tissue rather than true inflammation. Acute inflammation of the endometrium is usually a temporary event because the cyclic sloughing of the tissue prevents chronicity. In addition, the endometrium seems to have some bactericidal properties against certain organisms, the gonococcus, for example.

Intrauterine contraceptive devices are usually associated with a local low-grade inflammatory reaction, but clinical infection of the endometrium is an uncommon complication (page 240).

Nonspecific Acute and Chronic Endometritis. Endometritis may be caused by organisms that reach the endometrium in an ascending or descending fashion. Acute puerperal endometritis may result from invasion and growth of pathogens in an incompletely involuted uterus (subinvolution). In an induced septic abortion, infection may result from faulty technique, lack of asepsis, or the introduction of foreign bodies or chemicals. The organisms most frequently found are coliform bacilli, staphylococci, clostridia, and occasionally streptococci. The result may be a purulent discharge from the uterine cavity.

Nonspecific chronic endometritis with fibrotic changes may occasionally be seen in association with salpingitis or as a result of intrauterine radiation.

Symptoms: Characteristic complaints are menometrorrhagia with purulent discharge when there is no bleeding. Temperature, sedimentation rate, and leukocyte count may be slightly elevated.

Diagnosis: Tenderness and moderate enlargement of the uterus indicate involvement of the myometrium, or an endomyometritis. The clinical diagnosis is confirmed by curettage and histological examination of the specimen.

Treatment: Acute endometritis initially is treated conservatively. High doses of estrogens stimulate the regeneration of the mucosa. Application of progestins results in secretory change and sloughing. Differential diagnosis requires curettage, which should be performed after the acute inflammatory reaction has subsided. Prophylactic antibiotics may be required. Adnexal involvement makes delay of the curettage advisable.

Pyometra

Pyometra is seen most often in the postmenopausal woman. Atrophic changes of the mucosa in association with adhesions resulting from acute infection may result in occlusion of the internal os and collection of pus above the obstruction in the uterine cavity. Pyometra is frequently found in association with a carcinoma of the endometrium.

Another cause of pyometra is intrauterine radiation, which may create adhesions and obstruction of the internal os, resulting in collection of intrauterine fluid that is easily infected.

Symptoms: Pyometra may cause pain in the lower abdomen. Purulent or serosanguineous discharge may be produced and the temperature, leukocyte count, and sedimentation rate are usually elevated.

Diagnosis: On pelvic examination the overdistended uterus feels like a cystic tumor. Discharge from the cervical os may be seen unless the os is completely obstructed.

Treatment: The first step must be dilatation of the cervical canal to provide drainage from the cavity. Curettage is delayed for at least a few days.

Asherman's Syndrome

This syndrome is characterized by replacement of the endometrium by adhesions and scar tissue. The uterine cavity may be partially or totally obliterated. The history usually includes a postpartum or postabortal curettage. Trauma to the basal layer of the endometrium and the underlying myometrium may result in formation of granulation tissue and later connective tissue synechiae that connect the opposite walls of the uterine cavity. Signs and symptoms depend on the extent of the synechiae and the amount of remaining normal endometrium. Amenorrhea or oligomenorrhea and infertility are common sequelae.

Diagnosis: The history is most important. The amenorrhea is accompanied by normal hormonal and cytologic findings. The basal body temperature curve is biphasic, suggesting ovulatory cycles. No specific finding is obtained on pelvic examination. Either the uterus cannot be sounded or a discrepancy between length of the canal and the size of the uterus on palpation is noted. Location of the synechiae may be demonstrated on hysterogram. Differential diagnosis includes tuberculosis of the endometrium.

Treatment: The intrauterine adhesions can occasionally be broken up simply by a probe. Usually, however, the scars must be removed by sharp and blunt dissection. The cavity is then kept patent by means of a Foley catheter or an intrauterine device. High doses of estrogen may stimulate proliferation of nests of endometrial tissue. Complete recovery and even pregnancy are possible after successful treatment of the Asherman syndrome.

Cervical Stenosis. More common than obliteration of the uterine cavity is stenosis of the upper third of the cervical canal and the internal os. Cervical stenosis is caused by faulty technique at curettage, especially for abortion. The incidence of the complication is lower after suction curettage.

Secondary amenorrhea and infertility are common but hematometra is not a consistent finding.

Treatment: The stenosis is relieved by careful dilatation of the cervical canal.

Pelvic Inflammatory Disease

Infection of the fallopian tube has potentially serious consequences in the young woman, particularly chronic disability and infertility.

Etiology: Infection is caused mainly by organisms ascending from the lower genital tract. Infection may also occur by vascular or lymphatic dissemination. Occasionally adnexal infection may result from direct contact with inflammatory processes in surrounding organs, such as appendical abscesses, ileitis, and diverticulitis.

Salpingitis caused by ascending spread of organisms is mainly bilateral and is seen almost exclusively in women of reproductive age. Salpingitis in virgins is suggestive of genital tuberculosis. In a large proportion of cases of pelvic inflammatory disease, staphylococci, streptococci, or *Escherichia coli* may be identified.

Typically, in the acute stage of adnexal infection, only the fallopian tubes are involved (salpingitis). In the subchronic and chronic phases the process often involves the ovary as well.

Mucosal edema and leukocytic infiltration of the tubal stroma are occasionally seen in response to reflux of menstrual blood. This is not a true infection, although it may explain the increased likelihood of ascending infection during or immediately after menstruation.

Acute Salpingitis

The microscopic features of acute salpingitis are edema and infiltration by polymorphonuclear leukocytes of the stroma of the tubal plicae. Later, edema of the epithelium itself may be seen. Seropurulent and fibrinous or fibrinopurulent exudates then collect in the tubal lumen (Fig. 110). Fibrinous adhesions and edema result in obliteration of the lumen and fimbrial adhesions, with blockage of the tube. The fallopian

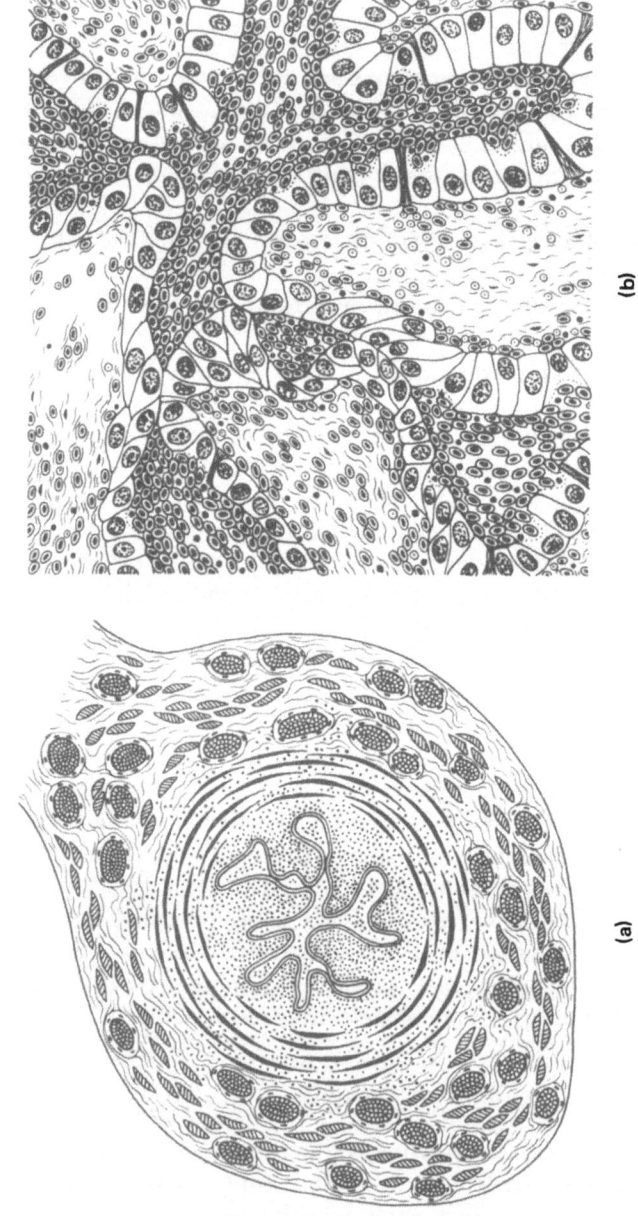

(b)

(a)

Fig. 110. Acute salpingitis. (a) Cross-section through the isthmic portion of the tube; pronounced hyperemia, edematous plicae, and leukocytic infiltration; (b) inflamed tubal mucosa, edema of the epithelium, purulent exudate in the lumen, fibrinous adhesions of the plicae of the mucosa and leukocytic infiltration of the stroma.

tubes are not enlarged in the initial stage of inflammation, but as exudate collects in the lumen, they become distended and retort-shaped. A collection of pus forms a pyosalpinx, whereas a tube filled with clear fluid is called a hydrosalpinx. An extravasation of blood into the tubal lumen results in a hematosalpinx.

Before the fimbriated end of the tube is sealed off by adhesions, infected exudate may leak into the peritoneal cavity producing an acute pelvic peritonitis. A collection of pus in the deepest portion of the peritoneal cavity results in a cul-de-sac abscess. Spread of infection to the ovary results, successively, in perioophoritis, oophoritis, and finally tubo-ovarian abscess.

Symptoms: The acute phase of pelvic inflammatory disease is associated with pain in the lower abdomen, dyspareunia, and fever. Vaginal spotting and purulent discharge from the cervical os may be noted as well.

Diagnosis: Localized bilateral rebound tenderness characterizes the acute phase. Because the tubes initially are soft and only slightly swollen they are not ordinarily palpable. The diagnosis is based on history and confirmed by exclusion of causes of acute abdominal pain such as appendicitis and ectopic pregnancy. Maximal tenderness in acute appendicitis is located higher in the abdomen, around McBurney's point, and gastrointestinal signs and symptoms such as nausea and vomiting are more prominent. Ectopic pregnancy is suggested by amenorrhea and a positive pregnancy test, although neither is required for the diagnosis. In ectopic pregnancy, furthermore, the leukocyte count, sedimentation rate, and temperature are more nearly normal.

The diagnosis is generally clear when a pyosalpinx or hydrosalpinx is palpated. The adnexa are felt as tense, club-shaped masses of limited mobility. Gentleness is required to prevent rupture of the cystic structures with exacerbation or generalized peritonitis. An abscess in the cul-de-sac bulges into the posterior vaginal fornix and rectum. It is felt as a tense fluctuant mass on rectal or rectovaginal examination and may be mistaken for a tuboovarian abscess that has extended into the posterior cul-de-sac. Diagnostic work-up includes cultures of cervical discharge to exclude gonorrhea and tuberculosis. Fluid from hydrosalpinges is usually sterile on culture. A high sedimentation rate and white count are found with early complications.

Treatment: Acute salpingitis is treated conservatively. Bedrest, fluids, and high doses of antibiotics such as ampicillin are prescribed. The use of corticosteroids to promote resorption of exudates and prevent formation of adhesions is controversial. Corticosteroids should not, of course, be used without antibiotics since suppression of the immune response might result in dissemination of the infection.

Cul-de-sac abscesses are surgically incised and drained through the posterior vaginal fornix. The pus is cultured and the organism tested for sensitivity.

Prognosis: Early diagnosis and immediate administration of antibiotics may preserve function of the fallopian tube. Once exudation has begun, formation of adhesions is almost unavoidable. Destruction of the fimbriae carries a poor prognosis with respect to subsequent fertility. Plastic surgical reconstruction of the tube is attempted only after the acute inflammation has subsided (page 187).

Chronic Inflammation of the Adnexa

If the acute stage is not properly treated, it usually develops into a chronic stage. The chronic course is characterized by the formation of fibrous adhesions between tube, ovary, and surrounding structures (Fig. 111).

Closure of the fimbriated end of the tube in the acute stage is an effective protection against spread of purulent material to the peritoneal cavity, but it may destroy tubal function. In primarily endosalpingeal infections the inflammatory reaction spreads from the mucosa outward to the serosa. Fibrinous exudation (perisalpingitis) produces extensive adhesions to surrounding tissues.

The tubal wall in chronic salpingitis is partially replaced by connective tissue and mucosal plicae are destroyed.

The tubal mucosa is finally destroyed and the exudate is slowly resorbed. The involved adnexa are thus transformed into a rigid connective tissue mass that is adherent to the sigmoid, rectum, uterus, and occasionally the bladder. The mass is frequently located in the posterior cul-de-sac and fixed to the posterior aspect of the broad ligament and the pelvic side wall.

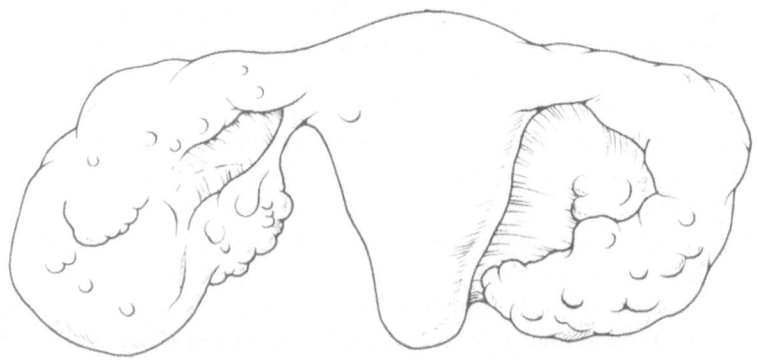

Fig. 111. Bilateral pyosalpinges. Occlusion of the fimbriated ends. Distention of the tubes, perisalpingitis, and inflammatory adhesions to the ovary.

Symptomatology: Chronic salpingitis and its sequelae are associated with persistent pain, unilaterally or bilaterally, in the lower abdomen. Vaginal spotting, chronic discharge, and low grade fever commonly persist.

Diagnosis: The lower abdomen is tender to palpation, one side usually more than the other. Rebound tenderness is absent in the chronic stage. Pelvic examination reveals unilateral or bilateral firm adnexal masses of varying size, often adherent to the uterus and pelvic side walls or fixed in the cul-de-sac. These masses are often softer than the adnexal masses of tuberculosis or ovarian carcinoma.

Treatment and prognosis: As long as signs of acute inflammation are present, treatment is conservative.

In the chronic stage, infertility may be treated surgically. The success rates vary between 10 and 15% (page 187) depending on the cause and the anatomical extent of the lesion.

The surgical results depend on the presence of residual tubal motility and remaining functional mucosa and fimbriae.

The rate of recurrence is high. Reinfection also must be considered. It may cause an acute exacerbation of a chronic process. Rupture of a tuboovarian abscess or a pyosalpinx creates a surgical emergency. Rupture may occur spontaneously or after intercourse.

Repeated bouts of infection usually require surgical intervention. Since most of these women are of reproductive age, an attempt is often made to preserve some functioning ovarian tissue, but the patient must be informed that the type and extent of procedure can only be decided at the operation. When both tubes are damaged beyond repair, it is usually best to remove the uterus and ovaries as well. In the young patient, this procedure should be followed by exogenous estrogen therapy.

Ovarian Abscess

An isolated ovarian abscess usually caused by hematogenous spread of pyogenic organisms is generally unilateral. The ovary becomes fixed to neighboring structures by periovarian adhesions. The usual pathogens are gram-negative bacilli. Tuberculosis (page 235) must also be considered in the differential diagnosis.

Symptoms: Septic temperature spikes persist after resolution of the underlying infectious process.

Diagnosis: A unilateral adnexal mass and isolated tenderness of the ovary are detected. The diagnosis may be confirmed by laparoscopy. Multiple small abscesses inside the ovary, however, may not be detected by palpation or inspection. Salpingitis, the rare ovarian pregnancy, and acute appendicitis may cause similar symptoms.

Treatment: The only effective treatment is immediate removal of the affected ovary.

Salpingitis Isthmica Nodosa

Salpingitis isthmica nodosa is a tubal disease of unknown cause. The isthmic portion of the tube is thickened to 1 to 2 cm in diameter by several firm, brownish-yellow nodules. On microscopic examination tubal mucosa is found scattered throughout the myosalpinx, which is hypertrophic and hyperplastic. In some cases a connection between the ectopic tubal epithelium and the lumen can be identified. It bears a certain morphologic similarity to adenomyosis of the uterus.

Etiology: The lesion is sometimes regarded as a form of chronic salpingitis. Tuberculosis also has been suggested as a possible cause.

Salpingitis isthmica nodosa is seen between the ages of 25 and 50 years. In more than one-third of all cases both tubes are involved. This condition gives rise to no specific symptoms.

Since it frequently interferes with tubal motility, it may be the cause of infertility or may lead to tubal pregnancy. The diagnosis is usually made during laparoscopy or laparotomy for infertility. Tubal endometriosis can be differentiated microscopically from salpingitis isthmica nodosa on the basis of the histologic structure of the mucosal implants in the tubal wall.

Parametritis

Acute or chronic parametritis complicating vaginal gynecological and obstetrical procedures has become unusual. The decrease may reflect improved asepsis and the availability of antibiotics.

Etiology: Inflammation of the parametrial tissue is usually preceded by trauma to the cervix during delivery or abortion (rough dilatation of the cervix or perforation of the uterus). Pathogens, usually hemolytic streptococci, thus gain access to the parametrial tissues. Parametritis may also be seen after injury to the vagina. Conization of the cervix (page 303) is often complicated by parametritis, but in that case the inflammatory reaction usually subsides spontaneously in a few weeks. Parametritis may develop as a unilateral suppurative process along the lymphatics of the parametrial connective tissue. The adnexa may be involved as a result of parametritis, or the parametritis may develop by spread from salpingo-oophoritis.

Symptoms: Unilateral persistent severe pains deep in the pelvis are most characteristic. Septic temperature curves are noted. The leukocyte count and sedimentation rate are elevated.

Diagnosis: Diagnosis is based on the history and the findings on pelvic examination. Deep abdominal palpation above the symphysis is painful. The infiltrated parametrium is palpable on vaginal or, prefer-

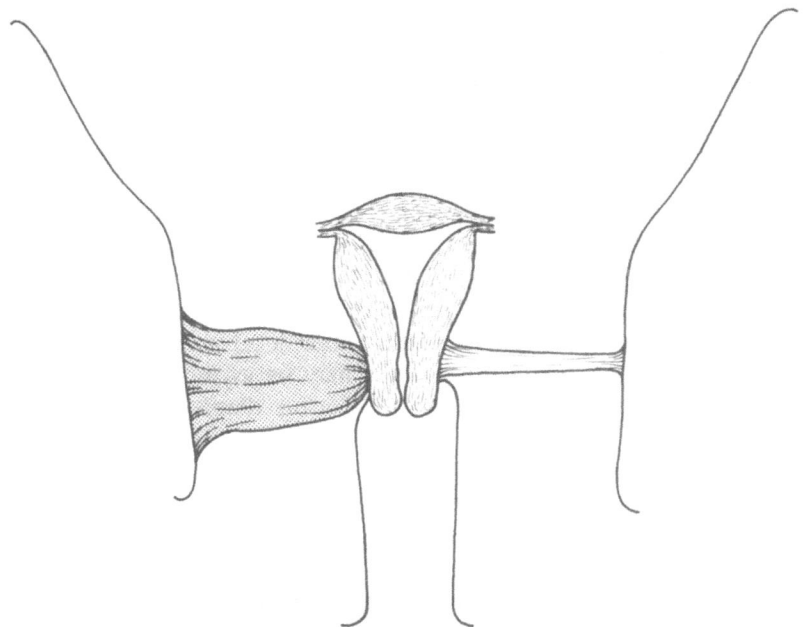

Fig. 112. Unilateral parametritis. Thickening of the parametrium with shortening of the fornix on that side; conical extension of the parametrium with thickened attachment to the pelvic wall.

ably, rectovaginal examination. The parametrium can be felt as a triangular soft (later firm) mass extending to the pelvic side wall (Fig. 112). The mobility of the uterus may be restricted with extensive parametritis. Malignant infiltration of the parametrium may produce a similar, though usually firmer, lesion.

Treatment: Antibiotics in high doses as well as anti-inflammatory measures are used for the treatment of acute parametritis. A parametrial abscess may be drained through the lateral vaginal fornix. For anatomic reasons, the infected parametrium cannot be removed surgically.

Specific Infections of the Female Genitalia

Tuberculosis. Depending on the prevalence of tuberculosis in general and the socioeconomic status of the population, up to 25% of all cases of female infertility may stem from pelvic tuberculosis. Approximately 10% of all cases of pelvic inflammatory disease are of tuberculous origin in some countries, although the percentage is lower and dropping in the United States and other industrialized countries.

Most of the patients are in the age group of 20 to 35. As a result of early diagnosis (population screening), there has been a shift in age-specific incidence. Whereas primary infection used to occur most often during childhood, it is now more frequently the young adult who develops tuberculosis. Genital tuberculosis was formely a major cause of primary infertility, but it is a much less important etiological factor today. Genital tuberculosis is more frequently seen today in the premenopausal, menopausal, and even the postmenopausal woman. It is still true, however, that an inflammatory adnexal mass in the young girl, especially the virgin, must be considered as possibly of tuberculous origin. In about 10% of cases genital tuberculosis is accompanied by a specific infection of the urinary tract. These are probably separate, hematogenously spread lesions rather than local extension from one organ to the other. Genital tuberculosis may be asymptomatic. In many cases the condition is first detected during diagnostic work-up for infertility.

Tuberculous infection of the genital organs occurs almost exclusively by hematogenous spread orginating from an infection in the lungs, pleura, peritoneum, or skeletal system. Only in very rare instances may peritoneal or mesenteric lymph node tuberculosis result from lymphatic spread. The primary site of hematogenous spread is, in 90% of cases, the mucosa of the tubes. Tuberculous salpingitis is almost without exception bilateral. The process usually starts in the ampullary part of the fallopian tube. From there the infection extends into the isthmic and interstitial portions and finally reaches the endometrium. Thus, tuberculous endometritis is almost always associated with salpingitis. Secondary involvement of the ovaries is seen in about 10% of cases. Extension to cervix, vagina, or vulva is unusual. Pelvic tuberculosis is frequently associated with ascites. In tuberculous salpingitis, the fimbriated ends of the tubes remain patent in about 50% of cases.

Genital tuberculosis is characterized by its chronic course. The typical tubercles with central necrosis develop in the tubal mucosa following lymphocytic infiltration and stromal hyperplasia of the mucosal plicae. The tubercles are surrounded by epithelioid cells and Langhans giant cells and a peripheral layer of lymphocytes (Fig. 113). The characteristic caseous material is sometimes absent. Obstruction of the fimbriated end of the tube produces findings on pelvic examination similar to those of a pyosalpinx of gonorrheal origin. Peritubal adhesions together with the enlarged tube and the ovary form a very firm aggregate tumor of variable size (Fig. 114). On laparoscopy or culdoscopy, tuberculous nodules may be seen on the tubal surface, on the peritoneum, in the cul-de-sac, on the sigmoid and rectum, and on the ileum. Tuberculous infection results in extensive adhesions between the pelvic organs and the pelvic side walls.

Tuberculous endometritis is characterized by granulomas in the functional layers of the endometrium. The cyclic sloughing provides material

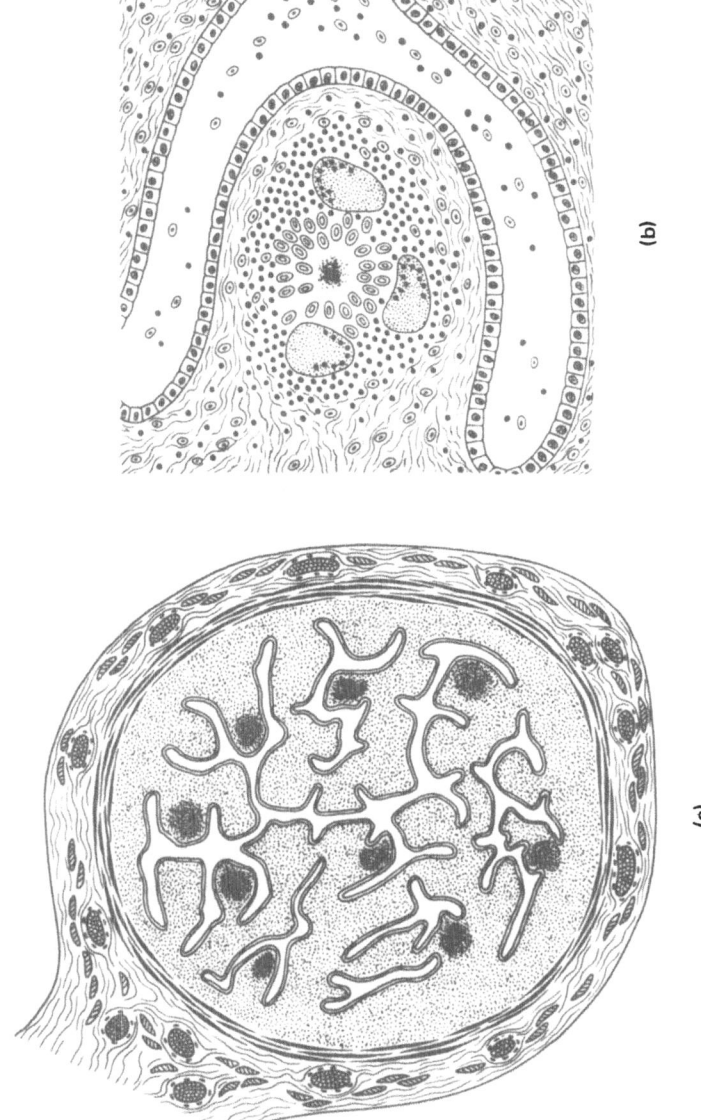

(b)

(a)

Fig. 113. Tuberculous salpingitis. (a) Cross-section through the ampullary portion of the tube; caseous granulation with tubercules replacing the stroma; lumen of the tube is largely obliterated; (b) Granuloma (tubercule) in the tubal mucosa; central necrosis surrounded by epithelial cells and Langhans giant cells (at 3, 6, and 9 o'clock); peripherally a collection of leukocytes is found.

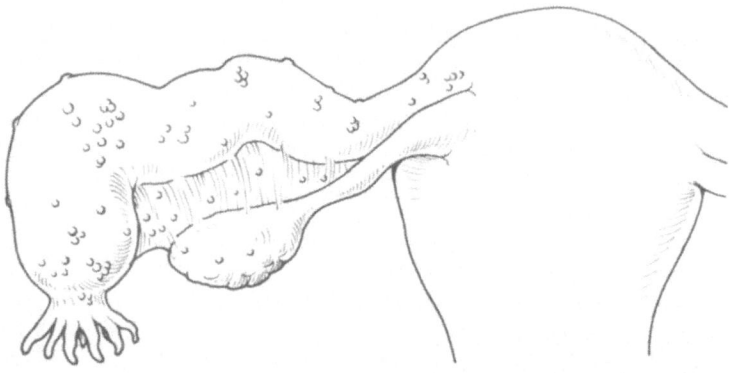

Fig. 114. Tuberculous salpingo-oophoritis. Thickening of the tube, abdominal ostium narrowed and fixed, but not closed; tuberculous perisalpingitis and perioophoritis.

for identifying the organism. Only in advanced cases does the inflammation extend to the basal layer and then cause amenorrhea.

Symptoms: Symptoms may be completely absent. Often the patient seeks medical advice because of infertility. Two-thirds of the patients complain of dysmenorrhea, which usually is of the secondary type. In approximately 20 to 50% of patients there is a menstrual disturbance: hypomenorrhea and oligomenorrhea or polymenorrhea. In about 25% of this group, secondary amenorrhea develops. General well-being may be affected. Easy fatigue, nocturnal perspiration, and slightly elevated temperatures are occasionally reported by patients with genital tuberculosis.

Diagnosis: Bilateral firm masses with extensive adhesions in the absence of symptoms of acute pelvic inflammatory disease suggest tuberculosis. Diagnosis may be confirmed by feeling multiple nodules in the area of the tubes as well as in cul-de-sac. Important clues are obtained from the history. In many patients the roentgenogram of the chest shows a lesion of old pulmonary tuberculosis. A hysterosalpingogram performed during infertility work-up may show enlarged club-shaped tubal ampullae with or without obstruction of the fimbriated end and with multiple filling defects in the lumen.

Clinical diagnosis must be confirmed by identification of the organism from menstrual blood and the typical histopathologic lesions in the affected tissue. Five to 10 ml of menstrual blood are required for bacteriological culture and animal inoculations. Two to three weeks are required for a result from the culture. The diagnostically more reliable animal inoculation test requires six to ten weeks. Curettage is performed in the premenstrual phase of the following cycle. Histological diagnosis is occasionally negative even in the presence of disease because of cyclic sloughing of the functional layer and its tubercles. Part of the material obtained by endometrial biopsy may be used for culture and animal inoculation.

A urologic work-up is indicated because of the occasional associated tuberculosis of the kidney or bladder.

Treatment: Administration of chemotherapy should be continued for nine months or longer. Since isonicotinic acid hydrazide (INH), para-aminosalicylic acid (PAS) and streptomycin affect different steps in the metabolism of the *Mycobacterium,* combined chemotherapy is often used. It also prevents the development of resistance and as a rule is better tolerated. Pelvic examinations every six months for at least three years are suggested.

The chances of medical cure are between 70 and 90%. Symptoms may remain in the presence of extensive pelvic adhesions. It is often difficult to distinguish between cure and temporary inactivation. Prognosis with regard to infertility is unfavorable. Surgical treatment is reserved for resistance to medical therapy and for adnexal masses, which must be differentiated from ovarian neoplasms.

Venereal Disease

Gonorrhea. The incidence of gonorrhea, especially among adolescents, has increased significantly in recent years. It is the most common venereal disease. Changes in sexual behavior of adolescents (especially earlier intercourse), promiscuity, and homosexual relations are considered the main reasons for this increase in frequency.

Gonorrhea is caused by a gram-negative diplococcus, first described by Neisser in 1879. Gonococci invade columnar secretory epithelium, but the squamous epithelium of the vagina is relatively resistant. The infantile vagina, however, is susceptible to gonococcal inflammation.

Infection occurs almost exclusively during intercourse. Dissemination by nongenital contact can be considered only in children. In the male, the ejaculate in acute gonorrhea contains gonococci. In women the initial sites of infection are the urethra (95%), the Bartholin gland ducts (20%), and the cervical mucosa (80%). The rectal mucosa may also be an important site of primary infection. The organs covered by squamous epithelium usually remain free. Infection of the cervix results in acute cervicitis with infectious purulent discharge. Bartholin gland abscesses (page 218) are typical manifestations of gonorrhea. The persistent cervical discharge may predispose to condylomata acuminata (page 216), which may spread over the entire vulva, vagina, and portio vaginalis of the cervix.

Gonorrhea may spread to the endometrium during or after menstruation, abortion, or delivery.

Gonorrheal endometritis is usually only transient because of the cyclic sloughing of the endometrium. The organism spreads easily from the endometrium to the fallopian tubes, however, causing an acute salpingitis.

Symptomatology: At first, slight dysuria is often the only symptom. With infection of the cervical glands, a purulent greenish-yellow discharge appears. Infection of Bartholin's glands is described on page 218). Acute and chronic gonorrheal salpingitis present the signs and symptoms described under pelvic inflammatory disease (page 229).

Infantile gonorrheal vaginitis causes a purulent discharge, severe dysuria, and a secondary vulvitis of acute onset.

Diagnosis: Soon after initial infection, a reddening of the periurethral (Skene's) ducts and Bartholin gland ducts is noticed. Purulent discharge from the urethra may be expressed by the examining finger by pressure against the anterior vaginal wall beneath the urethra. With cervicitis the vagina may contain copious purulent discharge and the exudate may be visualized at the external os.

Identification of the organism is essential for diagnosis. Cervical and urethral secretions are transferred by means of a sterile cotton-tipped applicator to a slide and then gram-stained. Intracellular gram-negative diplococci support the clinical diagnosis (Fig. 115), but are not absolutely diagnostic. The failure rate of identifying organisms on stained smears is considerable; specific culture media are preferable.

Biochemical identification of the organism is diagnostic. Material for culture is obtained from cervix, urethra, and rectum. In young girls with suspicious vaginitis, material for bacteriological culture may be obtained without damaging the hymen. In the postmenopausal woman a specimen from the posterior vaginal vault is usually obtained. Immunofluorescent and complement fixation techniques may be diagnostically helpful. If all tests are negative but the clinical impression is strongly suggestive for gonorrhea, appropriate treatment should be started anyway.

Treatment: Penicillin is the drug of choice. Resistance is still uncommon. An initial dose of 4.8 million units is recommended. Failure in treatment is usually caused by either a mixed infection with penicillinase-producing organisms or, more frequently, inadequate dosage. In patients with allergy to penicillin, broad-spectrum antibiotics such as tetracycline, oxytetracycline, erythromycin, or kanamycin may be used. Recent experience with spectinomycin has been encouraging.

Syphilis. Syphilis also has become more frequent in recent years. *Treponema pallidum* readily penetrates tissue through small epithelial lesions. It cannot, however, traverse the barrier of intact epithelium. The disease is usually transmitted during intercourse. The syphilitic ulcer (chancre), or primary lesion, is infectious, as are the moist mucous

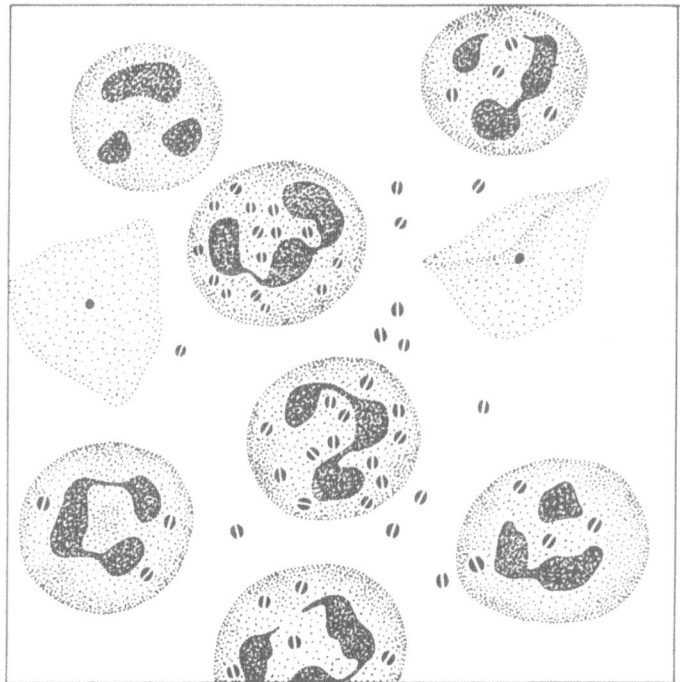

Fig. 115. Microscopic evidence of gonococci; gram-negative intracellular diplococci.

patches of the secondary stage. The primary syphilitic lesion in the female genitalia usually involves the vulva, vaginal wall (especially the posterior vaginal vault), or cervix. A hard chancre with sharp edges forms after an incubation period of three weeks. Multiple lesions and granulomas may be present. The inguinal lymph nodes enlarge significantly and become hard and tender.

The primary lesion of Stage I heals spontaneously in four to six weeks, but the spirochetes enter the circulation during this time. The second stage develops nine weeks after exposure. A generalized macular or papular rash and typical condylomata lata are secondary manifestations. The condylomata are usually limited to the vulva but may spread to the vaginal mucosa and to the surface of the cervix. The condylomata are flat, pale grayish granulomas. They have a central indentation and a watery secretion from their surfaces. The third stage is characterized by neurological symptoms. The ulcerating granulomas (gummata) of this stage may be located on the vulva and must be differentiated from carcinoma.

Symptoms: The primary lesion causes no symptoms and is frequently noticed accidentally. The patient may seek medical attention because of swelling of the lymph nodes. In the secondary lesion increased secretion from the moist patches occurs. The generalized rash usually directs the patient to the dermatologist.

Diagnosis: An indurated ulcer of the vulva, vagina, or cervix raises suspicions of a primary lesion of syphilis. Identification of *Treponema pallidum* in dark-field examination of secretion from the primary lesion confirms the diagnosis. Serological tests for syphilis are not yet positive at this stage. Condylomata lata of the second stage are also highly infectious; spirochetes are easily seen in darkfield microscopy of these lesions. Serological tests are positive during stages two and three.

Treatment: Pencillin is used in high dosages. Abstinence from intercourse until the disease is cured is essential to prevent spread.

Other Venereal Diseases. In some parts of the United States chancroid, lymphopathia venereum, and granuloma inguinale pose problems in differential diagnosis of venereal infections.

1. *Chancroid.* This disease, often called soft chancre to distinguish it from the "hard" chancre of syphilis, is caused by the Ducrey bacillus *Hemophilus ducreyi*, a gram-negative *coccobacillus*. It usually presents as a painful ulcerated maculopapular lesion of the vulva from three to ten days after sexual exposure. Shortly thereafter inguinal lymphadenopathy develops. The primary lesion as well as the lymph nodes may suppurate and produce a foul-smelling discharge.

Diagnosis: Diagnosis is often made by gram-stain and culture. Syphilis and the other granulomatous venereal diseases should be ruled out.

Treatment: Fourteen days of sulfonamides or a broad-spectrum antibiotic are usually adequate. Secondary infection may require prolonged treatment and lead to delayed healing.

2. *Lymphopathia venereum.* This infection, caused by a large virus, is confined to warm climates and is more common in the black population. The infection involves the lymphatics of the genital, inguinal, and anal areas. The initial lesion is an inconspicuous transient vesicle on the external genitalia appearing one to three weeks after sexual exposure. The primary infection is usually accompanied by fever and malaise. Two to three weeks later progressive inguinal lymphadenitis develops. A large mass (the bubo) may form in the groin. Breakdown of tissues leads to fistulas and extensive fibrosis. In the late stages of the disease, rectal and anal strictures as well as vulvar elephantiasis may be produced. Squamous cell carcinoma has been known to develop in chronically inflamed vulvar and anal tissues.

Diagnosis: The Frei test, which usually becomes positive two to six weeks after the initial lesion, is specific. The other granulomatous infections as well as carcinoma must be ruled out by appropriate cultures and biopsies.

Treatment: A three-week course of sulfonamides or a broad-spectrum antibiotic is often curative. The inguinal buboes may require drainage, and severe strictures of the bowel may occasionally necessitate colostomy.

3. *Granuloma inguinale.* In this disorder, the inflammation spreads by the cutaneous rather than the lymphatic route. The chronic ulcerating granuloma is produced by *Donovania granulomatis,* a microorganism of doubtful classification that is usually recognized by its encapsulated inclusion bodies (Donovan bodies) in the mononuclear inflammatory cells. The disease is found in blacks. The initial lesion usually appears on the vulva, vagina, or perineum as a small localized granuloma. Ulceration, secondary infection, and malodorous discharge follow shortly. Secondary infection may lead to lymphadenitis and vulvar elephantiasis, but the primary process is an extensive formation of subcutaneous granulation tissue. The lesion tends to spread slowly and directly to adjacent areas. Pain in the involved genital and perianal regions may be severe.

Diagnosis: Diagnosis is made by smear or biopsy of the ulcer with identification of the Donovan bodies in the mononuclear phagocytes of the granulomatous lesion. The other more common venereal diseases, which may be coexistent, must be ruled out.

Treatment: Streptomycin in daily doses of 4 g for four weeks has proved to be effective antibiotic therapy. Surgical excision and coagulation of the infected tissues is occasionally required in resistant cases.

Selected Reading

Gompel, C., and Silverberg, S.: *Pathology in Gynecology and Obstetrics.* Philadelphia and Toronto: Lippincott (1969).

Novak, E. R., and Woodruff, J. D.: *Novak's Gynecologic and Obstetric Pathology, 6th ed.* Philadelphia and London: Saunders (1967).

Sharman, A.: *Genital tuberculosis.* In: *Progress in Gynecology, vol. 3.* J. V. Meigs and S. H. Sturgis, ed. New York: Grune & Stratton (1957).

CHAPTER 19

Displacements of the Pelvic Organs

The clinical significance of displacements of the pelvic organs requires an understanding of their normal position and supporting structures.

Normal Position of the Uterus

The normal position of the uterus is defined by three criteria:

(1) *Version* of the uterus is determined by direction of the axis of the cervix in relation to the axis of the vagina (Figs. 116, 117). The uterus is anteverted when there is deviation of the cervical axis, with respect to the vaginal axis, in the anterior direction. If the cervical and vaginal axes are parallel, the uterus is in the middle position. The uterus is retroverted if the deviation of the cervical axis is posterior to the vaginal axis. Right and left version are self-explanatory.

(2) *Flexion of the uterus.* Under normal conditions, the axis of the corpus is flexed 130° anteriorly with respect to the axis of the cervix (Fig. 118). The normal uterus is, therefore, *anteverted* and *anteflexed.* The uterus is retroflexed when the axis is deviated posteriorly with respect to the cervical axis. A retroverted-retroflexed uterus is considered with few exceptions normal.

(3) *Position.* This term describes the position of the uterus in relation to the midline of the pelvis. The anteflexed, anteverted uterus lies near the symphysis, whereas the retrovexed, retroflected uterus often lies near the sacrum (retrocessed).

In the adult the fundus of the uterus is located near the pelvic inlet. The external os is located at the level of the ischial spine. When the cervix lies below the spine it is partially prolapsed. Because of its mobility, the uterus can adapt easily to a full bladder or rectum, returning to its original position after the adjacent organs are emptied.

The Clinical Significance of the Retroflexed Uterus. A retroflexed, or "tipped," uterine position may be congenital or acquired. The uterus is frequently retroflexed in young children, later changing to an anteflexed, anteverted position after puberty (Fig. 119).

An acquired retroflexion may result from loss of support after delivery or in the menopause.

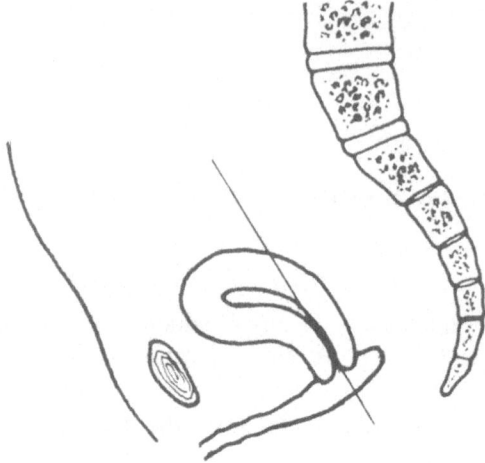

Fig. 116. Anteversion of the uterus (relation between the axis of the cervix cervix and axis of the vagina).

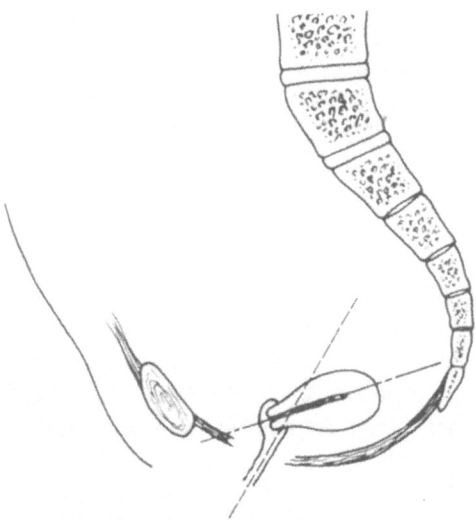

Fig. 117. Retroversion of the uterus (relation between the axis of the cervix and axis of the vagina).

Symptoms: A retroflexed uterus produces no symptoms and is of no pathologic significance. An occasional patient will complain of pressure in the rectum, constipation and back pain, premenstrual pain, or dysmenorrhea. Often these complaints are not related to the retroflexed uterus.

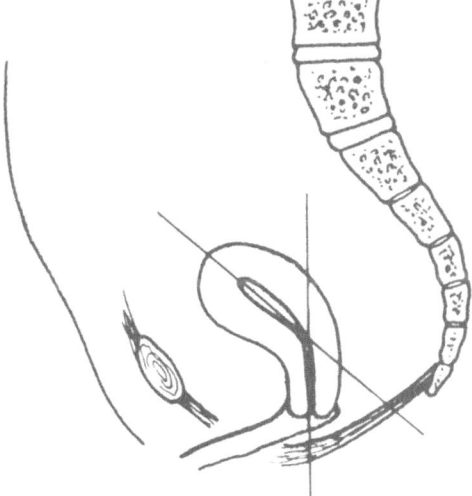

Fig. 118. Anteflexion of the uterus (relation between the axis of the corpus and that of the cervix).

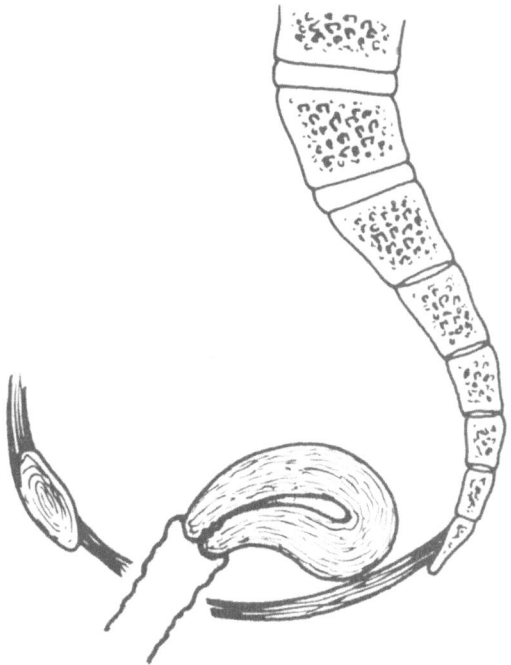

Fig. 119. Mobile retroflexion of the uterus.

Diagnosis: On speculum examination and on palpation, it becomes evident that the portio vaginalis is anterior and close to the symphysis. The corpus can be felt easily on rectovaginal examination close to the rectum. An attempt may be made to test its mobility. The portio vaginalis is then pushed posteriorly and the uterus elevated. The external hand tries to grasp the uterus as it is moved toward the anterior position. Frequently the attempt has to be terminated because of pain.

Treatment: Treatment is not required. Many gynecologists do not even inform the patient of retroflexion. If she is told, it should be explained that the retroflexed uterus is a normal finding that does not interfere with conception. In most instances the pregnant uterus rises in the anterior position spontaneously by the twelfth week of gestation. Occasionally, however, the uterus is incarcerated beneath the promontory. The uterus in that case must then be elevated under anesthesia and a pessary inserted. The incarcerated uterus may occasionally produce urinary retention if the urethra is pressed against the symphysis. Manual elevation under anesthesia is also performed if there is a large myoma or a uterus that is congested and enlarged as a result of venous obstruction. Very rarely, ventral suspension is indicated, as in long-standing infertility or habitual abortion. Many operations have in common the shortening of the round ligaments.

If the posterior wall of the uterus is adherent to the rectum, the condition is called *fixed retroversion* (Fig. 120). This is, in most instances, the result of pelvic inflammatory disease or endometriosis.

Symptoms: Retroversion is asymptomatic in most instances. Only occasionally are back pain or pressure in the rectum mentioned.

Diagnosis: The uterus on rectovaginal examination is immobile and cannot be brought into an anterior position, even under anesthesia.

Treatment: Asymptomatic fixed retroversion requires no treatment. Surgical correction may be indicated in cases of endometriosis and infertility without any other explanation.

Pelvic Relaxation

The descent of the uterus is called *prolapse.* If only the portio vaginalis protrudes through the introitus the prolapse is partial. A total prolapse means that the entire uterus protrudes through the vulva.

Often associated with prolapse of the uterus is descent of the anterior or posterior vaginal wall. The former produces a *cystocele* (prolapse of bladder through anterior vaginal wall) and the latter a *rectocele* (prolapse of rectum through posterior vaginal wall). The lower part of the cul-de-sac can herniate to produce an enterocele.

Descensus of the uterus is closely associated with that of the vaginal

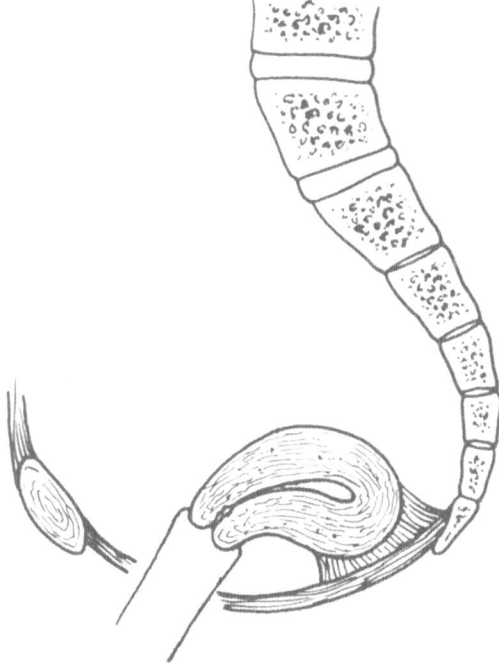

Fig. 120. Fixed retroversion of the uterus. The posterior wall of the uterus is fixed by scar tissue in the pelvis; the pouch of Douglas and the rectum are not shown.

walls. As has been mentioned (page 23), the levator plate of the pelvic floor is the most important supporting structure of the uterus. Insufficiency of the rectus musculature and an obese abdominal wall, in addition to a defective pelvic floor, contribute to prolapse (Fig. 121).

If the corpus is larger than the diameter of the hiatus in the pelvic floor, only the anterior and posterior walls of the vagina may descend. Congestion of the musculature and tension on the cardinal ligaments may result in an increase in connective tissue and the development of an elongated cervix (Fig. 122). This elongated cervix may be for a long time the only sign of insufficiency of the pelvic floor. The elongated cervix is not infrequently mistaken for a partial prolapse of the uterus.

Relaxation of supporting structures may be related to constitutional factors or injuries during deliveries. Usually both factors are involved. Uterine prolapse is common in Puerto Rican women, for example, but uncommon in blacks. Pelvic relaxation in nulliparous women results from a congenital weakness of connective tissue supports.

Pelvic relaxation usually becomes manifest in later years when hormonal stimulation ceases and elasticity of the supporting tissues decreases.

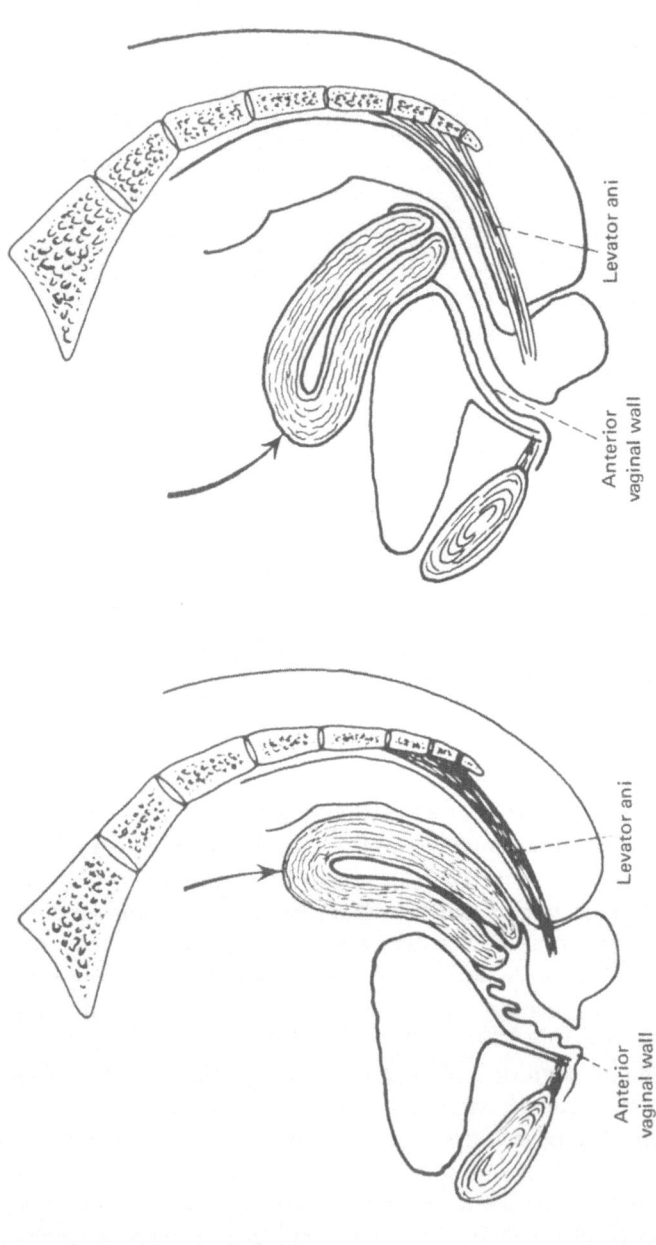

Levator ani

Anterior
vaginal wall

Levator ani

Anterior
vaginal wall

Fig. 121. Effect of intraabdominal pressure on the uterus in midposition. There is already a cystocele. Note the obtunded posterior bladder angle.

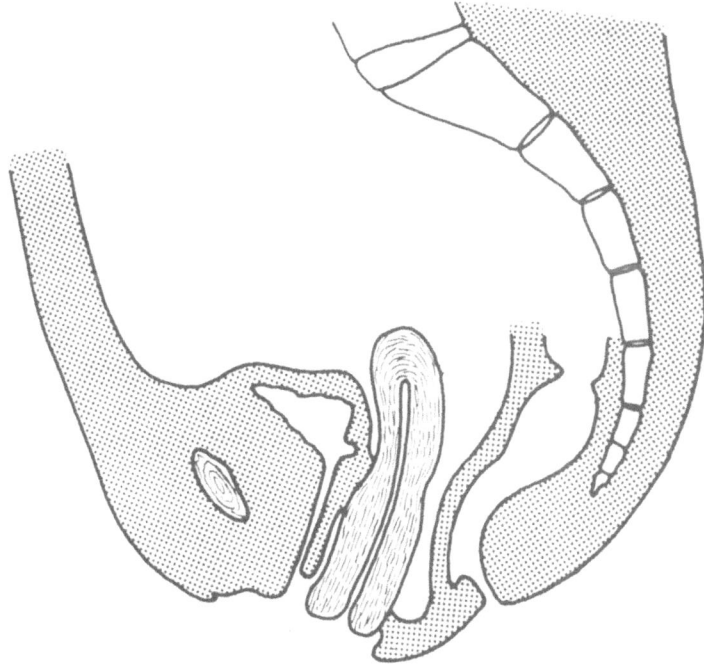

Fig. 122. Elongation of the cervix with uterine descensus.

Complete prolapse occurs in most instances after the menopause, when the atrophic uterus may protrude easily through the relaxed pelvic floor.

Symptoms: A slight relaxation of supporting structures may be observed in nearly all parous women. In the patient in whom pelvic relaxation becomes symptomatic, pressure in the vagina later leads to the feeling that something is "falling out". Back pain radiating to the groin is occasionally associated with prolapse of the uterus. Constipation is frequently associated with a rectocele.

The main complaint with relaxation of the anterior vaginal wall is stress incontinence, which is present in 70% of patients with extensive pelvic relaxation. The patient loses urine when intraabdominal pressure is increased, as with coughing or laughing. Stress incontinence must be differentiated from urgency, which is associated with the frequent urge to urinate but voiding of little urine. Urgency incontinence is as a rule not caused by a cystourethrocele, but rather is associated with urethrovesical calculi, irritation of the vesical mucosa, and sometimes psychogenic factors. When complete prolapse of the uterus is associated with urinary retention, voiding is possible only if the uterus is pushed back in the vagina.

Functional Disturbances in Stress Incontinence

The normal mechanism of voiding in women is incompletely understood. Continence and voiding in women require coordination of three functions.

(1) Function of the detrusor muscle. (Smooth muscle bundles surround the bladder and the upper third of the urethra. The urethra is supported by the pubovesical ligament originating from the symphysis pubis.)

(2) The posterior urethrovesical angle.

(3) Support by the pelvic floor. (The urethra ends in the bladder at an agle of 90%) (Fig. 123). Increasing amounts of urine in the bladder increase the pressure to 45 to 120 mm Hg. This increase in intravesical

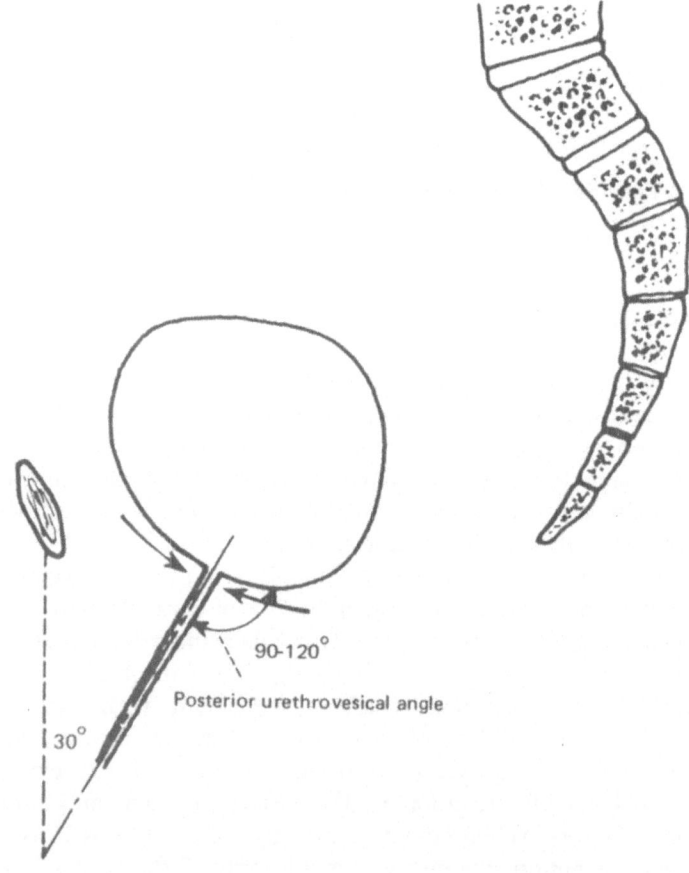

Fig. 123. Normal urethrovesical angle.

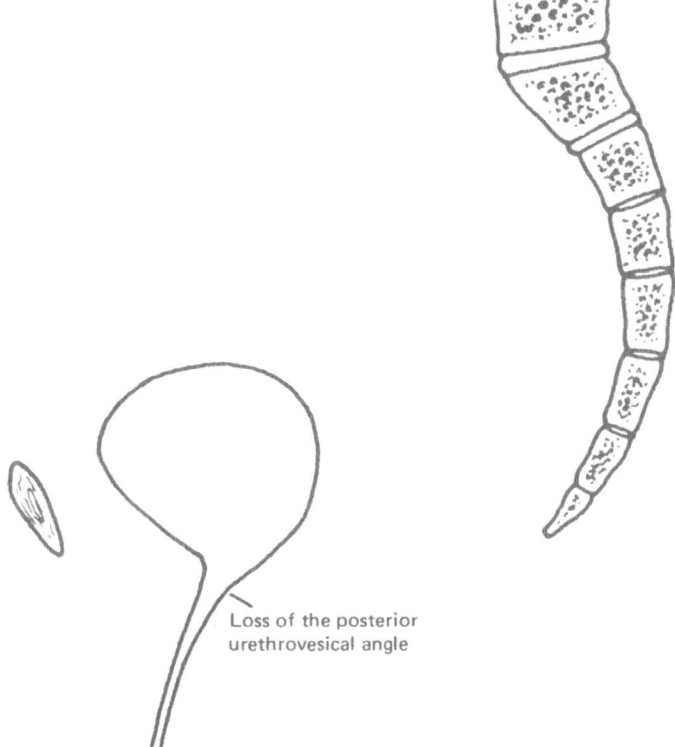

Loss of the posterior
urethrovesical angle

Fig. 124. Loss of the posterior urethrovesical angle with pelvic relaxation.

pressure creates the desire to void. Urination starts with contraction of the detrusor, which produces funneling at the trigone. At the same time the bladder pressure is voluntarily increased and the pelvic floor relaxed. As a result, voiding occurs.)

In the woman with stress incontinence, the posterior urethrovesical angle is blunted (Fig. 124) and the patient is constantly in the first stage of voiding. This condition is called stress incontinence Type I.

Descensus of the anterior wall of the vagina causes not only loss of the posterior urethrovesical angle but also descent of the base of the bladder and rotation of the urethral axis backward and downward (Fig. 125). This condition produces stress incontinence Type II.

Diagnosis: Complaints are characteristic. Descensus is associated with a patulous vagina and bulging of the posterior wall (rectocele) through the introitus (Fig. 126). The perineum is often small and scarred and the fourchette obliterated. On straining, the patient may lose a few drops of urine from the filled bladder (the bladder is filled before the examination with 150 ml of water) and the cystocele and rectocele are seen to bulge in the midline. These bulges are exaggerated by pressing

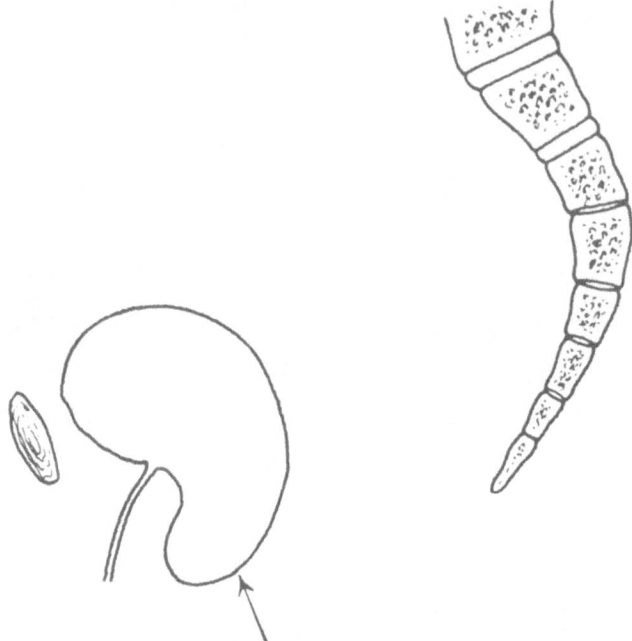

Fig. 125. Descensus of the floor of the bladder in connection with descensus of the vagina and uterus. Possibility of reconstruction of the urethrovesical angle.

down on the perineum with two fingers (Fig. 127). The Sims speculum is more helpful in making the diagnosis than is the Graves speculum. By asking the patient to contract the sphincter ani the levator tone can be evaluated. Stress incontinence can be diagnosed by a number of simple clinical tests. If supporting the periurethral connective tissue with an instrument relieves the incontinence, the patient probably has stress incontinence and a surgical correction is likely to be helpful.

The extent of the incontinence is demonstrated by cystourethrography (chain cystourethrogram) and by the cystometrogram (bladder pressure study).

The rectocele can be best appreciated by inserting a finger in the rectum and pushing the posterior vaginal wall upward (Fig. 128).

Treatment: Cystocele and rectocele require treatment if they are symptomatic or very large. An attempt can be made in mild cases of stress incontinence to strengthen the pelvic floor by exercises.

The principle of surgical correction is the replacement of the bladder and the restoration of the posterior urethrovesical angle. This can be accomplished vaginally by an anterior colporrhaphy and a mattress suture at the bladder neck (Kelly stitch). When possible, removal of the

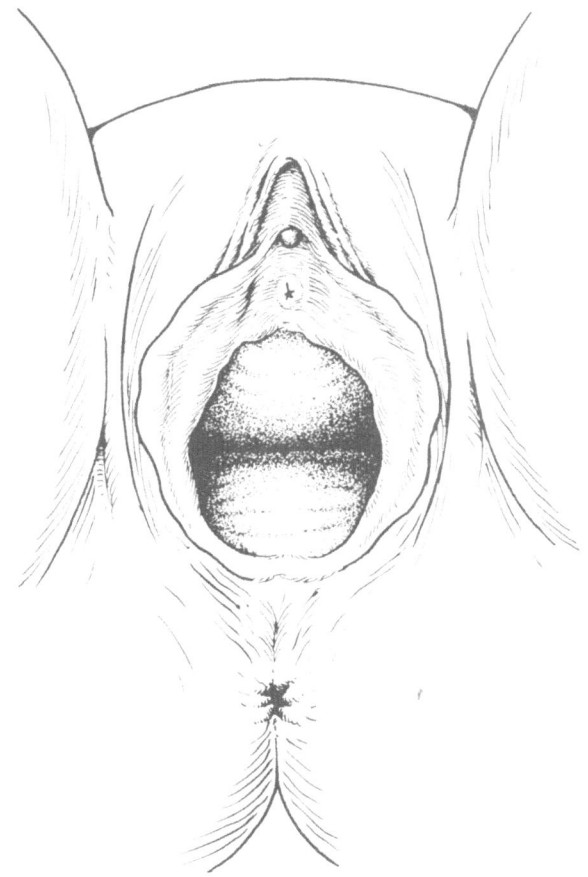

Fig. 126. Findings on inspection in the case of descensus: gaping vulva, bulging of the anterior and posterior vaginal walls.

uterus is desirable to prevent recurrence. A posterior colporrhaphy may narrow the hiatus and help restore the pelvic floor support. In patients who desire more children, only an anterior and posterior colporrhaphy is performed. Delivery usually requires a cesarean section, but even abdominal delivery does not eliminate recurrences. It is, therefore, preferable to delay the surgical correction until the patient's family is complete. The Manchester–Fothergill operation, consisting of amputation of the portio of the cervix and plication of the cardinal ligaments, is a procedure that is occasionally used to correct prolapse in women who desire to keep their uteri.

An abdominal approach to stress incontinence, especially for Type II, is the Marchall–Marchetti procedure. The connective tissue around

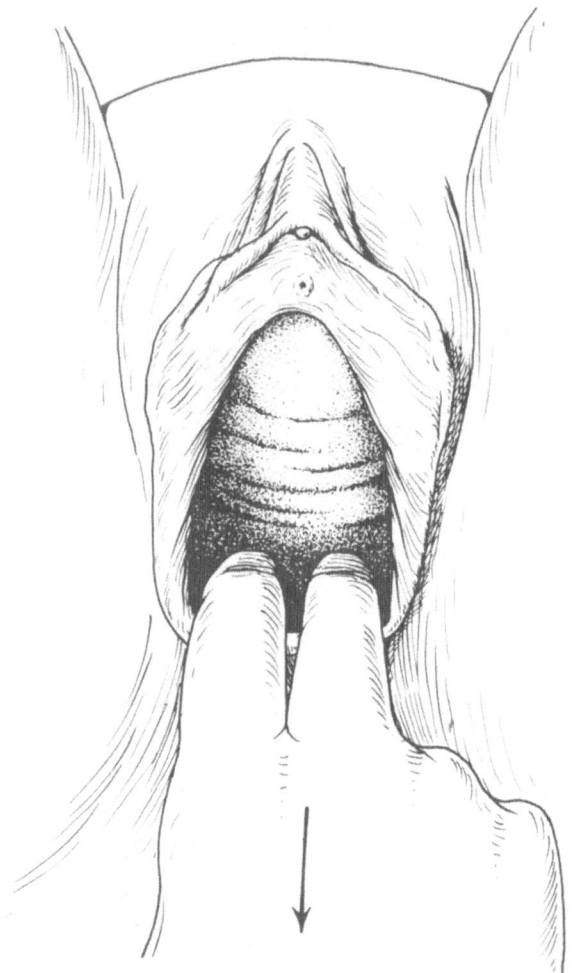

Fig. 127. Test of the relaxation of the pelvic floor by pressure on the perineum.

the bladder neck is attached to the periosteum of the symphysis, thereby restoring the urethrovesical angle. This procedure is occasionally indicated when laparotomy is performed for other reasons.

Prolapse of the uterus is best treated by vaginal hysterectomy (Fig. 129). Anterior and posterior colporrhaphy is often performed in addition. Colpocleisis (Le Fort procedure) is rarely performed today and then only in very old patients. With modern anesthesia the risk of vaginal hysterectomy is minimized and the time for its completion not greater than that of a colpocleisis.

Pessary: The conservative treatment using plastic or rubber pessaries has only a very small place in modern gynecology. The pessary must

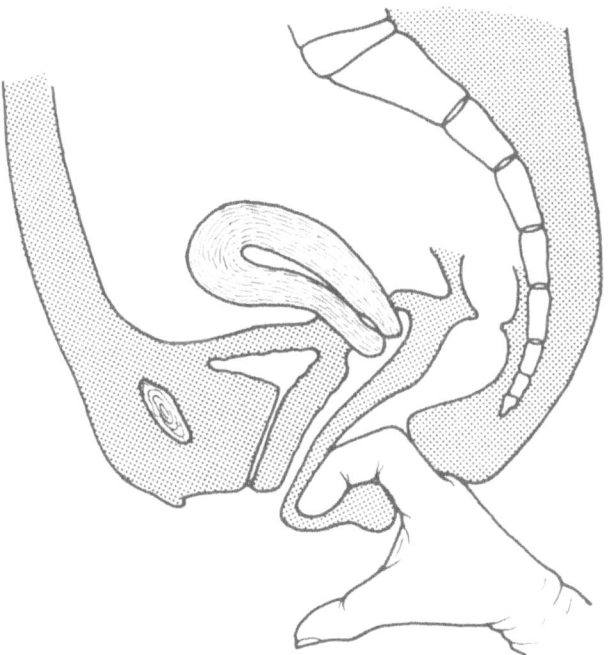

Fig. 128. Detection of the rectocele by rectal examination; by rectal palpation the extent of the rectocele can be ascertained.

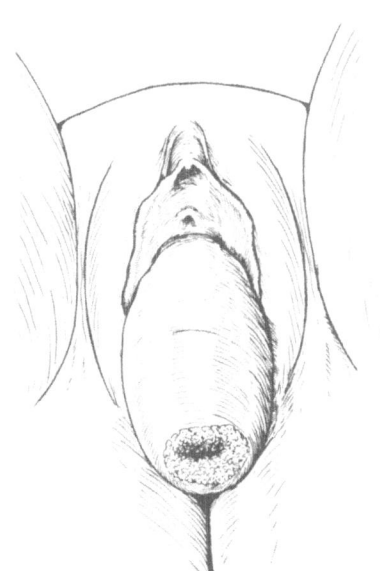

Fig. 129. Complete prolapse of the uterus with eversion of the vaginal wall; decubitus ulcer on the portio.

be removed at least every four weeks, and vaginitis and vaginal neo-plasms occur despite the administration of estrogenic vaginal creams.

Prophylaxis: Since prolapse of the uterus and cystourethrocele are frequently associated with previous pregnancies, the management of labor and delivery is of great prophylactic importance. Liberal use of episiotomy and outlet forceps prevent constant pressure on the pelvic floor and reduce the likelihood of relaxation. Exercises during pregnancy and the postpartum period further strengthen the supporting musculature.

Selected Reading

Green, T. H.: The problem of urinary stress incontinuence in the female. *Obst. Gyn. Surv.* **23**:603, 1968.

Hodgkinson, C. P., and Drukker, B. H.: Stress urinary incontinuence. R. M. Wynn, ed. *Obstetrics and Gynecology Annual,* Vol. 2, p. 367, 1973.

Zacharin, R. F.: *Stress Incontinuence of Urine.* New York: Evanston, San Francisco, and London: Harper & Row (1972).

CHAPTER 20

Urologic Problems in Gynecology

Gynecological diseases such as malformations, inflammations, and tumors are frequently associated with functional disturbances of the urological tract. Because of its close anatomical relation to the reproductive tract, the urinary tract is often a site of postoperative and postirradiation complications. Diagnostic urologic procedures are mandatory in the work-up of a variety of gynecological disorders. Although the more complex procedures and therapeutic measures are in the province of the urologist, an understanding of these procedures enables the gynecologist to advise his patient appropriately.

Malformations

Congenital malformations of the genital organs are frequently associated with anomalies of the urinary tract. Intravenous pyelography and cystoscopy are therefore part of the diagnostic investigation of genital malformations.

Aplasia of the kidney is observed with or without associated genital malformations in 1 to 2% of patients with urologic complaints. Malformations of the kidney and ureters are found in 3 to 4% of a similar group of patients. A double ureter unilaterally or bilaterally is the most frequent anomaly. The extra ureter may empty ectopically, for example, into the vagina and therefore produce urinary incontinence. Intravenous pyelography is diagnostic. Treatment consists in excision of the extra ureter and the associated kidney, which is usually nonfunctioning. In the rare situation of bilateral ectopic ureters, plastic procedures are attempted with implantation of the ureters into the bladder. The prognosis may be jeopardized by damage to the associated kidneys. Ectopy of one kidney is a not infrequent anomaly. The pelvic kidney may be located retroperitoneally in the pelvis and may frequently be taken for a gynecological tumor (Fig. 130). If this fused ectopic kidney is removed, the results are catastrophic. Hypospadias and epispadias are seen occasionally in patients with intersexuality. A urethra may be formed by means of plastic procedures using a skin flap.

A rare anomaly is exstrophy of the bladder. In this situation the bladder fails to close laterally and its mucosa and ureteral openings

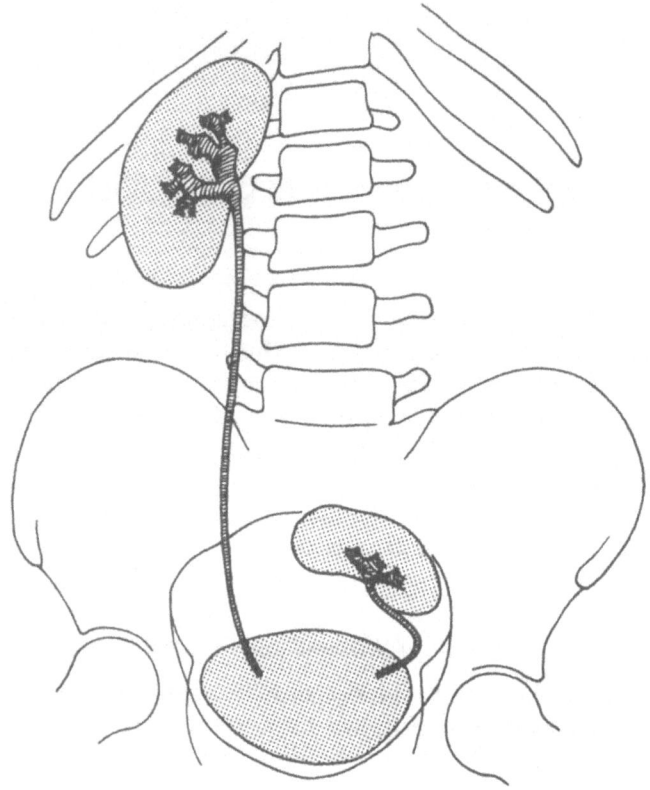

Fig. 130. Pelvic kidney on left (seen radiologically).

are external and visible. Treatment consists in surgical implantation of the trigone into the sigmoid with subsequent drainage of urine through the bowel. The defect is then closed by plastic surgical means.

Compression of the Ureters as a Result of Gynecologic Disease

Stenosis of the ureters by benign intraligamentous tumors rarely occurs, but compression and stenosis as a result of cervical carcinoma are not infrequent. (page 291). The ureters may be affected by scarring after radical hysterectomy or irradiation, or they may be occluded by direct spread of the carcinoma. Partial blockage of the ureters results in hydronephrosis and eventual loss of renal function. Bilateral occlusion of the ureters is the cause of uremia in many patients with untreated or unsuccessfully treated cervical carcinoma. Even after successful treatment, approximately 10 to 20% of all patients with carcinoma of the cervix eventually die with urologic complications. If the ureters are occluded by scar tissue as a result of surgical procedures or irradiation, and recurrence can be excluded, the ureters may be reimplanted into the bladder or an ileal or sigmoid conduit may be established.

Bladder Incontinence

Stress Incontinence. Stress incontinence of urine is often associated with pelvic relaxation (page 248). Incontinence of another variety may be the result of extensive dissection of the bladder or cervical tumors with resultant urinary retention.

The Irritated Bladder and Bladder Retention. An irritated bladder may be the result of cystitis, but it may also be psychogenic in the absence of inflammatory reactions and bacteremia. Urgency incontinence is characterized by the frequent desire to empty the bladder, which is, however, empty or almost empty. Since this symptom is only rarely related to cystocele, it will persist after colporrhaphy. Frequency may result from a bladder with reduced capacity as a result of scarring secondary to vesical tuberculosis or irradiation. Cystoscopy confirms the diagnosis.

Psychogenic urinary retention is poorly understood. Retention may develop in 60 to 80% of patients transiently after colporrhaphy as a result of postoperative edema, and may require suprapubic cystotomy or transurethral catheterization.

Urogenital Fistulas

Vesicovaginal fistulas, ureterovaginal fistulas, and urethrovaginal fistulas are usually complications of surgical procedures. Laceration of the bladder during surgical operation is followed by relatively few complications if it is recognized promptly and closed in layers. Unrecognized lacerations or placement of a suture through bladder and vagina may result in a vesicovaginal fistula, as after total hysterectomy. Such fistulas are usually located in the midline of the vagina high on the anterior wall. The fistula may be identified by instillation of methylene blue into the bladder. Treatment consists in vaginal closure by a partial colpocleisis (Latzko procedure). The success rate may be as high as 90%.

Ureterovaginal fistulas occur in from 2 to 10% of patients after radical hysterectomy, as a result of diminution in the blood supply to the ureters. Inadvertent ligation or transection of the ureter are occasional complications encountered in the removal of large intraligamentous tumors. The fistula in such cases is usually located laterally and can be identified by intravenous pyelography. Surgical treatment consists in implantation of the ureters into the bladder if ureteral catheterization fails to result in spontaneous healing.

Urethral fistulas produce incontinence only if they are located in the upper third of the urethra (page 252). They can result from colporrhaphy or as a direct spread of vaginal carcinoma. The reconstruction of the urethra may be quite difficult and correction of the incontinence may require transplantation of the ureters into the bowel.

Urogenital fistulas may be the result of irradiation or the direct spread of carcinoma. Fistulas after radiation therapy are not necessarily the result of overdosage. They may develop as a result of individual sensitivity or destruction of connective tissue by a large tumor. The surgical treatment requires complicated procedures, which are frequently unsuccessful.

Infection

Cystitis, pyelitis, and pyelonephritis are generally related. A urological infection may develop either by the ascending or the descending route. Only ascending infections are discussed here.

Latent infections of the bladder or kidney are much more common in women than in men. Bladder infections are seen frequently in the newborn and the adolescent girl because of the short female urethra. Cystitis is often unrecognized and ascending infection frequently follows (Fig. 131). Vesicoureteral reflux is more pronounced in young girls than in those of reproductive age. Recurrent urinary tract infections often begin in early childhood.

Predisposing factors in the reproductive age may be related to frequent intercourse ("honeymoon" cystitis) or anatomical and physiological changes related to pregancy and delivery. Partial obstruction of urinary flow from the kidney and irritation of the urethra increase the likelihood of spread of infection in the urogenital tract. It is therefore understandable that urinary tract infections are the most common secondary infections associated with gynecological disease. Urinary tract infection is observed in about half of all women with a cystocele, with or without other symptoms. In 7 to 10% of women with cervical cancer there is involvement of the urogenital tract. The frequency increases with the stage of the tumor. Fistulas are associated with infections of the urogenital tract.

A frequent source of infection is catheterization, which produces cystitis in 1 to 2% of cases, of which 5 to 6% progress to pyelonephritis. Atony of the bladder after major operations requires an indwelling catheter but microorganisms may make their way into the bladder by passing between the wall of the catheter and the urethra. Trauma to the bladder or urethra during catheterization predisposes to ascending infection. Without systemic antibiotics the incidence of bacteremia rises to 80 to 100% after 2 to 3 days of an indwelling catheter. Asymptomatic pyelonephritis occurs as a late complication in 15 to 25% of patients after radical surgical procedures.

Urinary tract infections are caused predominantly by gram-negative

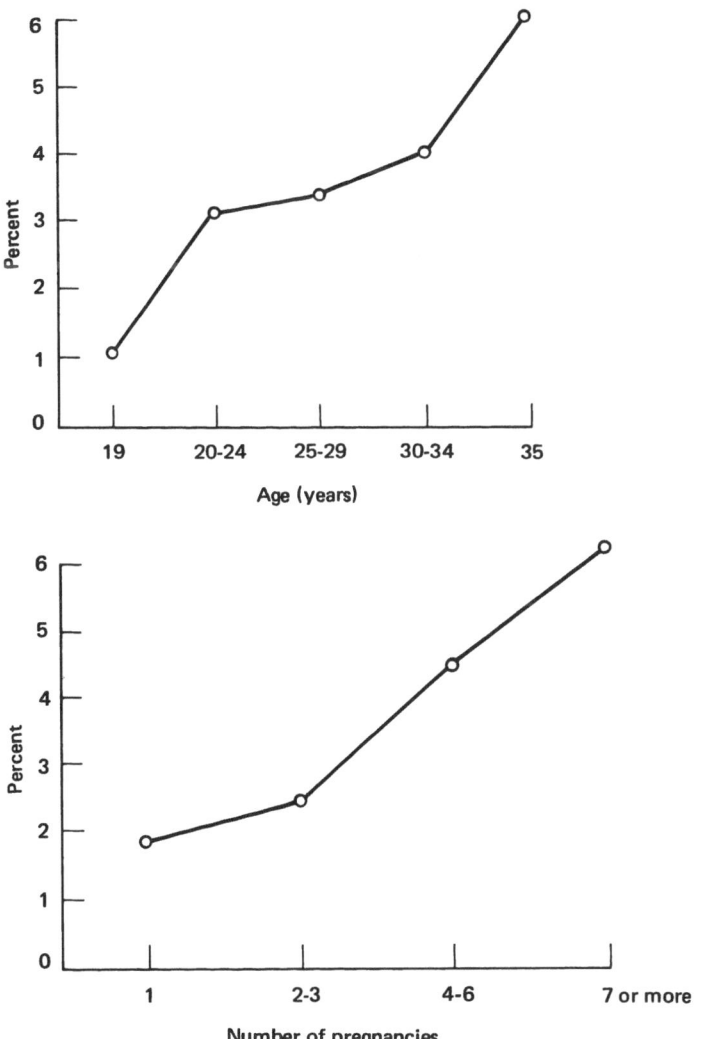

Fig. 131. Frequency of asymptomatic bacteriuria in women. *Top:* In relation to age. *Bottom:* In relation to the number of pregnancies (after E. H. Kass, 1960).

coliform organisms, *Proteus,* and *Psuedomonas.* These infections require careful selection of antibiotics. The causative organism is ascertained through cultures, which should be accompanied by sensitivity tests. Bacteria in excess of 100,000 per ml indicate latent infection. Less extensive growth may result from manipulation and contamination during catheterization.

Prophylaxis involves reduction in the use of catheterization and strict aseptic conditions when catheterization is necessary. For routine purposes, a so-called clean-catch midstream urine is preferable. To reduce the frequency of ascending infection suprapubic cystotomy has replaced routine indwelling catheters in many centers.

Acute and Chronic Cystitis. Acute cystitis is seen most frequently by the gynecologist as a complication of an atonic bladder or an indwelling catheter. The vesical mucosa may be diffusely or patchily erythematous and edematous. A cystocele is frequently associated with chronic cystitis because of the residual urine. The cystitis may be cured after a cystocele is repaired by colporrhaphy or the Marshall-Marchetti procedure (page 255). Postoperative recurrence as a result of indwelling catheters is observed in 80 to 90% of patients.

On cystoscopy, chronic cystitis appears as a whitish thickening of the mucosa with reduced vascularity. Chronic cystitis in women is frequently localized in the trigone. Chronic cystitis is often related to cervical carcinoma. In Stage I carcinoma, edema of the mucosa is frequently observed. Involvement of the vesicovaginal septum results in bullous edema and possible hematuria. Cystitis may also result from irradiation.

Symptomatology: Burning and pain on urination (dysuria) and urinary frequency are the main symptoms and signs. Constant pain in the bladder area and elevation of temperature occur less frequently. A sudden rise of temperature in connection with flank pain indicates pyelonephritis. In this situation the urine may be cloudy and bloody (hemorrhagic cystitis).

Diagnosis: The urinary sediment contains epithelial cells, leukocytes, bacteria, and erythrocytes.

Differential diagnosis: Straining on urination and erythrocytes in the urine may be the initial symptom and sign of a ureteral calculus. Tuberculosis of the bladder is another cause of recurrent cystitis. Polyps of the bladder cause symptoms similar to those of cystitis. Leukoplakia may be mistaken for cystitis of the trigone.

Treatment: Systemic antibiotics are administered on the basis of the sensitivity tests. Cystoscopy and ureteral catheterization are deferred until the acute phase has subsided. Radiation cystitis requires treatment by prolonged irrigation of the bladder. Prognosis of the acute form is good, but that of the chronic form depends on the primary disease and on the development (or lack of development) of pyelonephritis. The treatment of these complications is the province of the urologist.

Urethritis. Isolated infection of the urethra is rare except with gonorrhea (page 239). Foreign bodies may cause urethritis in small children. The so-called honeymoon cystitis is frequently associated with urethritis. The symptoms are dysuria and frequency. Urethritis may also be the result of chronic urethral diverticulitis; this diagnosis is suspected when pus is expressed from the external os of the urethra by pressure along the urethra and is confirmed by radiologic studies. The multiple tube urine test may help to differentiate cystitis from urethritis. The patient is asked to urinate successively into several test tubes. If the numbers of leukocytes and bacteria are larger in the first tubes and progressively smaller in subsequent tubes, urethritis is more likely than cystitis.

Selected Reading

Kass, E. H.: *Progress in Pyelonephritis*. Philadelphia: F. A. Davis Company (1960).

CHAPTER 21

Back Pain as a Presenting Symptom

Approximately 30 to 40% of patients in routine gynecologic practice complain of back pain. Back pain is four times more frequent in women than in men. The mechanical effects of pregnancy and childbearing may in part explain the difference. In addition, abnormalities in the position of pelvic organs or disease thereof may produce back pain as a result of involvement of the lumbosacral spine directly or its nerve supply. Organic disease of pelvic organs is estimated to be present in 10 to 20% of patients complaining of back pain in a gynecologist's office. The complaint is not etiologically related to gynecologic disorders in the majority of patients. It is the result of a functional disturbance of the musculoskeletal system (approximately two-thirds of patients) or local deformities of the skeletal system (approximately one-third of patients). Back pain is rarely a leading symptom of neurologic, urologic, or medical diseases. Back pain in women is therefore caused, with few exceptions, predominantly by orthopedic disorders, and only occasionally by gynecologic diseases.

Time of onset, intensity, localization, and radiation of pain provide diagnostic clues. Relations between "back pain" and gynecologic disease are summarized in Table 24. If organic gynecologic disease can be ruled out, the patient may be referred for orthopedic and, occasionally, medical or urologic evaluation. If no organic cause can be found, psychosomatic factors may be considered before the patient is treated symptomatically.

Injuries of the Genitalia

Coital Injuries. First coitus (defloration) usually causes single or multiple lacerations of the hymen and occasionally slight bleeding. Only rarely is hemorrhage profuse enough to require suturing. Injuries of the vaginal introitus or clitoral area require suturing more often because of the greater vascularity. Coital injuries of the vaginal wall have to be repaired after removal of the clotted blood. Temporary insertion of a drain and local application of antibiotics may reduce the risk of parametritis or ascending infection. These injuries are not infrequent after rape and they may occur in postmenopausal women with minimal

Table 24. Gynecological Causes of Back Pain

Disease	Special features
Primary dysmenorrhea (Uterine anomalies, genital hypoplasia)	Cyclic appearance
Secondary dysmenorrhea (Endometriosis, submucous myoma)	Cyclic appearance
Acute and chronic inflammation (Pyosalpinx, hydrosalpinx, cul-de-sac abscess, adhesions, endometriosis)	Mostly presacral deep in the pelvix
Malposition of the pelvic organs, especially descensus of the vagina and uterus	With radiation to the groin
Benign and malignant tumors of the genitalia, pelvic venous thrombosis	Depending on size and extent, with radiation to the groin and thigh
Osteoporosis	Primary amenorrhea, prolonged secondary amenorrhea, climacteric, menopause
Neurosis	Exaggerated description of complaints

trauma. The atrophic mucosa, in association with the reduced elasticity of the connective tissue structures, renders the vagina more vulnerable than in the reproductive age group. The possibility of extension of lacerations into the posterior cul-de-sac has to be considered when examining vaginal injuries. Injury brought about by falling astride some object is usually seen in children. Apart from lacerations visible in the introitus or anal region, the presence of paravaginal or parametrial hematomas has to be considered. Perforating injuries of bladder or rectum have to be explored by x-ray examination and urine analysis. Surgical repair requires evacuation of hematomas and proper drainage in order to prevent local abscesses. Antibiotics and tetanus prophylaxis are indicated.

External trauma to the lower abdominal region may involve the internal genitalia as a result of fractures of the pelvic bones. It is often difficult to evaluate the extent of internal injuries and bleeding. Injuries of bladder and ureters are identified by intravenous pyelogram and cystoscopy, and require immediate repair in order to prevent extravasation of urine and formation of fistulas. Residual pelvic adhesions have to be considered.

A detailed objective description of the findings without using words such as "rape" or "virgin" may be of great medicolegal value.

CHAPTER 22

Endometriosis and Adenomyosis

Endometriosis is defined as the occurrence of functioning endometrium outside the uterus. Adenomyosis refers to endometrial tissue deep within the myometrium.

Endometriosis

Endometriosis has been attributed to several factors. According to Sampson's theory (1921), endometrium may be regurgitated with the menstrual blood through the tubes into the peritoneal cavity. This tissue then implants elsewhere in the pelvis, primarily on the ovaries or in the posterior cul-de-sac. This theory has been supported by studies in monkeys as well as in women. Viability and proliferative capacity of endometrial cells obtained from menstrual blood have been demonstrated in tissue culture. Since endometrial implants are also found outside the true pelvis, vascular and lymphatic spread with subsequent ectopic implantation have been invoked as additional mechanisms, although they are at most of secondary importance. Mechanical transplantation of endometrial fragments during surgical procedures may explain the appearance and proliferation of endometrium in laparotomy scars.

Another theory attributes the development of endometriosis to metaplasia of celomic epithelium. It explains the appearance of endometriosis in most of its common sites.

The ectopic endometrial tissue sometimes responds to ovarian hormones as does the mucosa of the uterine cavity itself. The implants may be stimulated by endogenous as well as exogenous hormones. The functional zone may react with proliferative and subsequent secretory changes. Bleeding similar to menstruation may occur at the sites of endometriosis at the time of menstruation. Symptoms and progress of the disease are related mainly to the cyclic changes in the ectopic endometrial implants. Severity of the disease depends primarily on location and size of implants.

Common sites of endometriosis in decreasing order of occurrence are: ovaries, posterior cul-de-sac with involvement of uterosacral ligaments, rectovaginal septum, serosa of fallopian tubes, rectosigmoid, and bladder. Occasionally, endometrial implants are found on the cervix,

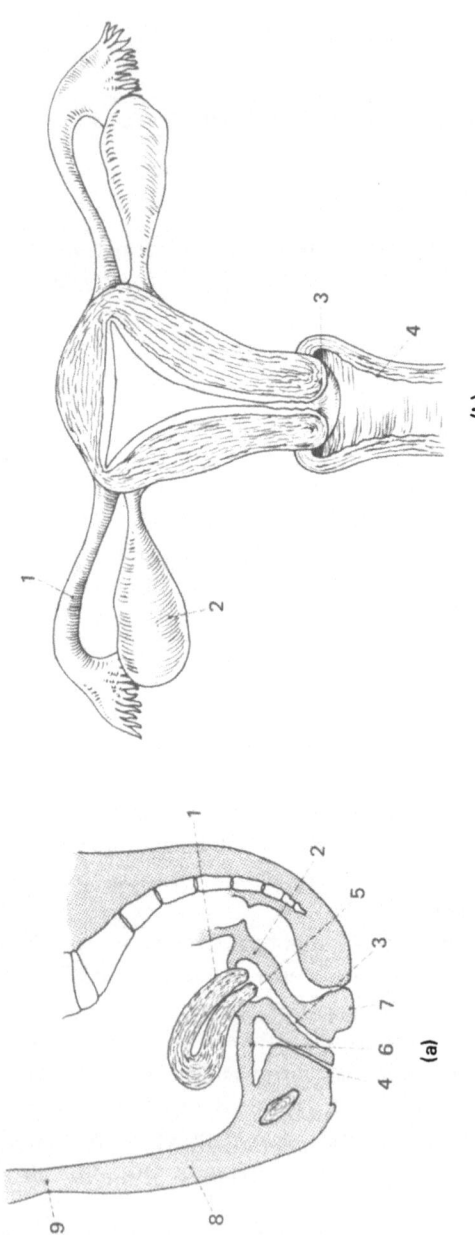

Fig. 132. Some of the locations of endometriotic implants. (a) 1. posterior wall of the cervix and pouch of Douglas; 2. rectovaginal septum; 3. vaginal wall; 4. vulva; 5. portio vaginalis; 6. bladder wall; 7. perineum; 8. laparotomy scars; 9. umbilicus. (b) 1. in or on the tube; 2. in the ovary (chocolate cyst); 3. portio vaginalis; 4. vaginal wall.

vaginal mucosa, inguinal nodes, and, less commonly, in distant regions such as the umbilicus, laparotomy scars, and appendix (Fig. 132).

Small implants of endometrial tissue on the surface of the ovary may develop into hemorrhagic cysts that rupture easily. Spillage of the contents of the cyst is followed by adhesions to neighboring peritoneum, leading to larger endometrical cysts (Fig. 133). Repeated bleeding into the free peritoneal cavity and result in hemorrhagic cysts of variable size. They resemble hematomas and are called chocolate cysts because of the color and consistency of their contents. With each successive menstruation, the pressure inside the cyst increases, often resulting in necrosis of the ectopic endometrium in the wall of the cyst. Since the process is self-limiting, in the late stages it may be difficult to recognize any endometrium histologically. Adenocarcinoma in the ovary may arise rarely from this ectopic endometrium.

Endometrial implants, in the posterior cul-de-sac with involvement of the uterosacral ligaments, on the posterior aspect of the cervix or the posterior vaginal vault, and in the rectovaginal septum, do not reach the size of ovarian endometrial cysts (Fig. 133). Repeated bleeding from multiple nodules soon leads to extensive adhesions of dense fibrous tissue on adjacent structures. In the end stage the whole pelvis may be occupied by a nodular, firm, fixed mass.

The peak incidence is between the ages of 25 and 35. There are no precise figures relating to true prevalence since the number of patients

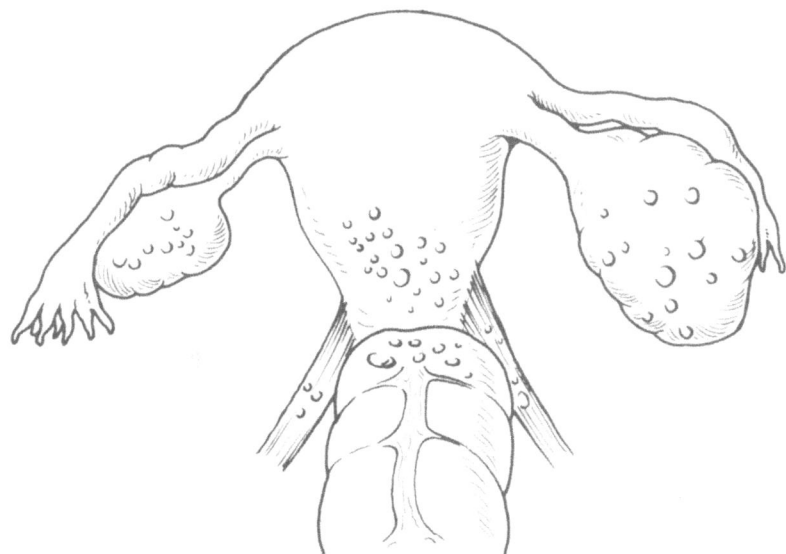

Fig. 133. Multiple endometriotic implants on the posterior wall of the cervix, in the pouch of Douglas, and on the ovaries.

requiring treatment does not reflect the total numer of patients with this disorder. Endometrial implants may be found, however, in from 5 to 15% of specimens obtained from unselected gynecological operations.

Symptoms: The number and size of endometrial implants quite often bear no relation to the degree of discomfort. Depending on location small implants may produce severe pains, whereas large masses may cause only slight discomfort. The leading symptom is secondary (acquired) dysmenorrhea. The pains usually start before the actual onset of menstruation, reaching their peak of severity on the first day of bleeding, and then gradually decrease. They are usually localized deep within the pelvis. The uterus is frequently retroverted and fixed. The endometrial implants located in the rectovaginal connective tissue are felt on rectal and rectovaginal examination as firm nodular infiltrates. Diagnostic accuracy can be improved by laparoscopy, although biopsy is necessary for proof. When endometriosis of the bladder or rectum is suspected, cystoscopy and proctoscopy become part of the diagnostic work-up.

Differential diagnosis: Chronic pelvic inflammatory disease, especially tuberculosis, must be considered. Ovarian endometriosis must be differentiated from benign and malignant tumors. Since mobile chocolate cysts cannot be differentiated on pelvic examination from ovarian cysts of other origins, the diagnosis is usually made during operation and by histological examination. Small implants in the uterosacral ligaments may be difficult to differentiate from the so-called pelvic congestion syndrome. Implants in the uterosacral ligaments may cause dyspareunia, and endometrial tissue in the rectovaginal septum and rectum may cause pain on defecation. Endometriosis of the wall of the bladder is a cause of dysuria, and involvement of the mucosa may appear as cyclic hematuria. Because of the early formation of adhesions, endometriosis frequently results in infertility. It is in some series reported to be the most common cause of secondary infertility. In women who marry late, endometriosis may be a cause of primary infertility.

Diagnosis: Diagnosis is suggested by history, especially when an adnexal mass is felt. Endometriosis, unlike pelvic inflammatory disease, is not accompanied by signs of inflammation. A normal or only slightly increased leukocyte count and a normal or only slightly elevated erythrocyte sedimentation rate favor the diagnosis of endometriosis. Nodules in the uterosacral ligaments as well as firm infiltrates in the rectovaginal septum suggest endometriosis.

Treatment: Early diagnosis and treatment may prevent the formation of extensive adhesions and scars and consequent infertility. Hormonal and surgical treatments, separately or combined, are available.

Hormonal treatment is based on the observation that endometriosis often improves during pregnancy, when decidual changes may be seen in the ectopic endometrium as well. In favorable circumstances, atrophy and necrosis of the ectopic endometrial tissue follow. These so-called burnt out endometriotic sites no longer respond to hormonal stimulation and the progress of the disease is halted. Attempts to reproduce these gestational changes by administration of exogenous steroids seemed logical. To achieve "pseudopregnancy," estrogen-progesterone combinations (with low estrogen and high progestational content) are given. Depending on the reaction of the implants, the treatment should be continued for several months but not longer than one year. The occasional breakthrough bleeding can be controlled by a temporary increase in steroids, usually a doubling of the daily dose. A modification in treatment is the continuous use of long-term synthetic progestins (without estrogens), which may result in atrophy of the endometrium. The medication may be given daily, weekly, semimonthly, or monthly.

Once extensive adhesions and scarring have formed and the endometrial tissue is no longer responsive to hormones, surgery becomes the treatment of choice. The operative procedure should be tailored to the age and reproductive history of the patient. In young women who want to become pregnant, conservative operations are appropriate. Sharp excision or electrocauterization of isolated implants, lysis of adhesions, and salpingolysis may be performed. Since the required surgical procedure is often unpredictable, the patient should be informed that the type of operation can be decided only during the procedure. If it is impossible to preserve any functioning ovarian tissue, the adnexa should be removed regardless of age. In patients in whom reproductive function is not desired and in whom extensive endometrial implants are found at laparotomy, total hysterectomy with bilateral salpingo-oophorectomy may be the procedure of choice. Elimination of ovarian function by oophorectomy guarantees permanent cure. This radical approach may be indicated if hormonal treatment fails and if removal of implants is impossible without jeopardizing adjacent organs such as bladder and rectum.

In summary, it is evident that treatment of endometriosis should be individualized according to location of the implants, severity of signs and symptoms, age of the patient, and desire for pregnancy. The rate of success of hormonal treatment varies widely. Some authors report rates as high as 80%. Infertility on the basis of endometriosis is treated successfully in at least 20 to 30% of cases. Surgical procedures may offer greater chances of success to the infertile patient but may be less effective in relieving symptoms.

Adenomyosis

Adenomyosis involves extension of endometrium into the myometrium with simultaneous thickening of adjacent myometrium. The endo-

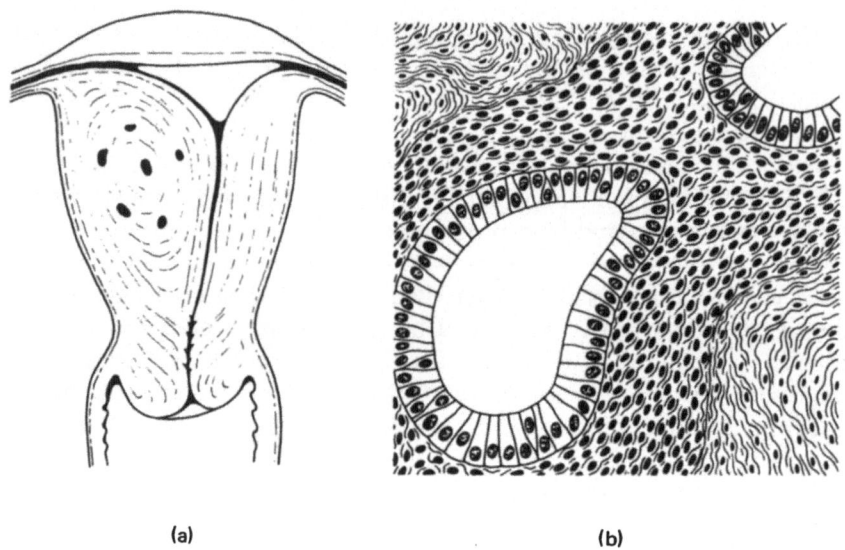

(a)   (b)

Fig. 134. Adenomyosis. (a) Endometrial islands in the myometrium surrounded by circular hypertrophic muscle bundles; (b) cystically dilated endometrial glands and stroma are seen embedded in the myometrium.

metrium should be at least one low-power field distant from the basal
layer of the endometrium. Although adenomyosis and endometriosis are
often discussed together, they are different diseases with respect to etiology and clinical features. Unknown local factors may stimulate endometrial growth and penetration of myometrium. Growth of the surrounding myometrium is a typical finding. Ovarian dysfunction related to
a predominant estrogen effect may be an etiologic factor (Fig. 134)..

The uterus is usually moderately enlarged and globular. Macroscopically, pockets of endometrium may be seen, frequently filled with blood
and extending almost to the serosa.

Microscopically, the endometrial areas in the myometrium usually
show both glandular and stromal elements, although one or the other
may predominate. A fully developed secretory change is seen only occasionally. Usually the picture is that of the early proliferative phase.
The abnormally located endometrium is apparently not fully responsive
to progesterone. The endometrium in the myometrium may appear to
be in the proliferative phase even when the superficial endometrium
is secretory. Histologically the adenomyotic implants may resemble cystic
glandular hyperplasia even without similar changes in the endometrium
lining the cavity.

Adenomyosis is predominantly seen in multigravidas in the fourth
and fifth decades, thus differing from endometriosis. In about 50% of
cases it is associated with myomata of the uterus. Endometriosis and

adenomyosis are occasionally found in the same patient. Adenomyosis has been reported in one third of all uteri excised for adenocarcinoma of the endometrium.

Stromal adenomyosis in which the tissue penetrating the myometrium consists only of stromal elements must be distinguished from endometrial sarcoma.

Signs and symptoms: Major signs are menorrhagia and secondary dysmenorrhea. The condition usually progresses, resulting in increasing pain.

Diagnosis: The uterus is usually moderately enlarged and globular. When myomas are associated it may be larger and irregular. The uterus is particularly tender and further enlarged just before the onset of menstruation. Myomas of the uterus, endometrial polyps, and carcinoma of the corpus may present with some of the same signs and symptoms. The diagnosis is often not made until the time of operation.

Treatment: Hormonal treatment is not effective. When adenomyosis causes distressing signs and symptoms, hysterectomy is the treatment of choice.

Selected Reading

Gompel, C., and Silverberg, S. G.: *Pathology in Gynecology and Obstetrics.* Philadelphia and Toronto: J. B. Lippincott (1969).

Novak, E. R., and Woodruff, J. D.: *Novak's Gynecologic and Obstetric Pathology, 6th ed.* Philadelphia and London: W. B. Saunders (1967).

CHAPTER 23

Pelvic Congestion Syndrome

This entity is poorly defined and poorly understood and it has been described in the international literature under a variety of terms. It is believed to be of neuromuscular or neurovascular origin as an organic manifestation of psychological conflicts that result from a fear of pregnancy or inadequate sexual response.

The venous channels in the pelvis are congested. The uterosacral ligaments are indurated and may later become fibrotic. Congestion of the pelvic organs may result also from failure to achieve orgasm.

Symptoms

The patient usually complains of deep pelvic pain that increases just before menses. Hypersecretion of cervical mucus may result in discharge. The pain of pelvic congestion may be accompanied by pains elsewhere, for example, migraine, fullness of the breast, and obstipation. The patient tires easily and may be mildly depressed. Dyspareunia is characteristic and coitus interruptus and lack of orgasm are frequently elicited in the history.

Diagnosis

The multiple complaints and the description of the pains suggest a psychosomatic origin. On pelvic examination the patient is usually tense. The uterus is often retroflexed and enlarged, possibly up to 10 weeks' gestational size. Attempts to move the uterus produce pain. The portio of the cervix may be hypertrophied and elongated. The entire pelvis may be painful to touch, particularly the posterior aspect of the symphysis.

On differential diagnosis endometriosis involving the uterosacral ligaments must be excluded, but the history is different. The pelvic congestion syndrome differs also in that the contracted uterosacral ligaments may become much less tense under anesthesia, whereas the findings in endometriosis remain. Chronic pelvic inflammatory disease also must be ruled out.

Treatment

In the early stages of the syndrome, psychological support may be successful, especially if the patient understands the relation between her conflict and this manifestation. Proper advice about contraception and sexual counseling are important. In some instances psychotherapy may be advisable. Tranquilizers and adrenergic blocking agents may provide additional support. In the chronic stage, if fibrosis is superimposed on the pelvic congestion, hysterectomy may be indicated, particularly in the older patient. The syndrome is still dubious and its very existence is questioned by some gynecologists.

CHAPTER 24

Benign and Malignant Neoplasms of the Genitalia

The Vulva

Benign Tumors. Benign tumors of the vulva are not fundamentally different from skin tumors in general. *Papillomas, condylomas,* and *adenomas* are the principal solid tumors of epithelial origin.

Condylomata acuminata are considered to be inflammatory lesions that develop in association with infections and chronic irritation of the genitalia (p. 217) although they resemble benign neoplasms.

Cystic tumors of the vulva are common. They are often derived from remnants of Gartner's duct (page 16) or they may represent retention cysts of sebaceous glands or tumors of sweat glands (hidradenoma).

Abscesses and cysts of Bartholin's gland result from occlusion of the duct, which in turn forms a cyst. The gland itself may be uninvolved in the cyst.

Among the mesenchymal tumors of the vulva are lipomas and fibromas; fibromas are often peduculated. Myomas may be derived from muscular strands in the labia majora, perhaps related to the insertions of the round ligaments (page 32).

Rare tumors of the vulva are lymphangiomas and hemangiomas.

Malignant change rarely occurs in hidradenomas (tumors of sweat gland origin) or papillomas.

Symptomatology: With the exception of condylomata acuminata even large tumors are often asymptomatic.

Treatment: Surgical removal is indicated if the tumor causes any symptoms. Most important, a malignant lesion must be excluded by prior biopsy. Condylomata acuminata require treatment of the underlying cause (page 217).

Bartholin cysts are treated by marsupialization or excision, as described on page 218.

Malignant Tumors of the Vulva. Precursors of vulvar cancer are dysplastic lesions, which can develop into invasive carcinoma. This statement implies that cytologic and histologic features of malignant transformation occur long before the lesion has become invasive. The latent

279

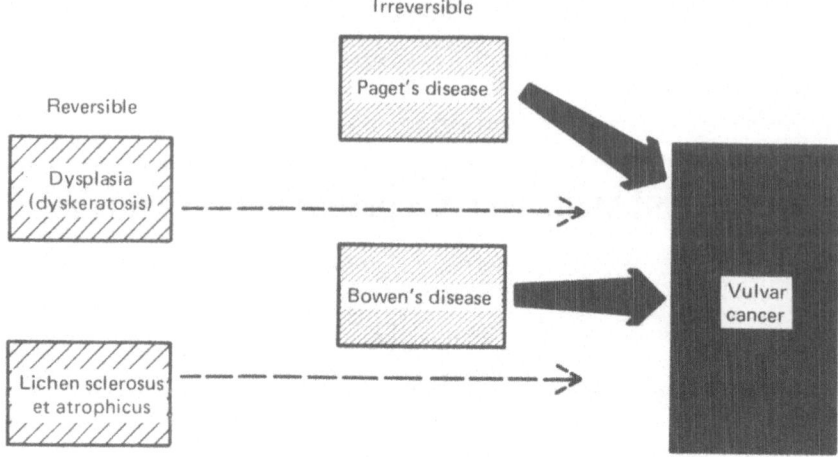

Fig. 135. Premalignant stages of cancer of the vulva.

period between premalignant lesions and invasive cancer may be years, but invasion may also occur within months (Fig. 135).

The most common dystrophic lesion of the vulva is lichen sclerosus et atrophicus. The designation implies a nonlocalized chronic degenerative process of the dermis. The cause is unknown. Only one-third of dystrophic areas are located on the external genitalia. The lesion is occasionally found in men but is five times more common in women. The lesion most commonly occurs postmenopausally, but it may be observed in the reproductive age group and occasionally in children. Lichen sclerosus et atrophicus of the vulva is progressive, involving the labia majora and minora, clitoris, and finally the introitus and perineum.

The lesion begins with degeneration and disappearance of collagenous and elastic fibers of the corium and degeneration of the peripheral nerve distribution. The epidermis is involved only secondarily, resulting in a small to moderate hyperkeratosis (Fig. 136). Loss of elasticity and atrophy of the external genitalia are obvious on inspection. A commonly used clinical term for this change is kraurosis vulvae. It should be stressed, however, that kraurosis is a gross description rather than a specific histologic diagnosis.

Symptomatology: Pruritus is the main complaint, with secondary injuries from scratching. Urinary symptoms and constipation, superimposed infections, and dyspareunia are additional complaints.

Diagnosis: The complaints vary depending on the stage of development of the lesion at which the patient is first seen. The initial lesion is small, localized, and flat or papillary with red edges. It can be found below the surface of the skin as a result of atrophy and sclerosis of

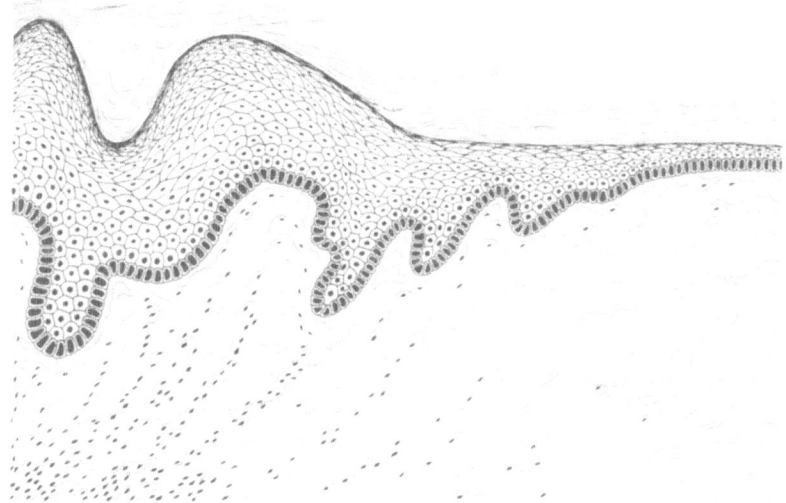

Fig. 136. Lichen sclerosus et atrophicus. From left to right: Progressive thinning of the epithelium with hyperkeratosis and loss of collagenous and elastic fibers.

the connective tissue. The surface is shiny and whitish (mother-of-pearl appearance). The vulva appears asymmetrical. In the final stages the entire vulva is atrophic; the labia minora and the clitoris are unidentifiable and the vestibule is rigid and small. Involvement of the perianal region results in sclerosis and atrophy around the posterior commissure.

Occasionally hyperkeratosis, which is often described clinically as leukoplakia, is associated with dysplastic changes. Cellular atypia in a leukoplakic area may be an early indication of a premalignant change.

Treatment: Estrogen creams, corticosteroids, and aluminum acetate solutions have all been tried, but the results are often disappointing. Patients must be checked carefully and biopsies obtained from suspicious areas frequently.

Dysplastic Changes of the Vulva. Dyskaryotic and dysplastic changes in the dermis are considered premalignant lesions, although not all of these lesions progress.

The cause of atypical dysplastic changes is unknown. Chronic inflammatory reactions are thought to predispose to dyskeratosis. Malignant transformation occurs in 10 to 20%. Basal cell carcinoma is the most frequent form of cancer developing from this lesion.

Another dysplastic lesion in the vulvar area is extramammary Paget's disease. The lesion usually affects patients in or after the menopause. The entire vulva, or only part of it, is involved with multiple whitish or red moist lesions separated by atrophic and eczematous tissue. His-

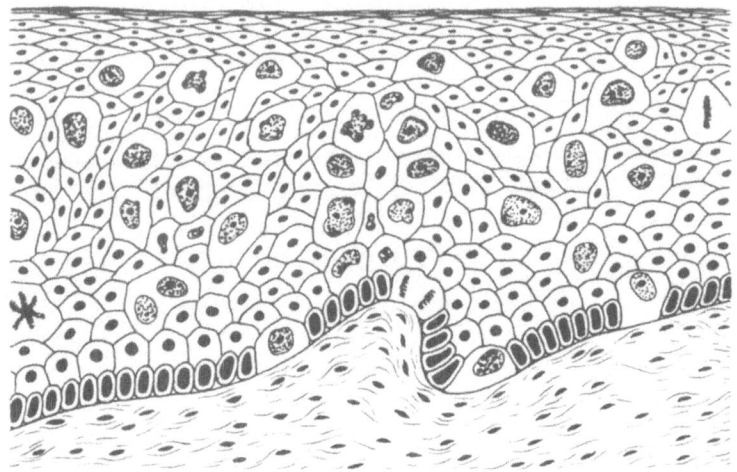

Fig. 137. Paget's disease. The epithelium lacks regular stratification and is occupied by polyhedral cells with a large amount of cytoplasm and pale vesicular nuclei, the so-called Paget cells.

tologically, the normal stratification of the squamous epithelium is lacking. It is replaced by large cells with pale cytoplasm and large nuclei, the so-called Paget cells (Fig. 137). The Paget cell contains acid and sulfated polysaccharides. The origin of the cells is not clear but histochemical reactions indicate that they may be derived from multipotential intermediary cells. Paget's disease is considered to be a form of carcinoma in situ of the vulva.

Another obligate precancerous lesion is Bowen's disease. This lesion is also generally considered to be an intraepithelial carcinoma, or carcinoma in situ, of the vulva. It is commonly found in postmenopausal patients. Bowen's disease is clinically characterized by single or multifocal large whitish lesions. The center is eroded, reddish-brown, and approximately one cm in diameter. Histologically, the epidermis contains dyskaryotic cells with atypical mitoses and lack of stratification. Progression to invasive cancer occurs in approximately 50% of cases.

Diagnosis: Localized leukoplakic or slightly elevated reddish papules in various locations, most frequently on the hairless part of the vulva, should arouse suspicion of premalignant or malignant lesions, especially in older women. Cytologic diagnosis often fails because of the marked hyperkeratosis of the superficial layers. The toluidine blue test (page 110) is positive. Final diagnosis is obtained by multiple biopsies, which may also reveal an in situ lesion or even early stromal invasion. Because of the multicentric lesions, the patient must undergo careful examination of the anus, urethra, cervix, and breasts.

Treatment: Since metastasis occurs in approximately 2% of cases, excision with a wide margin of normal tissue or simple vulvectomy is sufficient. Radical vulvectomy is required if histologic sections reveal areas of invasion. Recurrences are frequently seen in Paget's disease and careful follow-up of these patients is mandatory.

Again, clinical terms such as leukoplakia and kraurosis are purely descriptive. The only basis for therapy is a histological diagnosis.

Cancer of the Vulva. Primary cancer of the vulva accounts for approximately 5% of all cancers of the genitalia. In most cases, the lesion is squamous cell carcinoma. Adenocarcinoma of the Bartholin or sebaceous glands is rare (0.2% of all genital carcinomas). Melanomas, lymphomas, and malignant change in lipomas are extremely rare. Secondary cancer of the vulva (primary carcinomas of the rectum and cervix) more commonly involves the vagina than the vulva.

Squamous cell carcinoma arises from a premalignant lesion in 30 to 40% of cases. In 60 to 70%, the carcinomatous transformation appears suddenly without an antecedent premalignant stage. Cancer of the vulva is most common in patients between the ages of 60 and 70 years, although it is also found occasionally in younger women. Epidemiologic studies suggest that poor hygiene and chronic infections such as lymphopathia venereum are predisposing factors.

The clinical picture in the early stage includes localized, not infrequently multiple, small nodules with eroded surfaces. The labia majora are the most frequent location (40%), followed by the posterior commissure (28%), clitoris (17%), and labia minora (15%). Extension of the tumor occurs by exophytic papillomatous growth (Fig. 138). Another form of spread is endophytic. Large parts of the labia majora, the perineum, or the clitoris area may be involved in later stages. The margin of the carcinomatous area is involved in an inflammatory reaction. Localization of the tumor inside the introitus may be associated with a lesion on the opposite side.

One of the reasons that cancer of the vulva has a poor prognosis is the rich lymphatic supply of the vulva. The main lymphatic drainage involves the inguinal lymph nodes, the lymph nodes at the femoral ring (also called Cloquet's node), and finally, the external iliac lymph nodes. Some lymphatics cross behind the symphysis. The cancer finally spreads through the lymph nodes below Pouparts' ligament, extending to the deep inguinal, hypogastric, and iliac lymph nodes (Fig. 139). Thus, cancer of the vulva metastasizes early to the adjacent lymph nodes in the groin, which not infrequently undergo ulceration.

Symptoms: The tumor often causes no signs or symptoms. Pruritus and pain on urination or defecation are related to ulcerations. Foul-smelling, bloody, watery secretions are present in later stages. Loss of weight is a fairly constant sign.

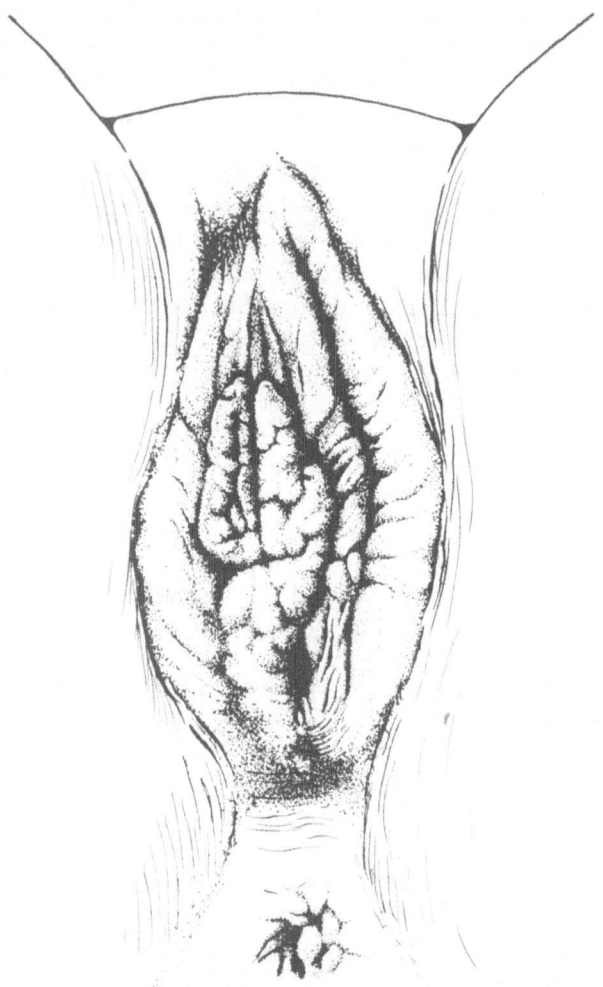

Fig. 138. Extensive exophytic carcinoma of the vulva in the region of the right labium minus with inflammatory swelling of the entire area.

Diagnosis: Single or multiple firm ulcerations with sharp edges and infiltration of the adjacent tissue are seen. Exophytic cauliflower-like ulcerations are also common. The tumor may involve part or all of the vulva, perineum, and rectum. Palpable lymph nodes fixed to the skin suggest metastasis. Proctoscopy may reveal involvement of adjacent organs. Prolapse of the uretha or a paraurethral cyst is occasionally mistaken for vulvar cancer, but both have smooth surfaces and are soft. A primary luetic lesion in older women or lymphopathia venereum may

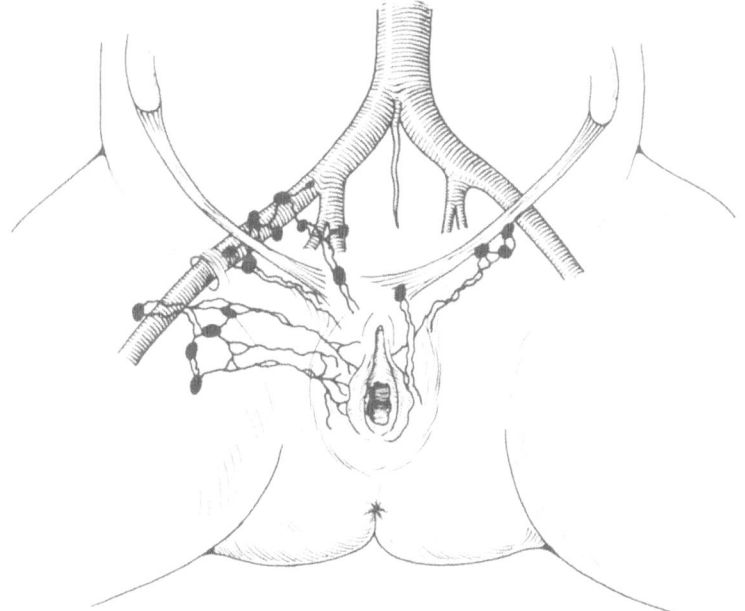

Fig. 139. Regional lymphatics and lymph nodes of the vulva. The lymph channels of the vulva drain into the superficial and deep inguinal lymph nodes, to Cloquet's node in the femoral ring, and to the hypogastric and iliac lymph nodes.

also simulate vulvar carcinoma. Biopsy provides the diagnosis in these cases.

Staging of cancer of the vulva: The cancer committee of the International Federation of Gynecology and Obstetrics (FIGO) proposed a classification according to the TNM system. (T stands for primary tumor, N for lymph nodes, and M for metastasis.) This system allows classification of the size of the tumor, relation of the tumor to adjacent organs, lymph node involvement, and distant metastasis. The TNM type of classification may eventually be applied to all malignant tumors.

Primary tumor—T:

Stage I. Tumor of the vulva less than 2 cm in diameter.

Stage II. Tumor more than 2 cm in diameter.

Stage III. Tumor of any size that involves surrounding structures such as urethra, vagina, anus, or rectum.

Stage IV. Tumor of any size with infiltration of the bladder mucosa, the upper part of the urethral mucosa, or the rectal mucosa, or extension to bones.

Lymph nodes—N:

Stage 0. No lymph nodes are palpable.

Stage I. Unilateral, not enlarged, mobile lymph nodes palpable (clinically not suspicious).

Stage II. Palpable unilateral or bilateral, firm, but mobile lymph nodes (clinically suspicious of malignancy).

Stage III. Fixed, confluent, or ulcerated lymph nodes.

Distant metastasis—M:

Stage 0. No metastasis observed.

Stage Ia. Infiltrated lymph nodes palpable in the small pelvis.

Stage Ib. Any distant metastases palpable. From this classification any combination can be derived, as given in the scheme.

Stage I. T_I, N_0, M_0
T_I, N_I, M_0

Stage II. T_{II}, N_0, M_0
T_{II}, N_I, M_0

Stage III. T_{III}, N_0, M_0
T_{III}, N_I, M_0
T_{III}, N_{II}, M_0
T_I, N_{III}, M_0
T_{II}, N_{III}, M_0
T_I, N_{II}, M_0
T_{II}, N_{II}, M_0

Stage IV. T_{III}, N_{III}, M_0
T_{IV}, N_0, M_0
T_{IV}, N_I, N_0
T_{IV}, N_{II}, M_0, and all cases with M_{Ia} or M_{Ib}
T_{IV}, N_{III}, M_0

Treatment and prognosis: Unfortunately, in approximately 50% of cases the inguinal lymph nodes are involved when the patient is seen for the first time. The treatment of choice is radical vulvectomy. Statistical evaluation is difficult because the number of cases in one medical center often does not allow the evaluation of different types of treatment and the patients treated are rarely comparable.

Radical surgery involves vulvectomy and en bloc resection of skin and fatty tissues (including lymphatics and lymph nodes) down to the external aponeurosis and fascia lata, with dissection of lymph nodes, of the femoral and inguinal canals. Whether resection of the deep iliac lymph nodes improves the 5-year survival and, therefore, whether this additional extensive dissection in older women is justified, is under discussion. The deep iliac lymph nodes are, with rare exception, involved only in the presence of positive nodes in the groin or femoral canal. Postoperative complications are frequent. In more than 50% the large skin defect breaks down and secondary healing requires prolonged hospitalization. In Stages I and II the 5-year survival is 50 to 80%, with an average of 75%. The figures decrease in Stage III (lymph node involvement) to 20 to 50% with an average of 41%, and in Stage IV it is close to 0. In some centers, radiation therapy with high-energy electrons has been used with success in vulvar carcinoma. Betatron external irradiation is not effective therapy, however. The mainstay of treatment of carcinoma of the vulva remains surgical operation.

Benign and Malignant Tumors of the Vagina

Benign Tumors. Solid benign tumors of the vagina are rare, with the exception of condylomata acuminata, which may be extensions of lesions of the vulva. Mesenchymal tumors such as myomas and fibromas in the vaginal wall usually do not exceed 1 cm in diameter.

Cysts in the vagina are more common, usually originating from Gartner's duct. They are located in the upper part of the vagina. Gartner's cysts are soft and they rarely cause symptoms. Inclusion cysts are caused by burying tags of epithelium (usually after repair of an episiotomy or vaginal laceration) and are located in the lower third of the vagina. They usually contain desquamated epithelial cells.

Symptoms: Symptoms are rare since the vagina is capable of great distension without causing pain. During delivery, Gartner's duct cysts are pushed aside and do not obstruct the birth canal.

Diagnosis: Small tumors and cysts are found coincidentally on pelvic examination. They can be confused occasionally with a cystocele or rarely with an endometriotic plaque in the vagina. Endometriosis, however, is usually associated with dysmenorrhea and dyspareunia.

Treatment: Cysts are treated by surgical excision. When Gartner's duct cysts extend into the parametrium, marsupialization may be preferable to the extensive dissection required for complete removal of the cyst.

Malignant Tumors of the Vagina. Premalignant lesions of the squamous epithelium of the vagina are similar to those of the cervix.

Symptoms and diagnosis: Dysplastic lesions and small areas of carcinoma in situ cause no symptoms. A suspicious or positive cytologic smear suggests a lesion of the vagina if cervical cancer is excluded especially if the uterus is absent. The entire vagina is then painted with Lugol's solution (page 111). The unstained areas are examined by colposcopy and subjected to biopsy.

Treatment: Simple excision of the premalignant lesions is sufficient provided that the margins of the specimen are free of atypical cells. Frequent follow-up is required.

Primary carcinoma of the vagina accounts for only 1.5 to 2% of all female genital cancers. Vaginal cancer is more common after the menopause. Predisposing factors are not clear, but prolonged use of pessaries may be etiologically related to carcinoma. In most cases the lesion is a squamous cell cancer.

Recently, clear-cell adenocarcinomas of the vagina have been reported in young girls (12 to 20 years of age) whose mothers had received diethylstilbestrol in early pregnancy. These cancers apparently arise from adenosis and spread readily. The number of new cases of this lesion is steadily increasing and the lesion has generated considerable interest.

Squamous cell vaginal cancer is located most commonly on the posterior wall of the upper third of the vagina. The primary lesion bleeds easily on touch. The tumors grow either exophytically, filling the vagina with a cauliflower tumor, or endophytically, progressing like a cuff along the vaginal vault. Local growth into the paravaginal and parametrial tissues results in a firm, fixed vaginal tube. Involvement of the adjacent bladder and rectum and lymphatic metastasis occur early. When the upper part of the vagina is involved, lymph node metastasis is similar to that in cancer of the cervix. Involvement of the lower third of the vagina results in extension similar to that in cancer of the vulva (Fig. 139). Not infrequently, both modes of spread occur simultaneously. Local extension finally causes obstruction of ureters, and uremia is the most frequent cause of death. Hematogenous metastasis to distant organs is rare.

Symptoms and diagnosis: Symptomatology is nearly identical with that in cancer of the cervix (page 291). Diagnosis is provided by speculum examination and biopsy. It is occasionally quite difficult to identify the vagina as the primary organ involved if the cancer extends into the cervix or the vulva. In case of doubt, the lesion is best regarded as a cervical or vulvar carcinoma with extension to the vagina.

Primary cancer of the vagina is staged in a manner similar to that employed for cancer of the cervix (page 309).

Staging of vaginal carcinoma:

    Stage 0: Carcinoma in situ.
    Stage I: Carcinoma limited to vaginal wall.
    Stage II: Carcinoma involves paravaginal connective tissue but has not reached pelvic wall.
    Stage IV: Carcinoma has penetrated mucosa of bladder or rectum, or has extended beyond true pelvis.

Treatment: Neither surgical nor radiation therapy has resulted in entirely satisfactory results. Radical surgical treatment requires careful selection of patients. The type of operation depends on the extent of the lesion. Colpectomy in addition to radical hysterectomy is often required (page 313). For some carcinomas of the lower vagina a vaginal radical hysterectomy (Schauta procedure) (page 371), combined with radical vulvectomy may be necessary. Occasionally, an anterior or posterior exenteration is required to remove the lesion. Radiation therapy requires local application of radium by vaginal mold or radium needles with additional external irradiation.

The 5-year survival rate after primary irradiation is approximately 32% and that of the combined approach is 38%.

Sarcomas of the vagina are rare and are generally sarcoma botryoides in young girls. The prognosis is very poor even after ultraradical surgical procedures.

Secondary Cancer of the Vagina. The vagina may be involved in cervical cancer Stages II, III, and IV by local extension of the tumor. Less frequently, vaginal cancer represents metastasis from a primary cancer of the ovary, fallopian tube, or choriocarcinoma. Occasionally, the vagina is involved by distant metastasis from mammary cancer or hypernephromas. Involvement of the vagina occurs by metastasis, direct extension, or recurrence in approximately 10% of adenocarcinomas of the endometrium. Treatment is discussed in relation to that of the primary tumor.

Benign and Malignant Neoplasms of the Cervix Uteri

Benign Neoplasms of the Cervix. Tumors such as condylomata acuminata, papillomas, and endometriomas occur occasionally.

Cervical polyps: Local hyperplastic lesions of the endocervical epithelium are common. They may be single or multiple lesions and are frequently pedunculated.

They are usually covered by squamous epithelium, often the result of epidermidalization (page 292). The surface is often ulcerated, especially in the case of large polyps. Malignant change occurs in less than 1%.

Hyperplastic lesions of the cervix have recently been observed in patients taking oral contraceptives. The polyps are indistinguishable macroscopically and colposcopically from ectopies or cervical polyps (Fig. 140). On histologic examination, the glands appear bizarre but the cells lack malignant features. It is believed that these lesions are derived from hyperplasia of reserve cells. They disappear after discon-

(a)

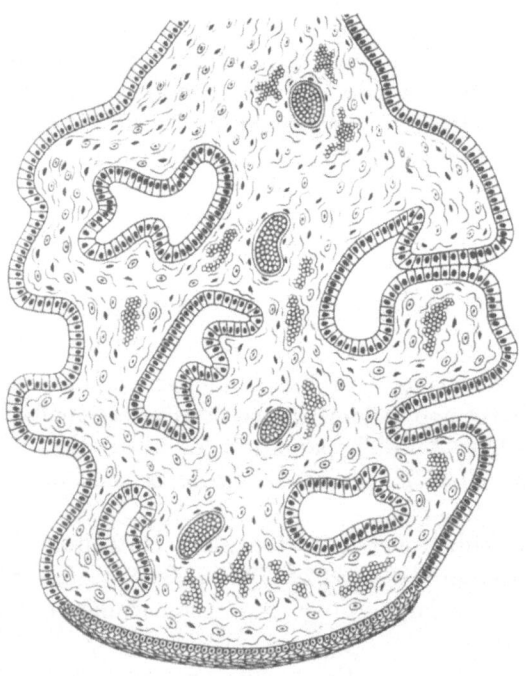

(b)

Fig. 140. Cervical polyp. (a) Pedunculated cervical polyp protruding through the external os; (b) histological section shows cervical glands surrounded by normal cervical epithelium. In the lower portion a stratified squamous epithelium is seen.

tinuance of the birth control pills. Cytologic smears are negative when other lesions are absent.

Symptomatology: Small polyps frequently produce no symptoms and are found coincidentally. Multiple and larger polyps may be associated with an increase in cervical mucus (mucorrhea). Contact (postcoital) bleeding is not uncommon.

Diagnosis: The polyps appear as small (up to 1 cm in diameter) red or bluish soft tumors around the external os with either smooth or ulcerated surfaces. Differential diagnosis includes a submucous myoma prolapsed through the cervical canal. Myomas, however, are firmer, larger, and often painful.

Treatment: Polyps should be removed and submitted for histologic examination. A curettage is desirable to remove the pedicle of the polyp as well as endometrial polyps, which may look grossly like cervical polyps.

Malignant Neoplasms of the Cervix. Cancer of the cervix is the most common malignant tumor of the female genitalia (50%). The success in treatment during the last decade is more the result of early detection of lesions by cytologic means than of improvements in treatment of advanced lesions. The need for routine cytologic examination of the adult cervix is thus obvious. Premalignant lesions observed below the age of 20, and the recent finding of adenocarcinoma of the vagina in young girls make speculum examination even of young girls necessary if there are any suspicious signs or symptoms.

The squamocolumnar junction: The area where columnar epithelium borders squamous epithelium is called the squamocolumnar junction. This border is not fixed, sharp, or stationary. At least 95% of all cases of cervical cancer begin in the area of the squamocolumnar junction.

The fluctuation of the squamocolumnar junction begins in fetal life. The squamous epithelium covers the lower part of the cervical canal as early as the sixth month of fetal life. In the newborn, columnar epithelium is found on the surface of the portio vaginalis. This glandular epithelium is subsequently replaced by squamous epithelium. Glandular epithelium and, therefore, the squamocolumnar junction are again present on the surface of the portio vaginalis in 75% of women in the reproductive years.

The presence of the squamocolumnar junction on the portio is frequently and incorrectly called an erosion. "Erosion" in pathologic terms is essentially an ulcer, or lack of surface epithelium. In cervical "erosions," however, one epithelium (squamous) is replaced by another (columnar). Colposcopically the "erosion" may be easily identified by its grapelike appearance as glandular epithelium. In colposcopic terms it is called an "ectopy." Extension of the glandular epithelium on the surface of the portio vaginalis produces an irritation, which stimulates the formation of squamous epithelium. The squamous epithelium may grow in from the periphery (epidermidalization), or may form from other epithelia by metaplasia, as the result of growth of reserve cells. These immature cells can develop into either glandular or squamous epithelium.

True reserve cells have the potential of differentiating into either squamous or columnar cells. The squamous epithelial tongues are at first very delicate. This stage of development can be recognized colposcopically and is termed an "open transformation zone." In time, the squamous epithelium becomes sturdier and covers the glandular ducts to form a "closed transformation zone." Production of mucus may result in formation of small nabothian cysts. Open and closed transformation zones may coexist and are considered physiological stages in the biology of the portio vaginalis. In the menopause, as a result of involution, the squamocolumnar junction moves up into the cervical canal. The portio is then covered by squamous epithelium, as in the prepuberal stage.

The movement of the squamocolumnar junction indicates cellular activity during the various phases of a woman's life. It is, therefore, not difficult to understand that in association with reproductive processes functional disturbances may develop.

Dysplasia and carcinoma in situ: It is generally believed that squamous cancer of the cervix, in the majority of cases, passes through precursor stages, characterized by progressive disturbances of maturation of the squamous epithelium (Fig. 141).

Dysplasia may originate from either basal cells of the squamous epithelium or reserve cells.

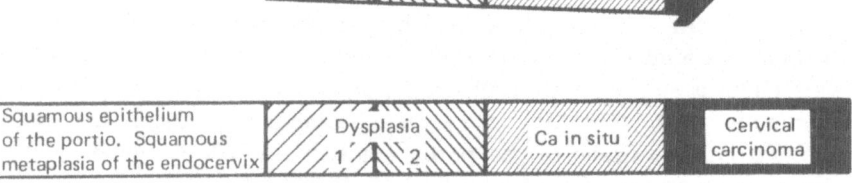

Fig. 141. The premalignant stages of cervical carcinoma (schema of malignant transformation) corresponding to facultative and obligate precancerous lesions. Mild dysplasia may be reversible (lower arrow), whereas severe dysplasia as well as carcinoma in situ can be regarded as truly precancerous.

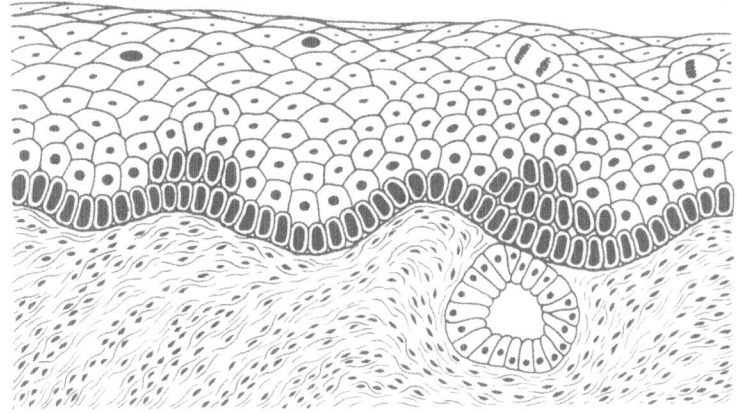

Fig. 142. Mild dysplasia. Extension of the basal layer, isolated mitoses, dyskaryotic cells in the superficial layer. The stratification of the epithelium is preserved. The superficial layer is narrow, but the intermediate and parabasal layers are thicker (slightly delayed maturation).

The term dysplasia (mild, moderate, or severe) refers to various stages of abnormal development that may end in carcinoma in situ The earliest dysplastic changes may regress spontaneously, but regression becomes increasingly unlikely as the dysplasia becomes more severe. On the contrary, progression to carcinoma in situ becomes increasingly likely (Fig. 142).

In mild dysplasia or basal cell hyperplasia, differentiation of the epithelium is preserved. Only delayed maturation and an increased rate of mitosis (Fig. 143) are obvious. Regression occurs in approximately

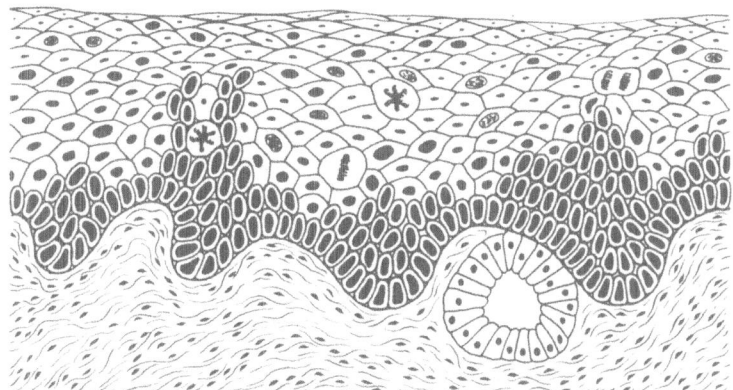

Fig. 143. Severe dysplasia. Pronounced thickening of the basal layer. From the basal layer plump pegs project toward the subepithelial connective tissue. Numerous mitoses, dyskaryotic cells in all cell layers, markedly delayed maturation with a very narrow superficial layer.

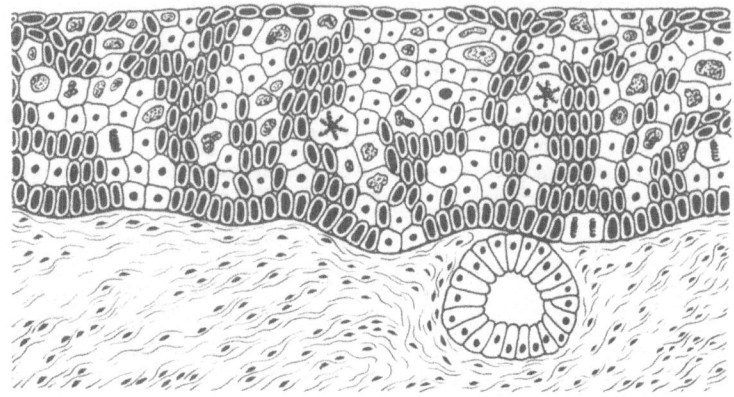

Fig. 144. Carcinoma in situ (simple replacement). Normal separation of the epithelium from the subepithelial connective tissue. Maturation is absent. The regular structure of the epithelium is no longer recognizable. Dyskaryosis, mitoses, and atypical cells occupy the entire thickness of the epithelium.

one-third of such lesions. As proliferation of basal cells increases, the stage of severe dysplasia is reached. The cells are then packed closely together; numerous mitoses are seen and atypical mitotic figures are obvious. The development into carcinoma in situ is obvious.

Dyskaryotic cells occur in which (by definition) nuclear atypia is evident. The cytoplasm of these cells matures, whereas the nucleus remains less than well differentiated, as indicated by abnormalities of size, shape, and chromatin distribution. Dyskaryotic cells are found in all layers of the epithelium. Polarization of such cells is absent, and differentiation is found only in the uppermost layer in severe dysplasia. The transition from severe dysplasia to carcinoma in situ is not sharp. In carcinoma in situ the number of cells with morphologic abnormalities is maximal and no differentiation toward normal stratification is found. The entire epithelium from basement membrane to the surface is involved. Mitoses and atypical basal cells can be observed in all layers (Figs. 144, 145). These changes are irreversible. The basement membrane, however, is still intact. Involvement of glands does not mean invasion, for the so-called cervical glands are part of the epithelium. Rootlike extensions of epithelium into the cervical connective tissue suggest early stromal invasion (Fig. 146).

At least 50% of severe dysplasias progress to invasive cancer. Statistically, the latent period is from several months to 10 years. The peak incidence of preinvasive carcinoma occurs about 10 years earlier than that of invasive cervical cancer. Cytogenetic studies agree with the cytologic and histologic data. In mild and moderate dysplasia most of the cells are diploid, but a few show numerical and structural chromo-

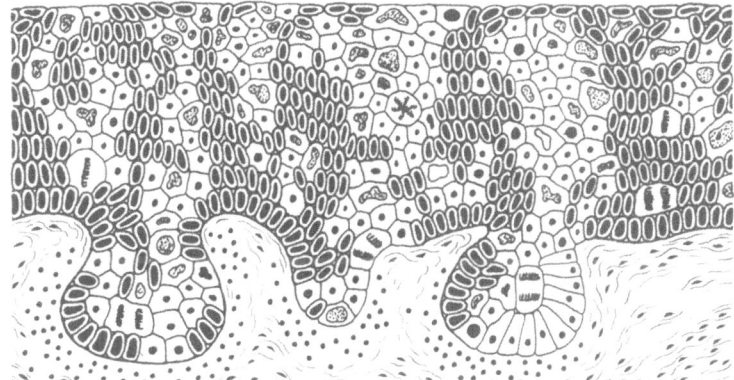

Fig. 145. Carcinoma in situ (plump projections). Marked, mainly plump projections toward the subepithelial connective tissue. Tendency to penetrate into cervical glands. Patchy subepithelial lymphocytic infiltrate. The regular structure of the epithelium is no longer present. Dyskaryosis, mitoses, and atypical cells occupy the entire thickness of the epithelium.

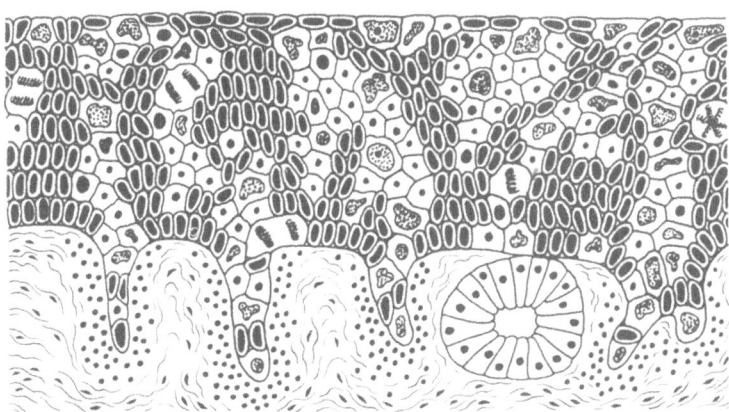

Fig. 146. Carcinoma in situ with early stromal invasion. Rootlike invasion by narrow pegs of atypical cells from the basal layer into the subepithelial connective tissue; plump projections are lacking; subepithelial lymphocytic infiltrates; basal membrane apparently intact.

somal abnormalities. In carcinoma in situ, an increased number of aneuploid cell lines seems to be established. So-called marker chromosomes indicate the evolution of new cell lines with structurally different karyotypes (Fig. 147). They increase proportionally as the dysplasia becomes more severe and develop through carcinoma in situ to invasive cancer.

Clinical considerations in premalignant lesions of the cervix: Premalignant lesions are asymptomatic. Dysplastic lesions are recognized by routine cytologic examination, which must accompany all pelvic ex-

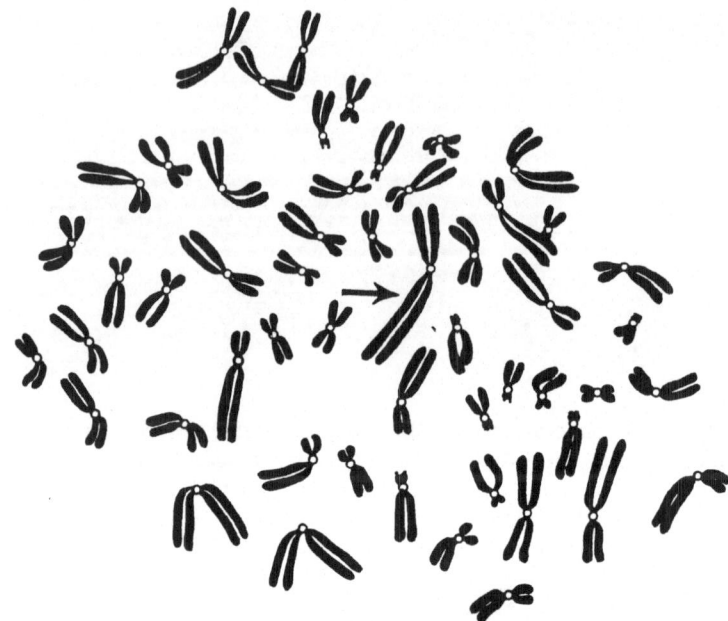

Fig. 147. Metaphase chromosomes of a carcinoma in situ. (This cell with 52 chromosomes belongs to one of the most commonly hyperdiploid cell lines.) In the middle of the picture (→) an abnormally large chromosome, the so-called marker chromosome, is indicative of neoplasia.

aminations, regardless of the age of the patient. In the United States, however, only 10% of the population receives regular check-ups including cytologic examination. The so-called irrigation smear has been suggested as a means of increasing the number of patients potentially benefiting from cytologic examinations. With the irrigation technique the patient herself aspirates cells from her vagina, using a prepackaged kit, and mails the slide to the laboratory. The method has increased the number of cytologic examinations, but the reliability of these smears is considerably less than that of those obtained by trained personnel. Furthermore, these patients do not have the benefit of pelvic eaxmination at the same time.

Cytodiagnostic procedures: The Papanicolaou smear (1941) aids diagnosis of premalignant or malignant lesions of the vagina, the portio vaginalis, and the cervical canal. Diagnostic reliability depends on the personnel who obtain the smear and those who examine the slide. Reliability should be around 95% if the smear is obtained and interpreted properly.

The Papanicolaou method is based on the shedding of cells from the proliferating squamous epithelium and mucous membranes. Dyskaryotic cells are identified amidst normal epithelial cells.

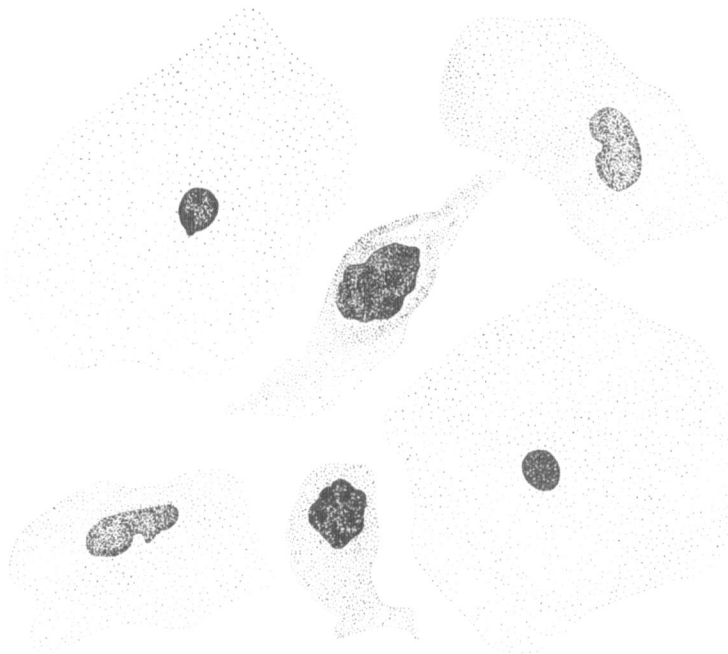

Fig. 148. Superficial cell dyskaryosis (between two normal superficial cells, upper left and lower right) with increase in the nuclear-cytoplasmic ratio and loss of rounding and hyperchromasia of the nucleus as well as variation in size and shape of the nuclei.

Mild dysplasia is characterized by superficial cells with dyskaryosis (Fig. 148). These cells have a slightly enlarged hyperchromatic nucleus. The cytoplasm is well differentiated.

Dyskaryotic cells from the intermediate and parabasal layers are observed in severe dysplasia and carcinoma in situ (Fig. 149). In addition, uniformly atypical cells are present in carcinoma in situ. The cells are similar in size to parabasal cells but contain multiple enlarged nuclei surrounded by a thin rim of cytoplasm, which is often vacuolated. Mitotic figures are not commonly seen in smears of premalignant lesions.

Polymorphic atypical cells (Figs. 150a and b) and occasional tumor giant cells are found predominantly in invasive cancer.

Qualitative changes in the individual cell indicate the degree of abnormality. No single cell is absolute proof of cancer. Malignant changes are classified according to the criteria originally suggested by Papanicolaou. Some cytologists prefer the following simplified scheme: normal (Classes I and II), suspicious (Class III), and malignant (Classes IV and V).

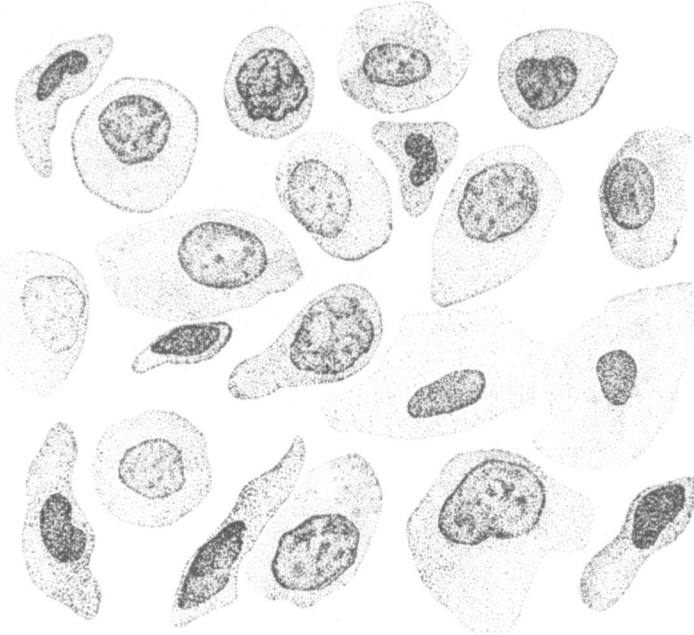

Fig. 149. Dyskaryotic cells from all layers of the epithelium. The cytoplasm is not altered. The assignment of individual cells to a definite cell layer is therefore possible.

If a Class III smear is obtained, it usually should be repeated before further investigation is carried out. A patient whose examination results in a Class IV or V smear is ordinarily subjected to histological confirmation without delay.

A difficult situation arises when a normal smear is reported after a suspicious or positive smear. This problem requires thorough investigation until the physician is convinced that the first smear was falsely positive. In case of doubt, further diagnostic methods (techniques in next paragraph) are mandatory.

Colposcopy and colpomicroscopy: By observation of the surface of the portio vaginalis under magnification (12×) and direct light (colposcope, page 103), physiologic changes on the surface of the portio can be differentiated from atypical lesions. This technique was developed by Hinselmann in Hamburg, Germany, in 1924. Colposcopy can be performed reliably only when the squamocolumnar junction or the lesion is located on the surface of the portio vaginalis. A lesion inside the cervical canal cannot be identified by colposcopy.

Colpomicroscopy is a different technique using higher magnification. The surface of the portio vaginalis is observed through a tube in direct

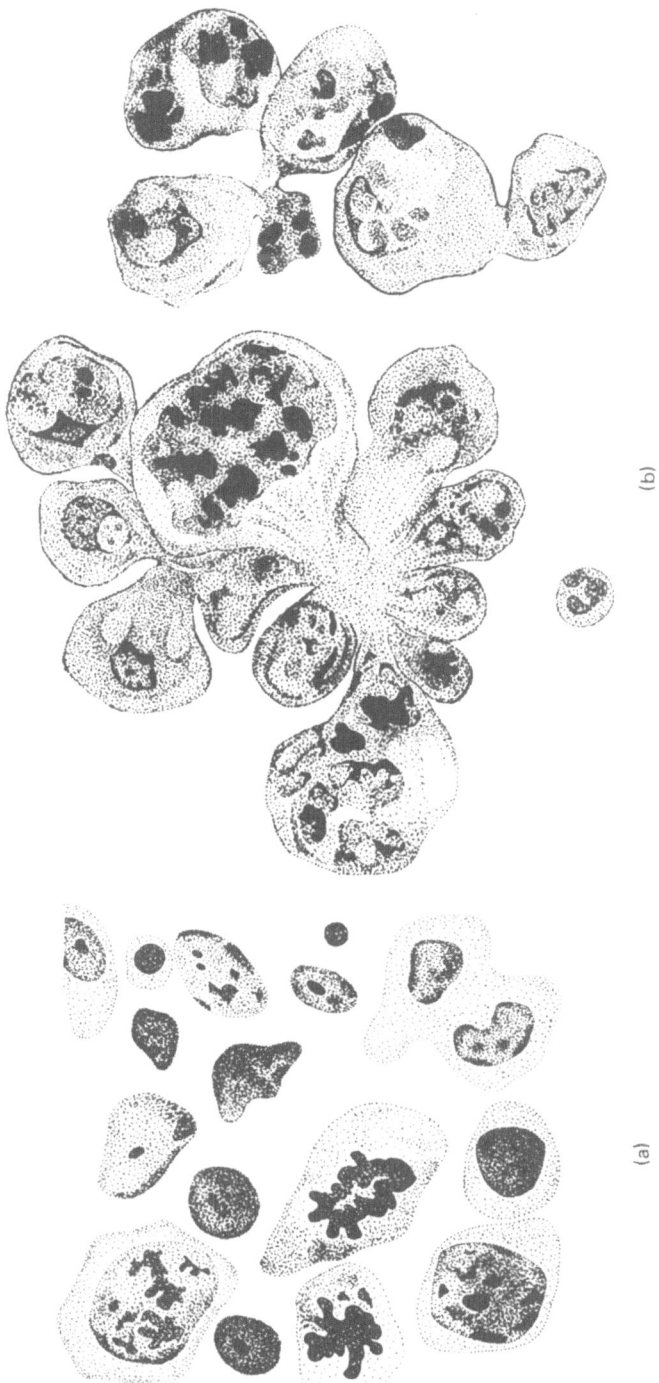

Fig. 150. Cytologic picture in frank carcinoma. (a) Polymorphic atypical cells (tumor cells). Irregularities in size and shape of nuclei and hyperchromasia of the nuclei, clumping of chromatin, degenerative changes of the cytoplasm, formation of so-called naked nuclei through extensive loss of cytoplasmic borders. (b) Tumor giant cell on left.

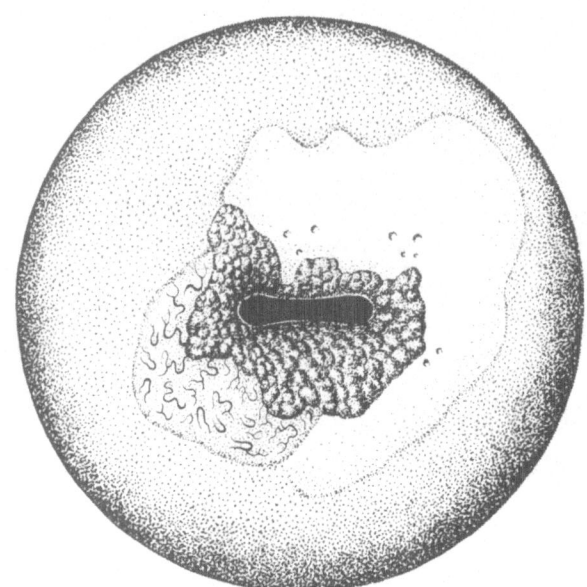

Fig. 151. Colposcopically suspicious finding on the surface of the portio. Leukoplakia between 12 and 6 o'clock. Atypical transformation zone with irregular vessels between 7 and 9 o'clock; circumferentially around the external os is a nonsuspicious ectopy.

contact with tissue that has been stained by toluidine blue or hematoxylin. A magnification of up to 350× is obtained, but the technique is time-consuming and is performed in only a few centers in the United States.

Colposcopically suspect areas with epithelial proliferation are termed "matrix areas." Typical findings include hyperkeratotic areas often described macroscopically as leukoplakia (white zone). If more detailed colposcopic observation reveals capillary sprouts, the finding is described as "punctation." The lesion has an irregular rough surface that contains innumerable capillaries that appear as red dots (Fig. 151). Punctation is often recognized after removal of the hyperkeratotic cellular layer. A "mosaic" is a whitish hyperkeratotic layer that is divided into small sections by capillaries (Figs. 152 and 153). In contrast to the physiologic "transformation zone," it is characterized by numerous bizarre and irregularly formed capillaries. Colposcopy aids in directing the punch biopsy to the abnormal areas of the cervix.

Punch biopsy: A good biopsy forceps should have sharp cutting edges (Fig. 154). Since the cervical lesion may be as small as 0.25 sq cm, it can easily be missed by taking too small a bite with the forceps.

Punch biopsy is diagnostic only when it reveals invasive cancer. In that case cone biopsy is unnecessary.

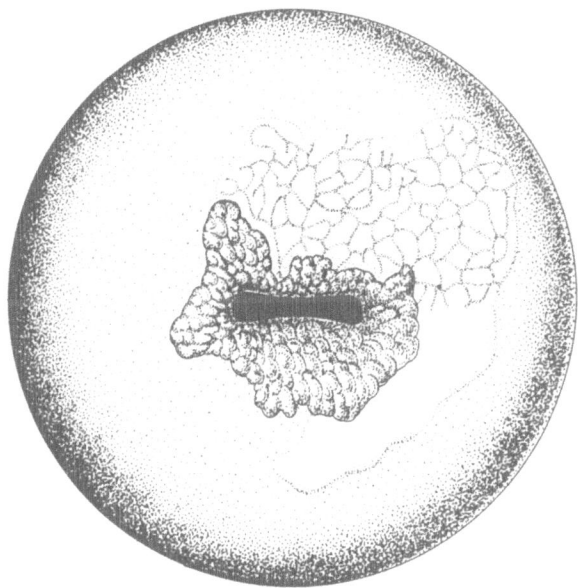

Fig. 152. Colposcopically suspicious finding on the surface of the portio. Leukoplakia between 3 and 6 o'clock. Mosaic subdivision of the leukoplakic zone by capillary sprouts; circumferential nonsuspicious ectopy around the external os with grape-like appearance.

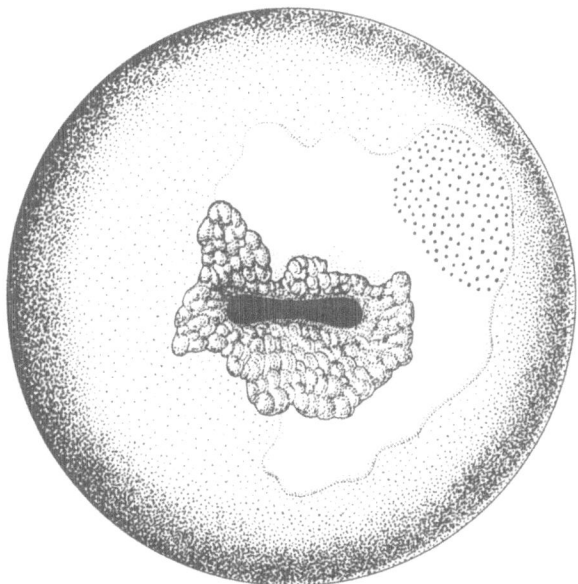

Fig. 153. Colposcopically nonsuspicious finding on the portio. Peripherally adjacent to a nonsuspicious ectopy a leukoplakic zone between 12 and 6 o'clock is apparent. Between 2 and 3 o'clock punctate capillaries in an area of irregular surface (ground).

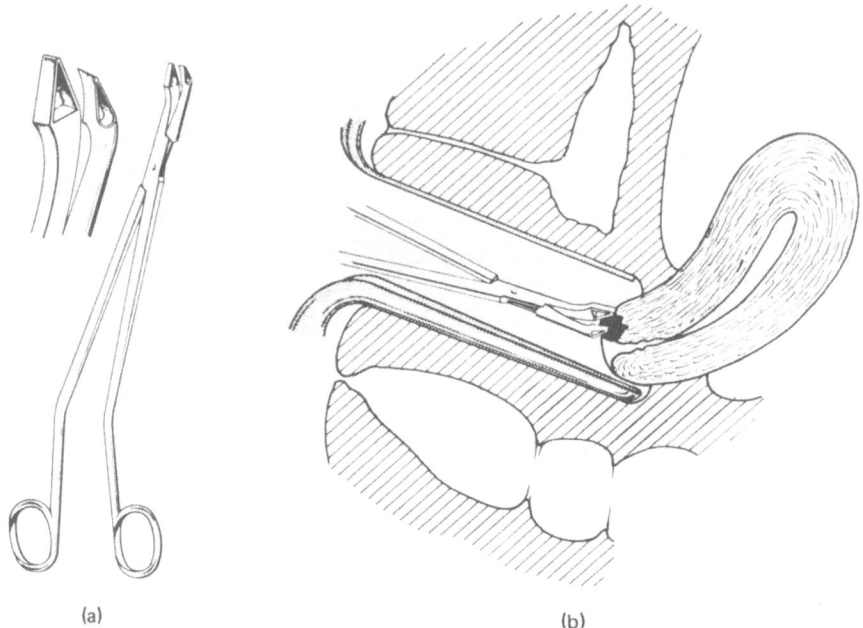

(a)   (b)

Fig. 154. Punch biopsy. (a) punch biopsy forceps; (b) Technique of removal of tissue from a site of carcinoma.

Cone biopsy: When punch biopsy and colposcopy fail to reveal invasive cancer in the presence of a positive cell smear, a cone biopsy is required for final diagnosis. The so-called sharp cone removes the squamocolumnar junction in a cone-shaped piece of tissue using a small knife. More than 95% of all premalignant and malignant lesions of the cervix are identified in the region of the squamocolumnar junction.

The shape of the cone depends on the location of the squamocolumnar junction. The cone biopsy must be broad-based but can be short if the squamocolumnar junction is located on the portio vaginalis. The extent of the lesion is identified by Schiller's test (page 111). If the squamocolumnar junction is located inside the cervical canal, the cone may be narrow-based but it must reach high into the cervical canal (Fig. 155). The cone is divided by the pathologist into multiple segments. The histologic diagnosis of the most malignant area is the final diagnosis. Disadvantages of the cone include bleeding, possible increased vascularity of the parametria for several weeks after the procedure, abortion, and scarring of the internal os that may lead to infertility.

The so-called four-quadrant biopsy does not significantly improve the diagnostic results of random biopsy, and the technique of taking multiple specimens (up to 25) is not popular. The following scheme

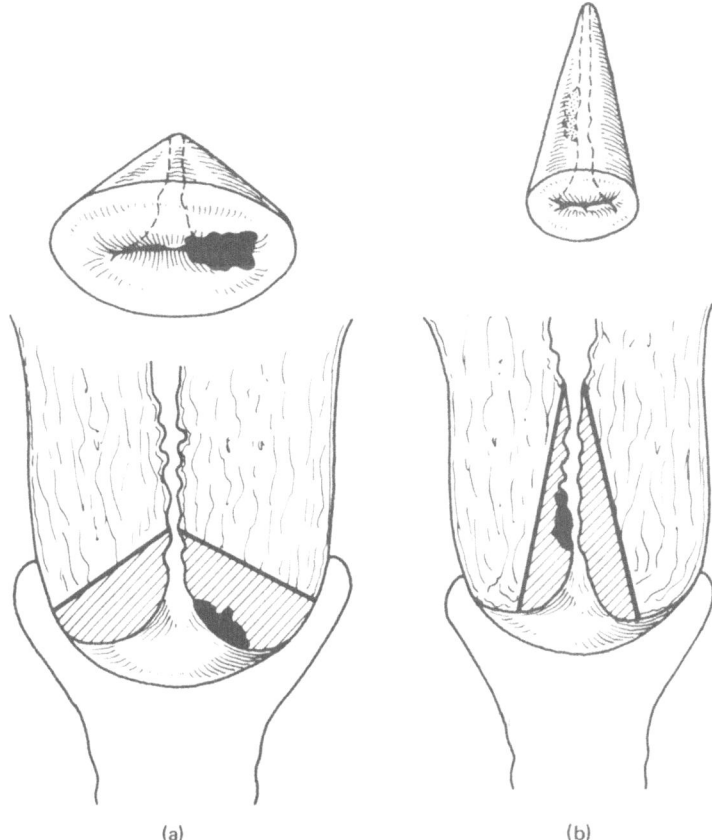

(a)   (b)

Fig. 155. Technique of conization. The shape of the cone depends on the location of the squamocolumnar junction. (a) In a young woman it generally has a broader base and is flatter; (b) in older women it has a smaller base but must include more of the cervical canal in order to reach the internal os.

is suggestive for diagnostic work-up of a patient with a suspicious or positive Papanicolaou test (Table 25).

Cervical Cancer.   More than 50% of all cancers of the female genitalia arise in the cervix. Only breast cancer is more common in women. The highest incidence of invasive cancer is between the fifth and sixth decades (Fig. 156).

Ninety-five percent of cervical cancers are squamous cell carcinomas. Five percent are adenocarcinomas. Only rarely does a cervical cancer arise from remnants of Gartner's ducts.

The cause is unknown. Evidence is mounting to suggest an association with infection by Herpesvirus Type I. Local inflammatory changes may also be predisposing factors. In general, cervical carcinoma is more

Table 25

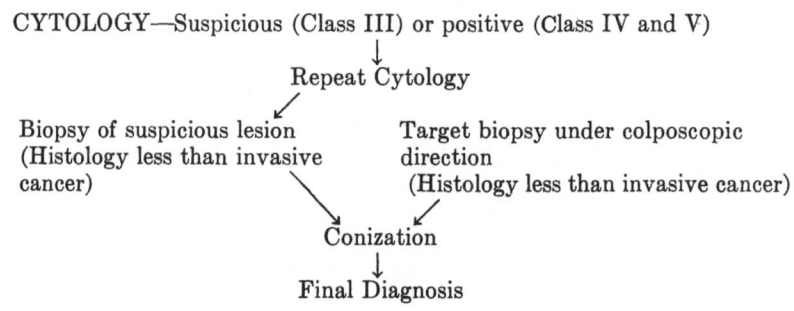

Target biopsy under the colposcope improves the probability of picking up invasion by punch biopsy. Colposcopy may, therefore, reduce the number of cone biopsies. In cases of clearly malignant smears, it may be preferable to proceed with histologic diagnosis without repeating the smear.

common in women with multiple sex partners (such as prostitutes), those who marry and have children early, and those with poor genital hygiene. That is, it is associated with low socioeconomic status, multiparity, and promiscuity. Poor penile hygiene has also been implicated. Whether smegma and lack of circumcision of the male partner cause carcinoma is doubtful except insofar as they predispose to poor genital hygiene. Familial factors may also be important, since it is known that the prevalance of cervical carcinoma is even lower in Jewish women than in Gentile women whose husbands are circumcised.

Microscopically invasive squamous cell cancer is composed of cell nests in the cervical tissue below the basement membrane. The cells are undifferentiated, with loss of polarity, dyskaryosis with nuclear hyperchromatism, and numerous atypical mitotic figures (Figs. 157, 158). Squamous cell cancer is often graded histologically with respect to differentiation. Five-year survival, however, is not well correlated with histologic grade.

Lesions that have penetrated the basement membrane minimally are called *microinvasive*. The exact definition of this lesion is still not fully agreed upon, but in general it is included in Stage Ia of the FIGO classification.

In younger women, cervical cancer usually begins on the surface of the portio, at or near the squamocolumnar junction. Growth is by extension, forming either a cauliflower-like exophytic tumor that fills the vagina or an endophytic crater-like pattern of growth (Figs. 159 and 160).

In patients of menopausal age, early cervical cancer may begin in the cervical canal. The cervix is distended without macroscopic changes

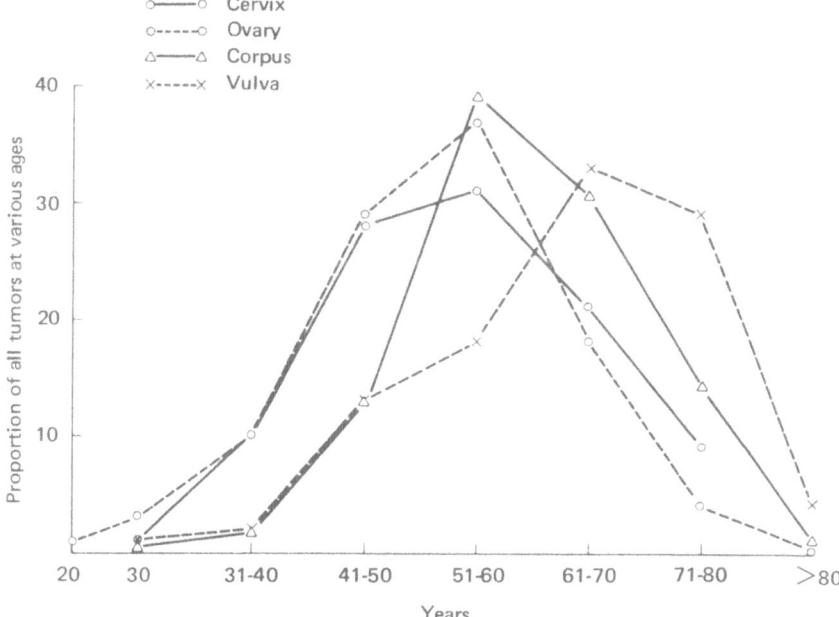

Fig. 156. Age-dependent frequency distribution of malignant tumors of the female genitalia, based on the total number of cases. The peak incidences of carcinoma of the vulva and corpus occur later than those of the cervix and ovary.

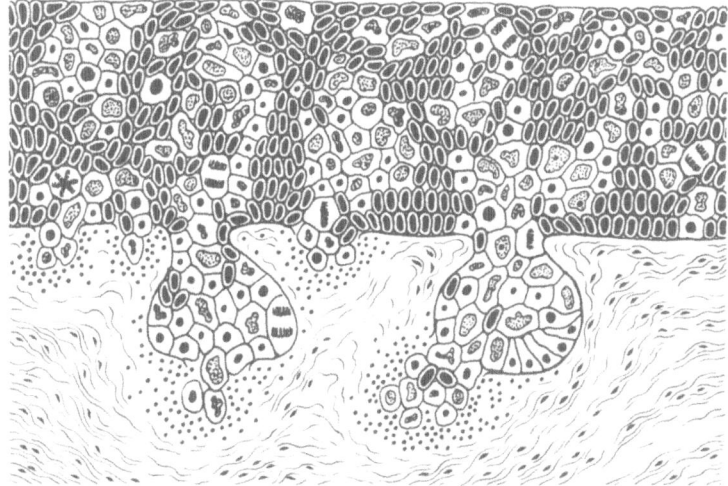

Fig. 157. Microcarcinoma. The basement membrane is penetrated in several places; infiltration of the stroma by narrow and wider pegs of atypical epithelium, plump infiltration of an involved cervical gland, marked lymphocytic infiltration. The regular structure of the epithelium is lacking. Dyskaryosis, mitoses, and atypical cells replace the entire thickness of the epithelium.

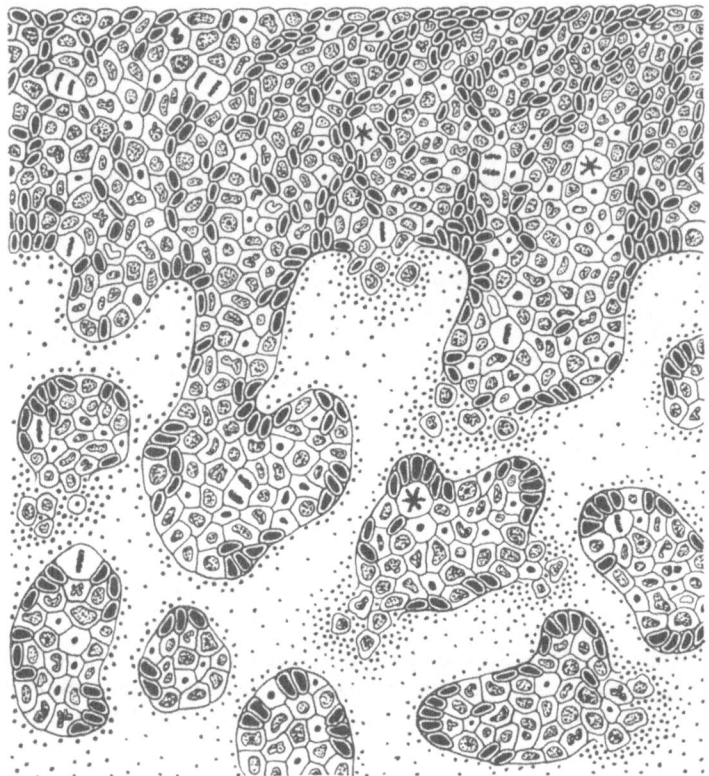

Fig. 158. Squamous cell carcinoma of the cervix. The atypical squamous epithelium grows by penetrating the basement membrane in completely disorganized wide fields, destroying the subepithelial tissue and replacing it completely in time. The narrow stromal bridges are extensively infiltrated by lymphocytes.

on the portio but necrosis and deep craters develop as a result of endophytic growth. Later, the surface becomes irregular and purulent and bleeds easily on touch. The edges and the surrounding tissue are firm and fixed. The, tumor may extend by continuous growth into the parametrial tissues or the vaginal wall, or both. Finally, the bladder or rectum is invaded with the formation of rectovaginal and vesicovaginal fistulas.

As a result of the rich lymphatic supply of the parametrial tissues, cervical cancer spreads rapidly. Lymph nodes in the parametrial tissues may or may not be involved. The obturator and deep iliac nodes may contain cancer even though the parametrial nodes are free of cancer. In terminal stages the entire pelvis may be filled with malignant tissue, forming a so-called frozen pelvis (Fig. 161).

Ureters are invaded by cancer, or obstructed by cancer or scar tissue, with resulting hydronephrosis and uremia the most common cause of

Fig. 159. Exophytic cervical carcinoma.

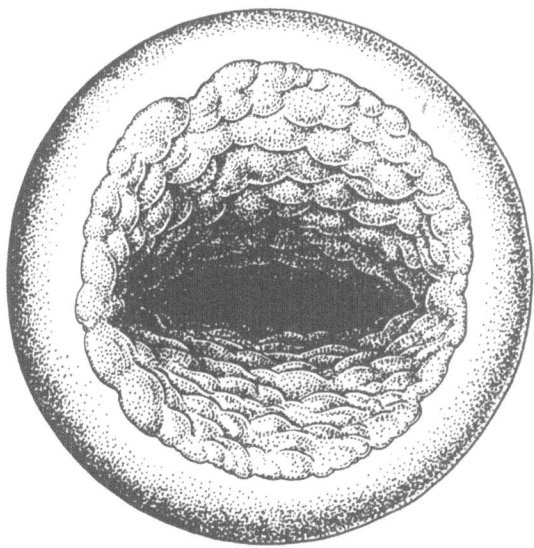

Fig. 160. Endophytic cervical carcinoma with formation of a deep crater.

death of patients with cervical cancer. Hematogenous metastasis to the liver, lungs, and skeletal system is a late development.

Symptoms: The earliest stages are asymptomatic. Only when the tumor begins to ulcerate is a brownish-yellow discharge seen, especially after intercourse. Physical signs thus indicate a lesion beyond the early

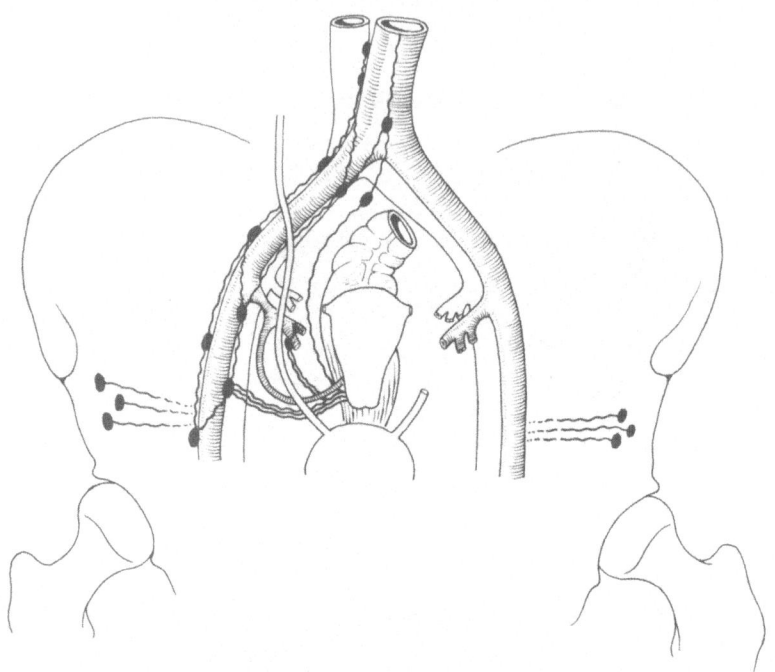

Fig. 161. Main lymphatics and regional lymph nodes of the cervix in the region of the parametria and the pelvic wall. The common iliac, obturator, external iliac, and hypogastric lymph nodes are the principal sites of lymphatic metastasis from cervical carcinoma.

stages. Pain in the pelvis and difficulty with urination and defecation are late symptoms. Edema of the lower extremities indicates extensive involvment of lymphatics and is a late manifestation.

Diagnosis: Early lesions are recognized only by cytology or colposcopy. Ulcerating lesions can be seen macroscopically. Endocervical lesions are identified by cytology and cervical curettage.

The rectovaginal examination identifies parametrial involvement, which must be distinguished from an inflammatory reaction (parametritis).

Lymphography, to identify involved lymph nodes by dorsopedal injection of a contrast medium, is popular in some centers, especially in Europe. Others consider this procedure of litle value in either diagnosis or planning of therapy.

Staging of Cervical Cancer. Cervical cancer is staged according to internationally established conventions. A TNM classification has not yet been accepted. The principal deficiency of the FIGO classification is its sole dependence on the results of pelvic examination. Stages

0 (carcinoma in situ) (Fig. 145) and Ia (microinvasion), (Fig. 157), however, are histologic diagnoses. During pelvic examination, inflammatory involvement is frequently confused with spread of tumor and vice versa. The only reliable staging, in addition to 0 and I, is in Stage IV, where cystoscopy and proctoscopy, with histologic confirmation of the lesion, prove involvement of adjacent organs.

Staging is important because it is the only method by which results of treatment from various medical centers may be compared. In addition, it allows comparison of the results of surgical and radiation therapy. For statistical purposes, the original staging must never be changed. When sufficiently large series are studied, however, the errors tend to cancel one another out and the results remain valid.

Stage 0: Carcinoma in situ (intraepithelial cancer, preinvasive carcinoma). This group is not included in the international annual report (Heyman's).

Stage I: The carcinoma is strictly confined to the cervix. Extension to the corpus is disregarded.

Stage Ia: Cases with microscopic invasion.

Stage Ib: All other cases of Stage I (Fig. 162).

Stage II: The carcinoma has extended beyond the cervix into one or both parametria, but has not reached the pelvic wall. The carcinoma has involved the vagina, but not the lower third.

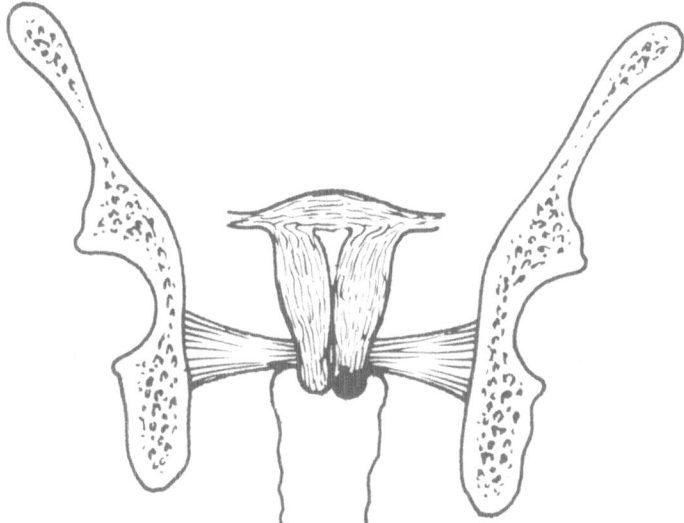

Fig. 162. Cervical carcinoma: stage Ib; the carcinoma is limited to the cervix.

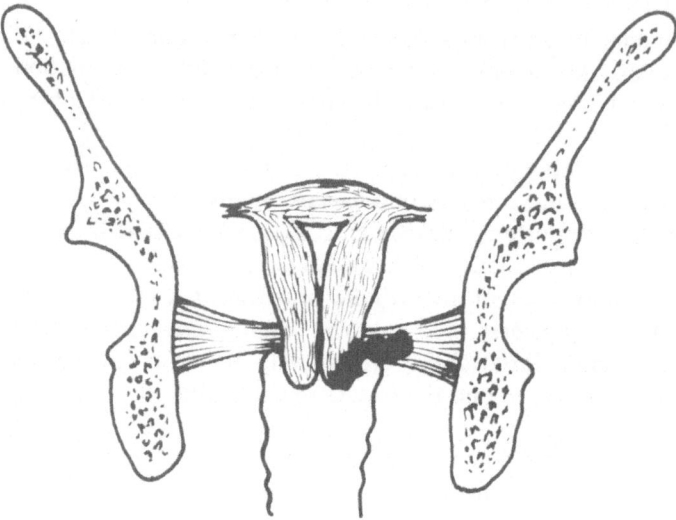

Fig. 163. Cervical carcinoma: stage IIa; the upper third of the vagina is involved with carcinoma.

Stage IIa: Involvement of the vagina but not the parametria (Fig. 163).

Stage IIb: Involvement of parametria, or parametria and vagina (Fig. 164).

Stage III: The carcinoma has extended onto the pelvic wall. On rectal examination there is no cancer-free space between

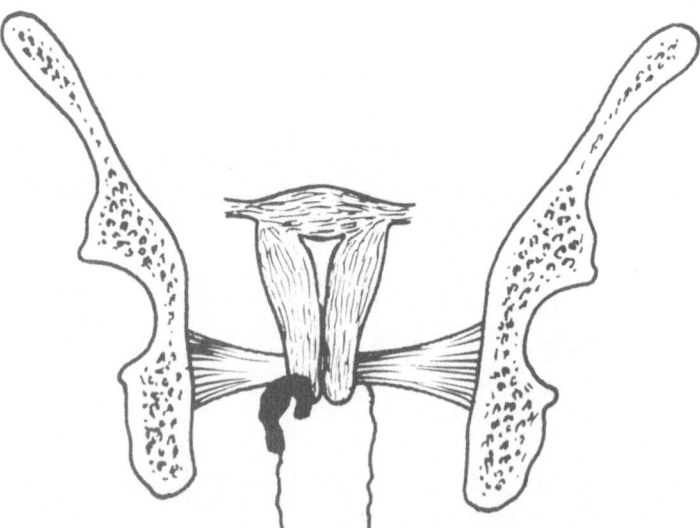

Fig. 164. Cervical carcinoma: stage IIb; the left parametrium is involved adjacent to the cervix.

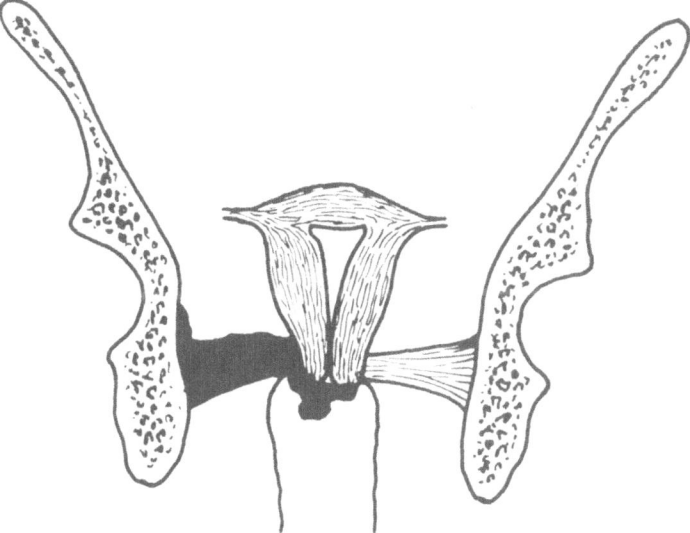

Fig. 165. Cervical carcinoma: stage III; the right parametrium is involved with carcinoma throughout its extent up to the pelvic side wall.

the tumor and the pelvic wall. The carcinoma involves the lower third of the vagina (Fig. 165).

Stage IV: The carcinoma has extended beyond the true pelvis (lymphatic or hematogenous distant metastasis) or has invaded the mucosa of the bladder or rectum. Bullous edema is not sufficient to classify a lesion as Stage IV. (Fig. 166).

Treatment: The treatment of severe dysplasia and carcinoma in situ depends primarily on the age and parity of the patient. In young patients

Fig. 166. Cervical carcinoma: stage IV; the carcinoma involves the anterior vaginal wall and has penetrated into the bladder.

who have not completed their families, the diagnostic cone may be therapeutic as well, provided the edges of the cone are free of tumor. Follow-up by cytologic smears is imperative. Repeatedly suspicious smears in younger patients requires considerable judgment in deciding whether to allow the patient to have at least one child. In patients beyond the age of 35, especially those with children, vaginal or abdominal hysterectomy is the treatment of choice.

Stage I: Treatment in Stage I is modified by the fact that the deep pelvic lymph nodes are involved in 15 to 20%. Radical treatment in Stage I is therefore required to cure lesions with lymphatic involvement. In a sense, 80 to 85% of patients with Stage I are overtreated by radical surgical or radiation therapy. To reduce the incidence of radical treatment, with its attendant mortality and morbidity, simple total abdominal hysterectomy is often now performed for some Stage Ia lesions. In Stage Ia (microinvasion) the lymph node involvement is less than 2%.

Stages Ib, IIa, and IIb: These stages may be treated by radical surgery, primary irradiation, or a combination of both. A properly performed radical hysterectomy entails removal of the uterus with an extended vaginal cuff (about one-third of the vagina), the parametria, and deep lymph node dissection.

Radiation therapy: This form of treatment often involves the insertion of radium. According to the inverse square law, radiation effect will decline very rapidly a few millimeters from the inserted radium. Therefore, additional external radiation is required, often in the form of megavoltage irradiation (cobalt 60). Numerous applicators have been developed for the insertion of radium into the cervical canal and uterine cavity. Currently the radiation dosage is expressed in rads. One mg radium in equilibrium with a 0.5-mm platinum filter gives 8.25 R per hour at a distance of 1 cm. In this terminology, a total tumor dose of 7,000 to 8,000 R is administered in treatment of cervical carcinoma (including radium and external radiation). The rectum and the bladder may be irradiated beyond their tolerance during therapy, which is approximately 6,000 R. The amount of radium in milligrams times the number of hours it remains in place is the dose in milligram-hours (mg-hr). A total of approximately 6,000 mg-hr is applied, usually delivered in two applications about two weeks apart (Fig. 167).

Preoperative insertion of radium, for reduction of the size of the tumor and prevention of spread, as treatment of cervical cancer is occasionally performed. Postoperative irradiation after radical surgical procedures for treatment of Stage I is also performed in some centers when lymph node involvement is detected.

The choice between surgery and irradiation is often difficult because the 5-year survivals are nearly identical. The choice depends more or less on preference and availability.

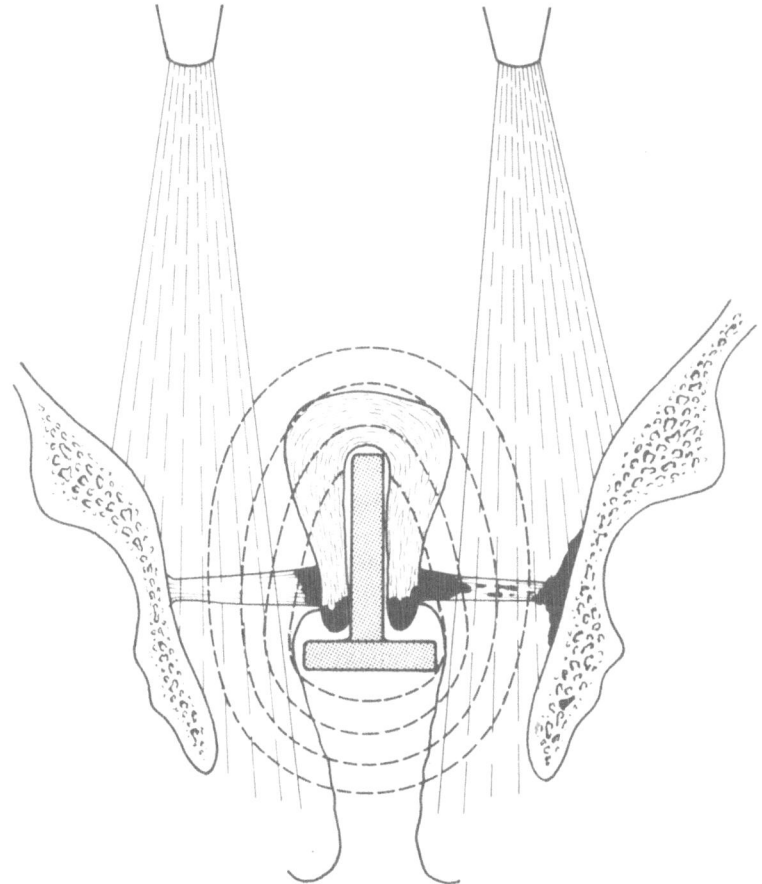

Fig. 167. Radiotherapy of cervical carcinoma. Schematic representation of local and external radiation. The local treatment permits the irradiation of the primary tumor and the immediately adjacent area. In this illustration the radium is applied with an intrauterine tandem attached to a colpostat. The infiltrate on the pelvic wall is reached by external radiation sources.

Irradiation has the disadvantage of adverse effects on sexual function because of stenosis of the vagina. In addition, the squamous epithelium becomes thin, irritated, and poorly vascularized.

Radical operations, because they involve removal of pelvic lymph nodes, provide prognostic information. Surgical treatment, however, is associated with a slightly higher mortality and a considerably higher morbidity. The incidence of ureterovaginal fistulas is approximately 2 to 10%. It thus seems that both forms of treatment are logical in selected situations and that neither surgical nor radiation therapy is preferable in all cases.

Stage II: Both radical surgery and irradiation are used; although most cases with more than minimal parametrial extension are probably better treated by radiation.

Stage III: Because it is illogical to cut through malignant tissue, primary irradiation is the only logical treatment for lesions that have reached the pelvic wall or the lower third of the vagina.

Stage IV: Irradiation is usually not effective, and ultraradical procedures such as anterior, posterior, or total exenteration are performed if the tumor has not metastasized beyond the pelvis, and can be removed without cutting through malignant tissue.

Treatment of the less common adenocarcinoma of the cervix is identical with that of squamous cell carcinoma.

Cervical Stump Carcinoma. This lesion should virtually disappear as the operation of supracervical hysterectomy becomes an unusual event. Many women, however, have had a supracervical hysterectomy 20 to 30 years ago and are now at risk. The incidence of cervical carcinoma in a stump is 4 to 8%.

Staging and treatment are the same as with cervical cancer in an intact uterus.

Cervical Carcinoma in Pregnancy. The diagnostic work-up in pregnancy is identical with that in the nonpregnant patient, including conization when the procedure is required. Severe dysplasia and carcinoma in situ do not require interruption of pregnancy if the pregnancy is desired. Vaginal delivery is attempted in the absence of obstetric contraindications, and definitive treatment is decided upon 6 weeks post partum.

Stage I and more advanced lesions require prompt treatment and termination of pregnancy. Radical hysterectomy following cesarean section is the treatment of choice when the fetus is capable of extrauterine survival. It may be possible to delay definitive treatment for a maximum of a few weeks to allow further critical development of the fetus. In early pregnancy, radical hysterectomy, removing the products of conception along with the uterus, is performed as soon as the diagnostic work-up is completed.

Postoperative care of cancer patients: A complete pelvic examination, including cytologic smear, is required quarterly for the first 3 years and semiannually thereafter. After 5 years, an annual check-up is sufficient. Weight, hematocrit, and urine analysis are required at each visit, as well as additional specifically pertinent laboratory tests.

In the presence of complaints referable to the urinary system or lower bowel, cystoscopy and proctoscopy, respectively, are required. Patients in the lower socioeconomic groups in particular need external support to improve their convalescence and rehabilitation.

Results of Treatment: By international agreement, a cervical cancer is considered to have been cured if the patient is free of recurrence

Table 26. Five-Year Survival in Carcinoma of
Cervix by Stage

Stage	5-year survival (%)
0 (carcinoma in situ)	100
Ia	95
Ib	75–85
II	55–70
III	30
IV	0–8

or metastasis five years after initial treatment. The cure rates for cervical cancer according to staging are shown in Table 26.

Recurrence: A recurrence is said to exist when a lesion is recognized after a period of from 6 months to 5 years in which the patient is apparently free of disease.

Many lesions develop as recurrences on the pelvic wall resulting from involvement of lymph nodes. They occur most frequently in the first year, and less frequently in the second year, after the original treatment.

Recurrences 5 years or more after treatment are called late recurrences.

The diagnosis can be very difficult since recurrences on the pelvic wall are often confused with scarring.

If the original treatment was surgical, recurrences can be treated by radiation. Further radiation after primary irradiation is limited by the tolerance of the tissues. A second course, when possible, requires a considerable reduction in dose of radium and external irradiation. Since a recurrence after radiation often indicates a low radiosensitivity of the tumor, prognosis is usually poor. An occasional patient with recurrence may be cured by ultraradical surgery.

The care of patients with incurable cancer requires considerable psychological skill. The patient is dependent on narcotics, often in extremely high doses, in the final stages. These women usually die from uremia, caused by obstruction of the ureters, and less often from hemorrhage or distant metastasis.

Benign and Malignant Neoplasms of the Corpus Uteri

Benign Neoplasms.

Endometrial polyp: Localized hyperplasia of the endometrium occasionally forms endometrial polyps, which often arise from the fundus and are solitary in 80% of cases. Rarely, the entire endometrium is involved to form the condition of polyposis uteri. Excessive endogenous or exogenous estrogen is considered an etiological factor by many, but by no means all, investigators.

Endometrial polyps are found in all age groups but occur most frequently in climacteric and postmenopausal women. Large single polyps are usually pedunculated and may be visible protruding through the external cervical os.

The endometrium of the polyp may respond to estrogen and progesterone, although usually only a response to stimulation by estrogens is seen. Cystic grandular hyperplasia is another possible pattern. In some polyps, the glands are cystic, dilated, and covered by an apparently inactive epithelium. Larger polyps, especially of the pedunculated variety, may undergo necrosis and ulceration.

Malignant change in an endometrial polyp is uncommon (0.36 to 1.12%), but 15% of adenocarcinomas of the endometrium arise near polyps.

Symptoms: Small polyps cause no signs or symptoms. Only the larger polyps may cause metrorrhagia and, if ulcerated, brownish mucous secretion.

Diagnosis: A polyp present at the external os may be of endometrial origin. Differential diagnosis may include endocervical polyps and pedunculated submucous myomas. Necrotic polyps may be confused with malignant lesions of the endometrium or cervical carcinoma. Menometrorrhagia, especially in older patients, requires fractional curettage, which may occasionally be therapeutic. Recurrent polyposis requires hysterectomy.

Myomas of the uterus (fibroids): Myomas arising from the myometrium are by far the most common neoplasm in women. After the age of 20 years about 20% of all women have myomas of smaller or larger size. The prevalence increases to 40% after age 35.

On histologic section whorls of smooth muscle bundles are seen, interspersed with strands of connective tissue. The peripheral layers appear compressed, forming a pseudocapsule, which provides a bloodless plane of cleavage and, therefore, easy enucleation of the tumor in most instances. There is some relation of estrogenic activity to the growth of myomas. The hormone dependence is suggested also by the fact that myomas are present in the reproductuve age group and undergo atrophy in the menopause. There is also an association of myomas with endometrial hyperplasia and occasionally with adenomyosis or endometriosis.

Myomas originate most commonly from the wall of the corpus. They are much less common (approximately 8%) in the cervix. Myomas vary in size from microscopic seedlings to extremely large tumors filling the abdomen (Fig. 168).

Intramural myomas may originate in the myometrium, enlarging the uterus without causing irregularities of the surface, or may develop secondarily from either a submucous or subserous location.

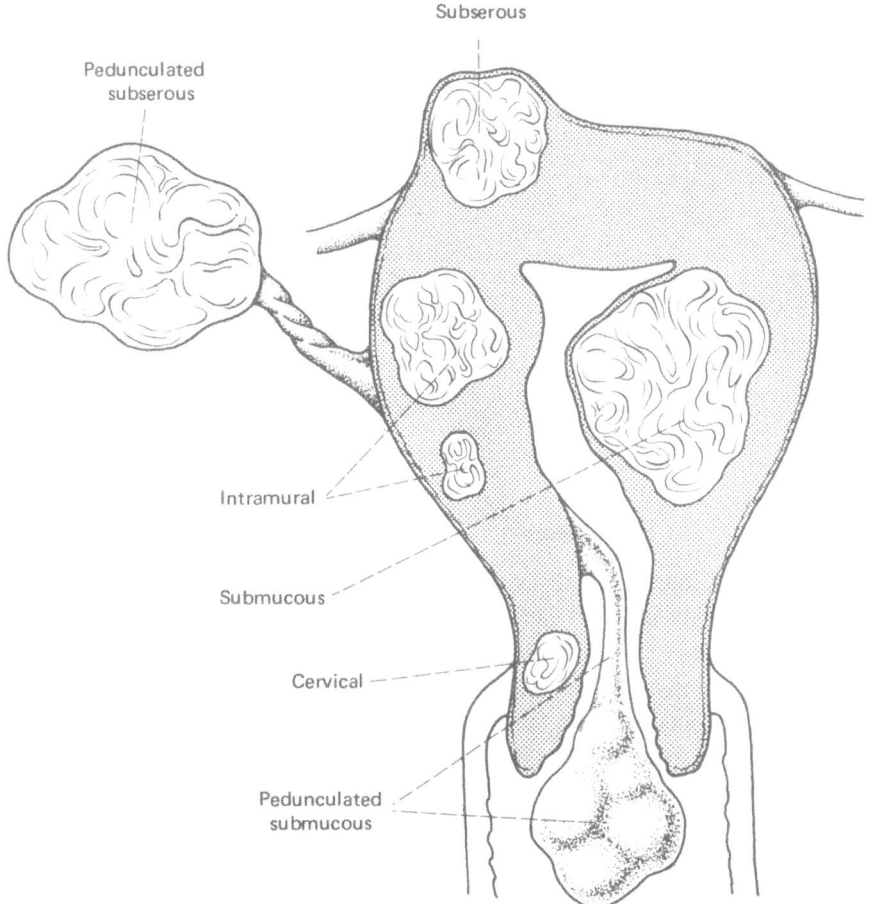

Fig. 168. Myomatous uterus. Schematic representation of the development and location of intramural, subserous, and submucous myomas. The subserous myoma in the upper left has a thick solid pedicle. The pedunculated submucous myoma has prolapsed through the external os.

Primary subserous myomas produce the typical irregularity of the uterine surface associated with these tumors. They may be sessile or freely mobile, pedunculated tumors.

Submucous myomas arise close to and extend into the endometrium. A pedunculated myoma may be "aborted" through the cervix.

Intraligamentous myomas arise as primarily subserous tumors and extend into the space between the sheets of the broad ligament. They are, thus, extraperitoneal (Fig. 169).

Myomas often undergo regressive and degenerative changes, largely as a result of poor vascularization that causes the tumor to outgrow its blood supply.

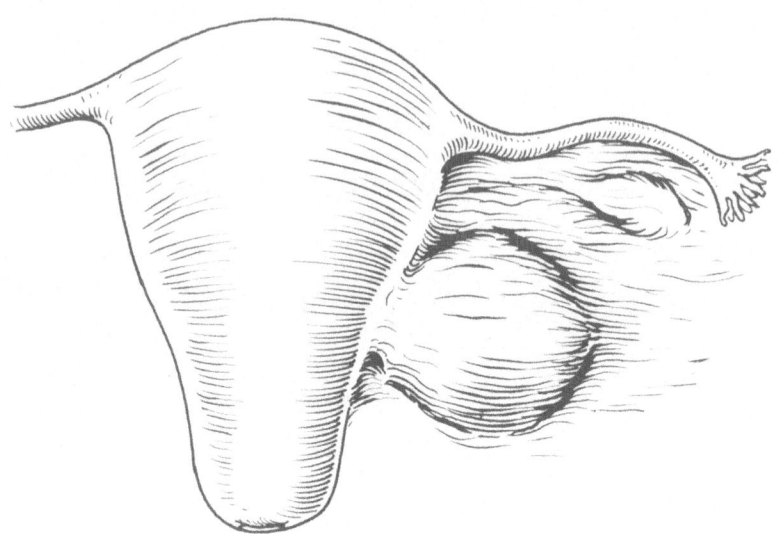

Fig. 169. Intraligamentous subserous myoma. Sessile origin from the side wall of the uterus. Adnexa can be felt separately.

After the menopause, myomas undergo atrophy of muscle layers and hyaline degeneration of the connective tissue. They may also undergo calcification, which can be identified on a scout film of the abdomen.

Occasionally a subserous pedunculated myoma may undergo torsion or detach and obtain its blood supply from another source, such as the anterior abdominal wall. Necrosis occurs in approximately 10% of myomas. Edematous myomas may undergo fatty degeneration, which in turn may result in cystic change.

Infection and abscess formation are rare; they are usually seen post partum or after infected abortions.

Sarcomatous change occurs in only a fraction of a percent of myomas and predominantly in older women.

Symptoms: The signs and symptoms depend on the size and location of the tumor. Small myomas are found incidentally in 15 to 20% of hysterectomy specimens as asymptomatic lesions. Subserous myomas, in particular, cause few symptoms even after having reached large size. Approximately 40 to 50% of patients with intramural myomas and nearly all patients with submucous myomas complain of menorrhagia and later metrorrhagia. The abnormal bleeding is explained in part by inadequate contraction of the myometrium and in part by the enlarged surface of the endometrium. Bleeding can be severe enough to result in an iron-deficiency anemia. In addition, there is frequently an associated endometrial hyperplasia (20 to 50%).

Pain in the lower abdomen is uncommon, even when the tumor is large, although bowel symptoms may occur. Submucous myomas "aborting" through the cervix are also associated with cramping pain. Pain in conjunction with myomas is often caused by another unrelated lesion, such as pelvic inflammatory disease.

Twisting of a subserous myoma may result in an acute abdomen (page 359). Cervical myomas may cause frequency or urinary retention.

Intraligamentous myomas rarely may compress the ureter, resulting in hydroureter and hydronephrosis.

Diagnosis: Myomas are diagnosed by pelvic examination. Pregnancy and ovarian tumors must be ruled out. Pregnancy can occur in a myomatous uterus. A pregnancy test is, therefore, a required part of the preoperative work-up.

The differentiation between myomas, especially of the pedunculated subserous variety, and ovarian tumors is sometimes impossible on pelvic examination alone.

Differential diagnosis should also include uterine anomalies, which may often be detected by hysterogram (page 133).

Submucous myomas are sometimes detected during curettage. Submucous myomas that protrude through the cervix may be confused with large polyps or a carcinoma.

Myomas may occasionally cause infertility by blocking the cornual part of the fallopian tubes. Early abortion may result from implantation on a submucous myoma. In the second trimester, failure of proper growth of the uterus may result in late abortion. The myoma usually enlarges as pregnancy proceeds and shrinks post partum.

Treatment: Asymptomatic small myomas do not require treatment. Large or symptomatic tumors in younger women who have not completed their families may be removed by myomectomy.

Laparotomy is indicated whenever ovarian tumors cannot be ruled out. Laparoscopy may occasionally be of diagnostic aid. Hysterectomy is usually performed for large or symptomatic tumors in women who have completed their families.

Irregular uterine bleeding requires a curettage. If the bleeding is caused by endometrial hyperplasia, the curettage may be at least temporarily therapeutic. Recurrent irregular bleeding is an indication for hysterectomy.

Whether high doses of progesterone or progestins can shrink myomas is not clear, but approximately one third of myomas grow if the patient is taking estrogen-containing birth control pills (page 203). Twisted or infected degenerating myomas require emergency hysterectomy.

Relation between Cystic Glandular Hyperplasia and Adenomatous Hyperplasia. Continuous unopposed estrogenic stimulation may produce hyperplasia of the endometrium, involving glandular structures

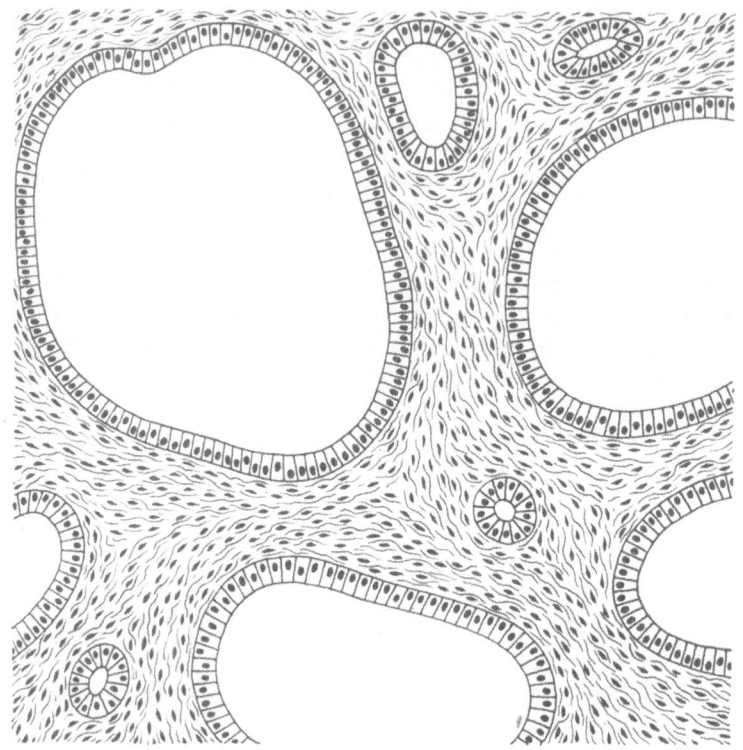

Fig. 170. Cystic glandular hyperplasia of the endometrium. The glandular lumina are cystically dilated in varying degrees. The epithelium is flattened in proportion to the extent of the dilation of the glands.

as well as stroma. In time the glands enlarge, creating the histological appearance of a "Swiss cheese" hyperplasia (Fig. 170). The epithelium is regular. Large, well-stained nuclei and occasional mitotic figures are seen. The stroma is characteristic of the proliferative phase of the endometrium. Glands and stroma are sharply separated. Endometrium and myometrium are clearly differentiated. Small hemorrhagic and necrotic areas may appear but they do not alter the generally regular pattern.

This type of hyperplasia is not premalignant, but cystic glandular hyperplasia may progress to adenomatous hyperplasia. Both forms can often be identified in one specimen.

Premalignant Changes in the Endometrium

Atypical adenomatous hyperplasia. Adenomatous hyperplasia is histologically characterized by hypertrophy and hyperplasia of glands and a reduction in stroma (Fig. 171). The glands are crowded back to back.

Fig. 171. So-called adenomatous hyperplasia of the endometrium. The glands push the stroma aside and are packed closely together in so-called back-to-back arrangement. The glandular epithelium may be multilayered and markedly flattened. Many mitoses are visible.

The epithelial cells are regular and may be stratified. Mitoses are common. There are occasional areas of benign squamous metaplasia. In from 1 to 10 years, 6 to 12% of adenomatous hyperplasias develop into carcinoma of the endometrium. The lesion must therefore be considered precancerous (Fig. 172).

Dysplasia of the Endometrium. Atypias of the endometrial epithelium that are premalignant are termed dysplasias. They are a stage between adenomatous hyperplasia and adenocarcinoma of the endometrium. Proliferation of cells into the glandular lumina is characteristic. The glands may be completely filled with cells. Individual cells may be poorly differentiated (polymorphic nuclei, abnormal mitoses, and edematous cytoplasm). The stroma is considerably reduced but the basement membranes of the glands are intact. The basic difference between adenomatous hyperplasia and dysplasia is the cellular atypia. Some pathologists use the term adenocarcinoma in situ for this

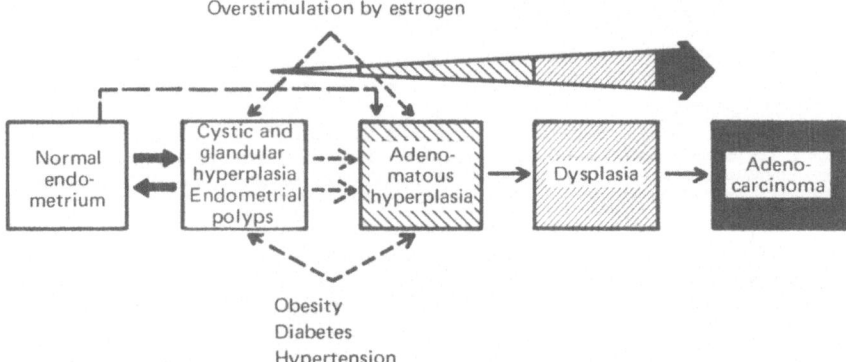

Fig. 172. Diagram of the genesis of endometrial carcinoma. Predisposing endocrine and constitutional endogenous factors. Reversible and irreversible premalignant stages.

lesion. More than 50% of such lesions progress to adenocarcinoma of the endometrium within 1 to 3 years. The change is considered irreversible and therefore precancerous. Similar patterns may be found in polyps of the endometrium (page 315).

Treatment: Adenomatous hyperplasia is responsive to progesterone and progestins and may be shed during menstruation (so-called medical curettage), as may some dysplastic lesions also. In general, prophylactic hysterectomy should be considered in women within and beyond the later reproductive years. An attempt at conservative therapy in younger patients (repeated curettage) may be justified.

Malignant Neoplasms of the Endometrium. Adenocarcinoma of the endometrium follows mammary and cervical cancer as the third most frequent carcinoma of the reproductive organs in women. The age distribution has a peak in the menopause and postmenopausal years (Fig. 156). Generally, as longevity increases, the prevalence of endometrial cancer increases. The ratio of frequency of cervical cancer to endometrial cancer was 10 to 1 a few decades ago, but it has decreased to 3 to 1 or less in the last decade. This change results in part from the reduction of cervical cancer by early cytologic detection.

Endometrial cancer is more common in patients in higher socioeconomic groups. According to some investigators there is a relation to unopposed estrogenic stimulation. The history frequently includes dysfunctional bleeding. Patients with estrogen-producing ovarian tumors and patients with Stein-Leventhal syndrome show a higher incidence of endometrial cancer than do patients in a normal control group with biphasic menstrual patterns. Not all investigators, however, accept the relation of estrogen to carcinoma of the endometrium.

Endometrial cancer also appears to be associated with obesity, hypertension, diabetes, and cardiovascular diseases.

One to two decades ago, an accepted treatment for preclimacteric, abnormal, uterine bleeding was intrauterine radium to destroy the endometrium. These patients are now completing the 10- to 20-year lag period required for the development of cancer; endometrial carcinoma seems to be increased in that group. Radiation may therefore be an additional etiological factor.

The main location for adenocarcinoma of the endometrium is the fundus and the area of insertion of the fallopian tubes (80%). The tumor spreads by superficial growth in the endometrium, polypoid extension, or exophytic growth into the uterine cavity. It may fill the entire endometrial cavity, extend downward into the cervical canal (1 to 20%), and be seen in the external os of the cervical canal. Another pattern of growth occurs when the fallopian tubes are involved. The tumor extends into the abdomen through the fimbriated ends. Infiltration of the myometrium is a later development.

Lymphatic spread occurs through the mesosalpinx and the infundibulopelvic ligament to the periaortic lymph nodes. Metastasis to the ovaries occurs in 5 to 12% of cases. Occasionally metastasis occurs via the round ligaments (Fig. 173).

Once the adenocarcinoma has penetrated into the cervix, the lymphatic spread is identical with that of cervical cancer, The inguinal lymph nodes are involved in approximately 28% of cases. Since the fundus is richly vascularized, hematogenous metatasis is more frequent than in cervical cancer. Distant metastasis occurs to the lungs, liver, skeletal system, and brain.

Pyometra may result from infection of necrotic tumor masses when the cervical canal is blocked or stenotic.

Various degrees of histologic differentiation may be seen, including the most undifferentiated patterns. Polymorphism and hyperchromasia of nuclei and abnormal mitoses are common (Fig. 174).

Stromal invasion or complete loss of stroma may be seen. In 15 to 20% of adenocarcinomas of the endometrium, squamous metaplasia may be identified. Such a tumor is termed an adenoacanthoma. Rarely a clear cell carcinoma that resembles hypernephroma is also found. In general, the more undifferentiated the pattern, the worse is the prognosis.

Signs: Endometrial cancer produces metrorrhagia in younger women and postmenopausal bleeding later in life. The discharge is frequently serosanguineous. Endometrial cancer accounts for up to 40% of cases of postmenopausal bleeding in some series.

Diagnosis: Cytology is an unreliable method of diagnosis for endometrial carcinoma. Cancer cells from the endometrium are identified in less than 40% of cases in the vaginal pool. Endometrial biopsies obtained with a small curette or a brush, and the recently employed jet washings improve the accuracy of diagnosis (page 118).

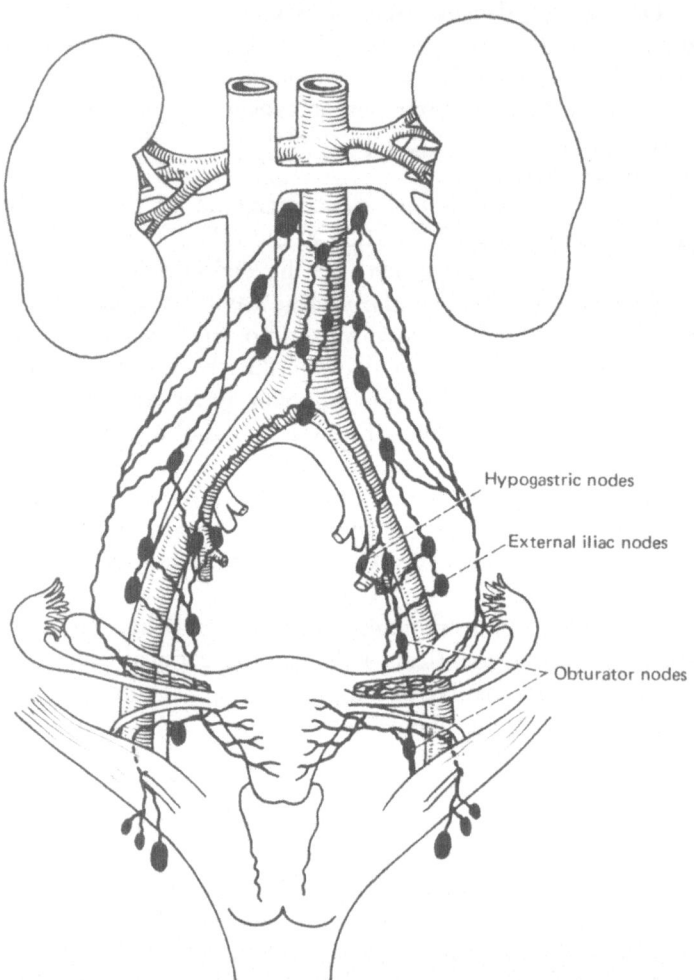

Fig. 173. Lymphatics and regional lymph nodes of the corpus. Carcinoma confined to the fundus spreads primarily by means of the lymphatics of the infundibulopelvic ligament to the periaortic lymph nodes. When endometrial carcinoma involves the lower uterine segment and cervix, the hypogastric iliac, and obturator nodes also may be involved.

Palpation of the uterus in later stages may reveal a slight increase (comparable to 6 to 8 weeks' gestation) in size and softness of the organ. Pyometra is usually associated with a tender, cystic, enlarged uterus.

The diagnosis is confirmed by a fractional curettage, in which the cervical canal is scraped by a small curette before dilatation. After dilatation, the endometrial cavity is scraped separately with a medium-sized curette. The specimens from the cervix and corpus are submitted to

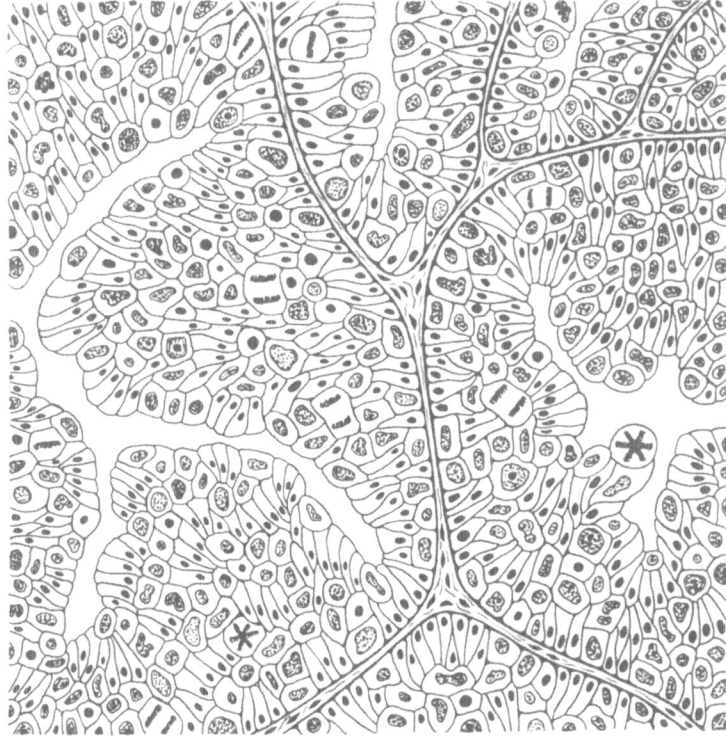

Fig. 174. Adenocarcinoma of the corpus. The glands are tightly packed in typical back-to-back arrangement. The stroma is compressed and the glandular epithelium is irregularly thickened. In places the glandular lumina are recognized as narrow clefts. The epithelial cells are polymorphic with marked irregularity of nuclear and cytoplasmic size and shape. Atypical mitoses are common.

the pathologist in separate containers. Perforation may readily occur in advanced stages of the disease when the myometrium is infiltrated by the tumor.

In case of pyometra, the cervical canal is dilated and the endometrial cavity drained. The uterus is scraped a few days later. Hysterosalpingography should be avoided because of the danger of spreading cancer from the uterine cavity.

Staging of adenocarcinoma of the endometrium: The staging proposed by the FIGO in 1962 is generally accepted and allows comparison of results on an international basis.

Stage 0: Suspicious but not clearly malignant areas histologically.

Stage I: The carcinoma is confined to the corpus uteri.

Stage II: The carcinoma has involved the corpus and the cervix.

If it cannot be decided on fractional curettage whether the cancer is primarily endocervical growing in the endometrium, or vice versa, then an adenocarcinoma is classified as carcinoma of the corpus.

Stage III: The carcinoma has extended outside the uterus but not outside the true pelvis.

Stage IV: The carcinoma has extended outside the true pelvis or has obviously involved the mucosa of the bladder or of the rectum.

The TNM system has not been standardized for the carcinoma of the endometrium.

Treatment: Three types of treatment are available:

(1) surgery
(2) radiation
(3) a combination of radiation and surgery

The treatment of choice in Stage I is total abdominal hysterectomy with bilateral salpingoophorectomy and resection of a large vaginal cuff. The operation is usually performed by the abdominal route but vaginal hysterectomy may be performed as well. The adnexa must be removed since the most common mode of spread is along the lymphatics of the ovarian vessesls. Preoperative radium is used by some medical centers (Heyman capsules) to shrink the tumor. Some investigators believe that preoperative radium reduces the likelihood of recurrences in the vaginal vault; but the exact role of preoperative radiation is not yet clear. Primary irradiation without hysterectomy is used only in patients presenting a very great operative risk.

The ideal treatment of Stage II is not established. A few centers treat the lesion as a carcinoma of the cervix by radical hysterectomy, but more recommend simple total hysterectomy with a full course of external radiation.

Radiation and various extensive surgical procedures are tried in Stages III and IV, but the therapeutic results are poor.

Recurrence: The definition is the same as that for cervical carcinoma (see page 315). Diagnosis is made by curettage, palpation, cytology, lymphangiography, venography, and roentgenography. Recurrences are observed in approximately one-fifth of patients in the vagina, parametria, uterus, and ovaries, in that order of frequency. Treatment is similar to that of recurrent carcinoma of the cervix (page 315).

Recurrence and distant metastases may be suppressed by treatment with high doses of progestational agents such as 17 α-hydroxyprogesterone caproate (Delalutin) or medroxyprogesterone acetate (Depo-Provera). Progestational drugs and other chemotherapeutic agents have, in general, provided only temporary remission.

Results of treatment: The total 5-year salvage of all stages is at present 62.6% (Annual Report of the Results of Treatment in Carcinoma of the Uterus and Vagina. Vol. 14, Stockholm, Ed. Office, 1967). The 5-year survival depends on size of the uterus and histologic grade of the tumor, among other factors. For Stage I the 5-year cure is 75 to 85%; for Stage II, 20 to 40%; for Stage III, less than 20%; and for Stage IV, less than 5%.

Sarcoma of the Uterus. Approximately 2% of malignant neoplasms of the uterus are sarcomas. They may be derived from the myometrium or from myomas (leiomyosarcoma) or, less frequently, from the stroma of the endometrium. They occur in all age groups. Sarcoma of the endometrium is also found in young women.

Leiomyosarcomas may grow rapidly. Endometrial sarcomas may occur in polypoid form projecting as far as the external os.

Sarcomas may cause pain in the lower abdomen and are often recognized by apparently rapid growth of "myomas." Irregular uterine bleeding is more commonly found with sarcomas of the endometrium. Diagnosis is made by histologic examination. The treatment of choice is abdominal hysterectomy with bilateral salpingo-oophorectomy and possibly postoperative irradiation. The prognosis is poor if any spread has occurred.

Benign and Malignant Neoplasms of the Fallopian Tubes

Benign Neoplasms. Myomas, lymphangiomas, and hemangiomas are uncommon lesions.

Malignant Neoplasms. Primary carcinomas of the fallopian tubes are the least common genital carcinomas (0.1 to 0.8%). They usually occur between the ages of 40 and 60 and only occasionally in younger women. These are adenocarcinomas and unilateral in 80% of cases. Primary carcinomas are more common in the ampullary than in the isthmic part of the tube.

Carcinoma of the fallopian tube is characterized by early involvement of the peritoneum. Metastasis occurs by lymphatics or by local spread to the uterus, ovaries, vagina, bladder, and rectum. Hematogenous metastasis is a late development.

Histologically the differentiated papillary or alveolar adenocarcinomas are more common than poorly differentiated medullary forms. Premalignant lesions have not been identified.

Symptoms and diagnosis: Since there are no pathognomonic signs, carcinomas of the fallopian tubes are in most instances accidentally found at laparotomy. Hydrorrhea occasionally provides a clue to diagnosis, but correct preoperative diagnosis is made only infrequently. Rarely, malignant cells have been identified in vaginal cytologic smears. The tumor is frequently mistaken for a hydrosalpinx or an ovarian mass. Since tubal carcinomas spread rapidly, peritoneal signs occur early. At laparotomy, brownish-red, papillary tumor masses are seen to have penetrated the wall of the fallopian tube.

Treatment: This is identical with that of ovarian tumors (page 352). Radical surgery plays little role in treatment, but irradiation and chemotherapy may be tried for advanced lesions.

Prognosis: Five-year survival is only about 10%. The poor prognosis reflects delayed diagnosis and resistance to irradiation and chemotherapy.

Metastatic tumors of the fallopian tubes are more common than primary cancers. They are most often secondary to ovarian, endometrial, or mammary carcinoma, in that order of frequency.

Benign and Malignant Tumors of the Ovary

The ovary comprises a variety of histological structures: the mesenchymal stroma around the cellular elements of the follicle (granulosa and theca cells), the external layer on the surface (the so-called germinal epithelium) and the androgen-producing hilus cells, as well as lymphatic and vascular structures. The ovary contains many different stem cells, which may, under the proper stimulus, proliferate and form a variety of tumors of various origins.

The development of some types of tumors can be explained only on the basis of pluripotential remnants of the gonadal anlage, such as mesenchyme or celomic epithelium. Others apparently develop from cell and tissue inclusions acquired during an early embryological stage.

Germ cells must be considered multipotential and may, for unknown reasons, form embryonal tumors containing structures of varying degrees of differentiation.

Maturation of a follicle requires the differentiation of surrounding stroma cells, or theca cells, which have the capacity to synthesize steroid hormones. The ovary thus remains in a "differentiating" stage throughout reproductive life. The functional changes of the ovary in different age groups may further explain the variety of ovarian tumors.

Malignant ovarian tumors account for 15 to 20% of all genital tumors. They are surpassed only by carcinomas of the cervix and endometrium. Approximately one-third of all ovarian tumors show malignant change. As a result of incomplete knowledge and various interpretations of ovarian embryology, a generally accepted classification is not yet available. Even the differentiation between benign and malignant tumors is often difficult, for many of the benign tumors must be considered potentially malignant.

One classification of ovarian tumors is suggested below.

I. Retention or functional cysts
   A. follicle cyst
   B. corpus luteum and theca lutein cysts
   C. endometrioid cysts
II. True neoplasms
   A. benign
      1. cystic (cystomas)
         a. serous cystadenoma
         b. mucinous cystadenoma
         c. benign teratoma (dermoid cyst)
      2. solid
         a. Brenner tumor
         b. fibroma
      3. hormonally active
         a. estrogen-producing
         b. androgen-producing
   B. malignant
      1. primary
         a. adenocarcinoma
         b. serous cystadenocarcinoma
         c. mucinous cystadenocarcinoma
         d. malignant teratoma
         e. dysgerminoma
      2. secondary (metastatic

Functional or Retention Cysts of the Ovary. Functional cysts of the ovaries are not true neoplasms. They are mentioned in this connection only because they present clinically as tumors and must be considered in differential diagnosis.

Follicle cysts may develop as a consequence of failure of ovulation and of subsequent transformation of the follicle into a corpus luteum (persistent graafian follicle) (page 59). The accumulating fluid may increase the diameter of the cyst to 6 cm. Involution with resorption of the liquid usually occurs in 1 to 2 months.

The patient may complain of lower abdominal pain on the side on which the cyst is found. The pain may be caused by tension on the ovarian capsule (tunica albuginea). Dysfunctional bleeding frequently follows a short-term oligomenorrhea. The basal body temperature curve is often monophasic (page 144).

On examination, a cystic, mobile mass with a smooth surface is felt. A patient in the reproductive years may be told to return for another pelvic examination in about 6 weeks. Increase in size during this period of time is inconsistent with the diagnosis of a functional cyst and requires further investigation without delay.

The small ovary with multiple cysts and the polycystic ovary of the Stein-Leventhal syndrome may be included in the group of retention and functional nonneoplastic cysts. They are related to permanent ovarian dysfunction and may result in characteristic morphological changes of the organ.

The ovary with multiple small cysts is often the result of inadequate maturation of the follicle (page 59) or perioophoritis. The surface of the ovary is irregular because of numerous small cysts. The ovary may attain a diameter of 6 to 8 cm. The follicular wall has a normal or atrophic epithelium. The cumulus oophorus may be present, but in most cases the oocytes are atretic.

The polycystic Stein-Leventhal ovaries differ in structure from the small cystic ovary. The subcapsular cortex shows marked fibrosis with thecal hyperplasia and luteinization of the theca interna. The remaining oocytes are degenerated. Corpora lutea and albicantia, indicative of previous ovulations, are lacking (page 118).

Corpus luteum and theca lutein cysts are less common than follicle cysts. A corpus luteum cyst may result from excessive bleeding of one of the numerous vessels that develop during the formation of the corpus luteum of menstruation or pregnancy. A tender cyst, 6 to 8 cm in diameter, may result. After resorption of hemoglobin and its breakdown products, the cystic fluid appears clear and yellowish. The wall is formed by luteinized granulosa cells. Continued synthesis of progesterone delays the onset of the next menstruation. The presence of unilateral lower abdominal pain in conjunction with an adnexal mass may lead to the erroneous diagnosis of tubal pregnancy. The corpus luteum cyst is thin-walled and easily ruptured.

Lutein cysts may also develop from a hemorrhagic corpus luteum. They are characterized by a wall formed by luteinized granulosa and theca interna cells. Multiple, bilateral, lutein cysts may be found in association with increased production of gonadotropin, as, for instance, in molar pregnancy or choriocarcinoma. The overstimulation induces atresia of the follicles and luteinization of theca cells with subsequent cystic change. Theca lutein cysts have recently received more attention because they are known to develop iatrogenically as a result of over-stimulation during an attempt at induction of ovulation (page 179). Spontaneous rupture with bleeding into the peritoneal cavity is infrequent but it may produce an acute abdomen. The very thin cyst wall is easily ruptured during an overly vigorous pelvic examination. After elimination of endogenous or exogenous gonadotropic stimulation, theca lutein cysts should disappear.

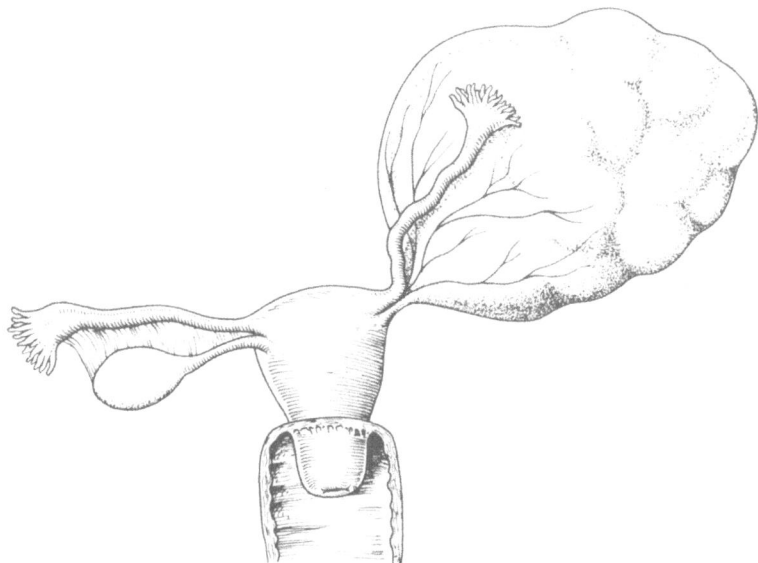

Fig. 175. Pedunculated ovarian cystoma. The blood vessels are clearly visible coursing from the pedicle into the tumor.

Chocolate cysts refer to any blood-filled cyst of the ovary, although the term is commonly, but inaccurately, used as a synonym for endometriotic cysts (page 269).

Cystic Ovarian Tumors. In contrast to functional or retention cysts, cystomas are true neoplasms of epithelial origin. Even the benign types carry some risk of malignant change (page 339). During growth of an ovarian tumor, the upper part of the broad ligament and the infundibulopelvic ligament are stretched, forming a pedicle that contains the vascular supply of the cystoma (Fig. 175).

The stalk increases the mobility of the tumor and predisposes to torsion.

Serous cystadenoma: Twenty to 25% of ovarian neoplasms are serous cystadenomas. They occur during the reproductive years as well as after the menopause. The cysts are unilateral in approximately 80% of cases. They are filled with a clear watery liquid that is sometimes stained slightly yellow. As a result of bleeding into the cyst, the fluid may be dark brown. The tumor may be so small that it is barely palpable or sufficiently large to fill the entire abdomen up to the umbilicus. The remaining normal ovarian tissue is usually destroyed by pressure from the expanding tumor. Many authorities believe that serous cystadenomas originate from the surface or "germinal" epithelium. When the lining

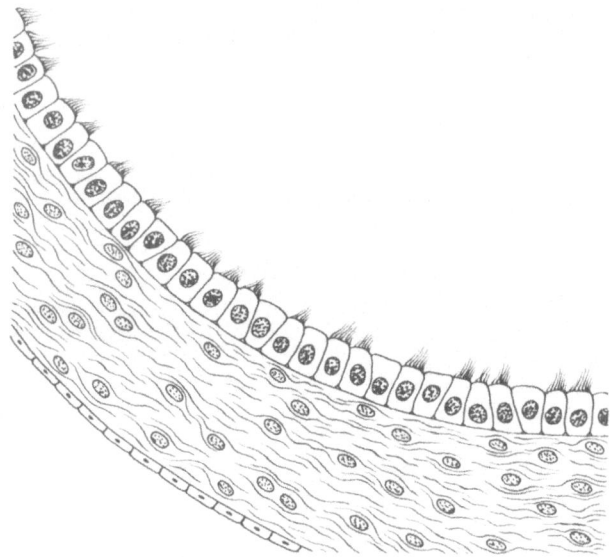

Fig. 176. Benign serous cystoma. The cyst wall is smooth. The epithelium consists of a single layer of flat to cuboidal cells, some of which are ciliated.

of the cyst wall is smooth and it consists of a single layer of cuboid or cylindrical cells, the tumor is clearly a benign serous cystoma (Fig. 176). Multilocular cysts often contain papillary projections covered by cuboidal epithelial cells. This type of tumor is called a *papillary serous cystadenoma* (Fig. 177). The papillary projections may penetrate the cyst wall and continue to grow on its outer surface. They may even implant on the peritoneum or the serosa of the intestine without being malignant. Such extension often leads to ascites, which may, however, regress after removal of the tumor and the implants.

Among ovarian tumors, the greatest likelihood of malignant change occurs in papillary cystadenomas. About half of all papillary serous cystadenomas are actually or potentially malignant (Table 29, pp. 348–349). Penetration of the cyst wall by papillary growth is particularly suspicious of malignancy. Bilateral cystadenomas should always arouse suspicion of malignancy.

Mucinous cystadenomas: The cystic tumors are so called because of their mucinous opalescent contents. Mucin is a glycoprotein formed of aminosugars. The older term "pseudomucin" has been abandoned as inaccurate and misleading. Mucinous cystomas account for 10 to 18% of all ovarian tumors. They are usually unilateral (95%) and may attain large size. The tumor is often multiloculated and its surface may be smooth or irregular. The histogenesis is not certain. Some authorities be-

Fig. 177. Papillary serous cystoma. The stroma is highly vascularized. The epithelial cells are cuboidal to columnar. Most of the cells are ciliated.

lieve that these tumors originate from transformed "germinal" epithelium. Since the epithelium resembles that of endocervical glands, which originate from müllerian structures, embryonic paramesonephric epithelial inclusions have also been considered a source. Another hypothesis is based on the similarity of the epithelium to that of the intestinal mucosa. Mucinous cystomas could, therefore, be considered teratomas in which one cell species predominates, as in the case of ovarian struma (Fig. 178).

The thin cyst wall may rupture spontaneously or during examination or laparotomy with the release of gelatinous material into the peritoneal cavity. This material is very slowly resorbed and it may result in a chronic peritonitis. The implantation of mucin-producing cells results in a peritoneal pseudomyxoma in approximately 7% of cases. This complication often leads to cachexia and death, although histologically malignant tissue is not always identifiable. Malignant change occurs in about 12 to 15% of these tumors.

Benign Cystic Teratoma (*Dermoid Cyst*). Approximately 10 to 18% of ovarian tumors are dermoid cysts. They are predominantly unilateral (75%). They grow slowly and rarely exceed 12 cm in diameter. They are of doughy consistency with a smooth, pearly-gray surface. The tumor is most commonly diagnosed during the third decade of life. Dermoid

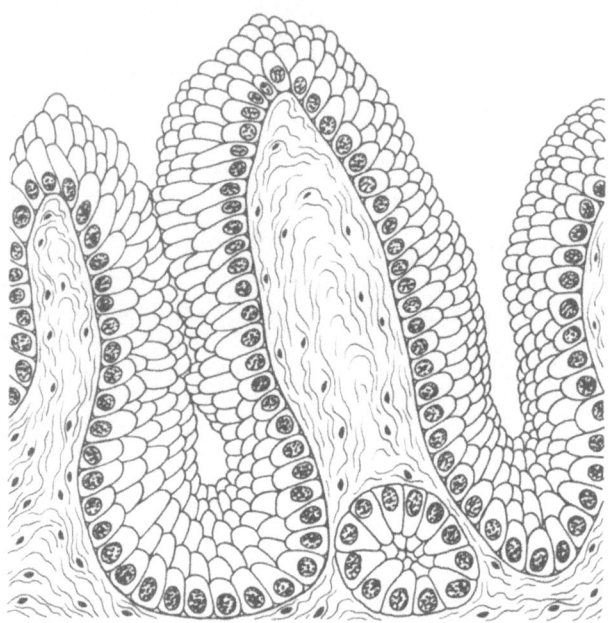

Fig. 178. Mucinous cystadenoma. The cyst wall consists of a single layer of columnar epithelium, the nuclei of which are round or flattened. A tangential section of the epithelium gives the impression of a multilayered epithelium.

cysts may have long pedicles. They tend to twist because of their mobility, limited size, and frequent location anterior to the broad ligament.

Dermoids are benign teratomas (page 349). One hypothesis holds that teratomas develop by parthenogenetic division of germ cells. In the benign dermoid cyst, the ectodermal elements (skin, hair, and sebaceous and sweat glands) predominate, although all three embryonic layers contribute to the tumor. Teeth in the tumor may often be recognized on a scout film of the abdomen. Structures of mesodermal or endodermal origin, such as cartilage, bone, thyroid, and intestines, are also found. In the benign dermoid, however, the tisues are generally well differentiated (Fig. 179). The contents of the cyst are derived from secretion of sebaceous glands. The tumor contains a solid tubercle containing the various tissues and organs.

Struma ovarii: A rather rare type of benign teratoma (1 to 3% of cases) is the struma ovarii. It consists almost exclusively of thyroidal tissue. The tumor is solid and soft. It is seen primarily during the reproductive years. Well-differentiated cells may synthesize thyroid hormone and cause hyperthyroidism. Malignant change is rare (5 to 10%). For benign tumors oophorectomy is sufficient treatment. Hyperthyroidism regresses rapidly after ovariectomy. Carcinomatous change must

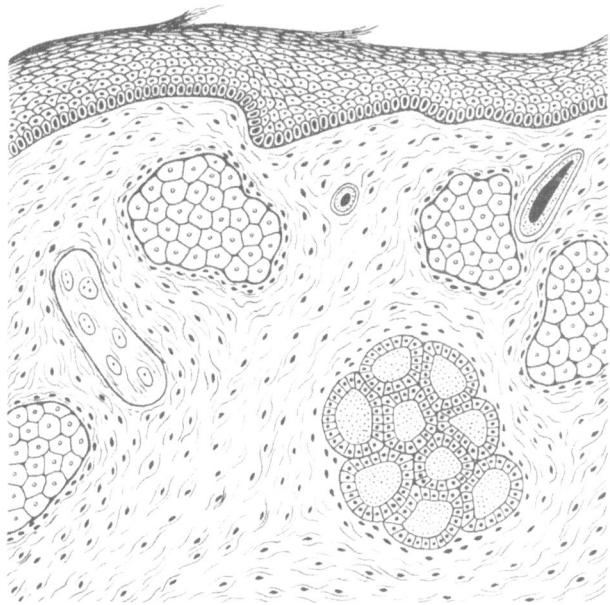

Fig. 179. Dermoid cyst (benign cystic teratoma). Segment of the cyst wall. Toward the luminal surface, the wall is lined by a cornified stratified squamous epithelium. In the underlying connective tissue cross-sections of hair (upper right), sebaceous glands (upper right, upper center, and lower left), sweat glands (lower right), and cartilage (middle left) are seen.

be treated according to the principles applicable to ovarian carcinoma in general.

Malignant change in a benign teratoma is uncommon (about 1%). It occurs mainly in older patients, whereas primary malignant teratomas usually affect children and young women.

Parovarian cysts: Parovarian cysts develop from rudiments of the wolffian ducts and the mesonephros. The normal ovary is closely attached to the parovarian cyst but can be clearly isolated from it. Since parovarian cysts are occasionally pedunculated, they sometimes undergo torsion. The diagnosis is almost always made at laparotomy. The parovarian cysts may be peeled away leaving a functioning ovary. The proximity of the ureter to intraligamentous cysts requires care in surgical removal. Other parovarian cysts are found outside the broad ligament in the region of the infundibulopelvic ligament. Both types are benign. The clinical picture is similar to that of ovarian cysts (page 331).

Solid Ovarian Tumors. Brenner tumors are solid neoplasms of the ovary with a characteristic epithelial structure. They account for about 1 to 2% of all ovarian tumors, and are most commonly seen in women beyond the age of 40. Microscopically, the tumor is composed of nests or cords of round or oval epithelial cells with abundant cytoplasm and surrounding connective tissue (Walthard rests). These nests of cells may originate from inclusion of celomic cells in embryonic life, or they may develop by metaplasia of the epithelium of the ovarian surface.

The tumor is almost exclusively unilateral, appearing as a firm lobular mass that is white on the outside and yellowish on section. It usually measures 10 to 15 cm in diameter. It cannot be differentiated from a uterine leiomyoma or an ovarian fibroma by palpation. Malignant change is so rare that unilateral oophorectomy is considered sufficient treatment.

Of the less-common ovarian tumors of mesenchymal origin, only the fibroma is of major clinical significance. Ovarian fibromas account for about 5% of all ovarian neoplasms. They are predominantly unilateral (90%) and are found more often in women after the menopause than in other age groups. The tumor grows slowly, measuring 15 to 20 cm in diameter, and is often pedunculated. Fibromas originate from stromal cells and are therefore formed of connective tissue. Larger tumors tend to undergo cystic degeneration. In about 25% of cases, Meigs' syndrome develops. This syndrome is characterized by unilateral or bilateral hydrothorax and ascites. The fibroma may cause no local symptoms even when hydrothorax and ascites cause dyspnea and increase in abdominal girth. Malignant change is rare (less than 1%) and unilateral oophorectomy is therefore adequate treatment. Ascites and hydrothorax regress after removal of the fibroma.

Hormone-producing Ovarian Tumors. There are several theories about the origin and development of hormone-producing ovarian tumors. These tumors may reflect the various modifications of the ovarian stroma that may be associated with differing endocrine effects, for it is known that ovarian stromal cells may differentiate along several histological lines and produce a variety of interconvertible steroid hormones. These tumors may also be termed mesenchymomas or gonadal stromal tumors. They may be classified histologically or according to their endocrine effects as feminizing, masculinizing, or inert tumors.

Estrogen-producing Ovarian Tumors. Granulosa cell tumors are estrogen-producing ovarian tumors that account for 1 to 3% of all ovarian tumors and are mainly unilateral (95%). Granulosa and theca cell tumors rarely reach large size. They may remain as small nodules inside the ovary, which maintains its normal size and shape. Larger tumors may be palpated as solid firm masses with a smooth or slightly irregular surface. The cells resemble granulosa cells and are arranged in either rosettes or columns supported by a connective tissue network (Fig. 180). Local pressure symptoms are rare and the clinical picture is more often related to the excessive production of estrogen.

Fig. 180. Granulosa cell tumor. The neoplastic tissue is traversed by strands of connective tissue. The tumor cells all look alike and form rosettes around a central lumen. The lumina contain albuminous detritus.

Granulosa cell tumors may be found in all age groups, but mainly during the postmenopausal years. Irregular menstrual bleeding is the predominant sign in the premenopausal woman, and postmenopausal bleeding in the older women. The bleeding is related to endometrial hyperplasia. Children with this tumor may develop precocious puberty (page 155).

Early diagnosis and treatment are important because of the possible relation of granulosa cell tumor to endometrial carcinoma. The significance of this relation, however, is questionable according to some investigators. The frequency of malignant transformation of the tumor itself varies between 10 and 30%. In addition, there is often a discrepancy between a histologically benign appearance and clinically malignant behavior. A primary malignant form is occasionally found. These tumors are generally resistant to radiation therapy.

Treatment in children and women of childbearing age who have not completed their families may be restricted to unilateral oophorectomy. Regular periodic gynecological examinations and measurements of estrogen are required to detect recurrences. In older women who have completed their families, total hysterectomy with bilateral salpingo-oophorectomy is the procedure of choice.

Thecomas are unilateral, solid or partially cystic tumors of soft or firm consistency. They usually grow at a rate slower than that of granulosa cell tumors. They remain smaller than granulosa cell tumors and produce larger amounts of estrogens. They occur three times as often as granulosa cell tumors and are found almost exclusively in older women (70% after the menopause). Histologically, the tumor is characterized by fusiform epithelioid cells densely arranged in a whorled pattern with little interlacing connective tissue. Cytoplasmic lipoid droplets, presumably indicating steroids, are demonstrated by fat stains. Degenerating areas with necrosis and hyalinization are common.

The clinical picture results from excessive production of estrogen, and is indistinguishable from that of granulosa cell tumors. The thecoma does not undergo malignant transformation, but is sometimes associated with endometrial carcinoma, according to many investigators.

These estrogen-producing tumors are frequently grouped as granulosa-theca cell tumors, since they may have both cell types, with one or the other component predominating.

Luteomas are luteinized granulosa-theca cell tumors, which produce predominantly estrogens and rarely progesterone or androgens.

Androgen-producing ovarian tumors are uncommon. Since they may cause defeminization and virilization, they are usually easily diagnosed.

Arrhenoblastoma: The arrhenoblastoma is unilateral in 95% of cases and is variable in size. It occurs most often between the ages of 20 and 30.

Histologically, tubular formations similar to those in the testis are found. The degree of differentiation varies considerably. They may arise from residual elements of the originally undifferentiated embryonic gonad (page 8). Malignant change occurs in 20 to 25% of cases.

Hilus cell tumors are quite uncommon. They are derived from hilus cells, which correspond to interstitial (Leydig) cells of the testis. The cases described in the literature are benign.

Adrenal rest tumors are also rare. They may arise as embryonic inclusions of adrenal cortical cells. Microscopically, this benign tumor resembles adrenal cortex, with hemorrhagic and cystic areas.

The production of androgen causes progressive defeminization with increasing signs of masculinization (amenorrhea, infertility, and involution of the genitalia). Later stages are characterized by hirsutism, deepening of the voice, and enlargement of the clitoris.

The signs, except hypertrophy of the clitoris, may gradually disappear after removal of the tumor. Virilization and a palpable adnexal mass are diagnostic. Hormonal analysis differentiates this condition from the adrenogenital syndrome, Cushing's disease, Stein-Leventhal syndrome, and adrenal carcinoma.

Oophorectomy is performed in younger women, and total abdominal hysterectomy with bilateral salpingo-oophorectomy after the menopause.

Malignant Neoplasms of the Ovary. Malignant ovarian tumors are classified as primary and secondary, or metastatic.

Primary malignant ovarian tumors: Primary ovarian carcinoma: Primary ovarian carcinoma accounts for about 15 to 20% of all gynecological cancer and approximately 5% of all malignant tumors in women. Recent statistics indicate a relative increase in incidence of ovarian carcinomas, compared with those of the cervix and endometrium. Mortality statistics indicate that death from ovarian cancer is exceeded only by that from cervical cancer. The mortality of ovarian cancer in the white population doubled between 1930 and 1960 while the mortality of cervical cancer was halved during the same period of time. The relative increase in mortality of ovarian cancer is explained by improved diagnosis of cervical cancer (page 299) and increased life expectancy. Although ovarian carcinoma may be seen in younger women and even children, 76% of all women with this lesion are older than 45. The peak incidence occurs in the 6th and 7th decade (Fig. 181).

Epidemiological factors are poorly understood. The estimated mortality rate for Europe is 7 to 8 per 100,000 women (especially high in Denmark—11.02). In Japan it is only 1.69. Socioeconomic studies reveal a higher incidence among unmarried women of a higher social status.

The majority of malignant ovarian tumors are of epithelial origin. They arise from "germinal" or surface epithelium of the ovary or inclusions of celomic epithelium acquired during embryonic development. There is some dispute about the origin of the mesonephroid ovarian tumors (mesonephromas) that look like clear-cell carcinomas. Rare forms include the malignant teratoma and in some classifications, the dysgerminoma.

Primary adenocarcinomas of the ovary are predominantly solid or cystic structures of variable size and consistency, with smooth and irregular surfaces and multiple morphologic patterns. Well-differentiated glands (Fig. 182) and papillary epithelial structures may be interspersed with areas of rather poorly differentiated adenocarcinoma (Fig. 183). The stroma varies from fibrotic (scirrhous type) to sparse (medullary type).

Serous and mucinous cystadenocarcinomas differ from benign cystomas in their exuberant proliferation of epithelium, marked polymorphism of nuclei, and invasion of neighboring tissues and organs (Fig. 184). Transformation into malignant tumors may be delayed for a long period of time and differentiation between malignant and benign areas is not always clear.

The endometrioid carcinoma of the ovary develops either from a benign variant or without apparent antecedent benign stages. Histologically, it resembles adenocarcinoma of the endometrium. Endometrioid

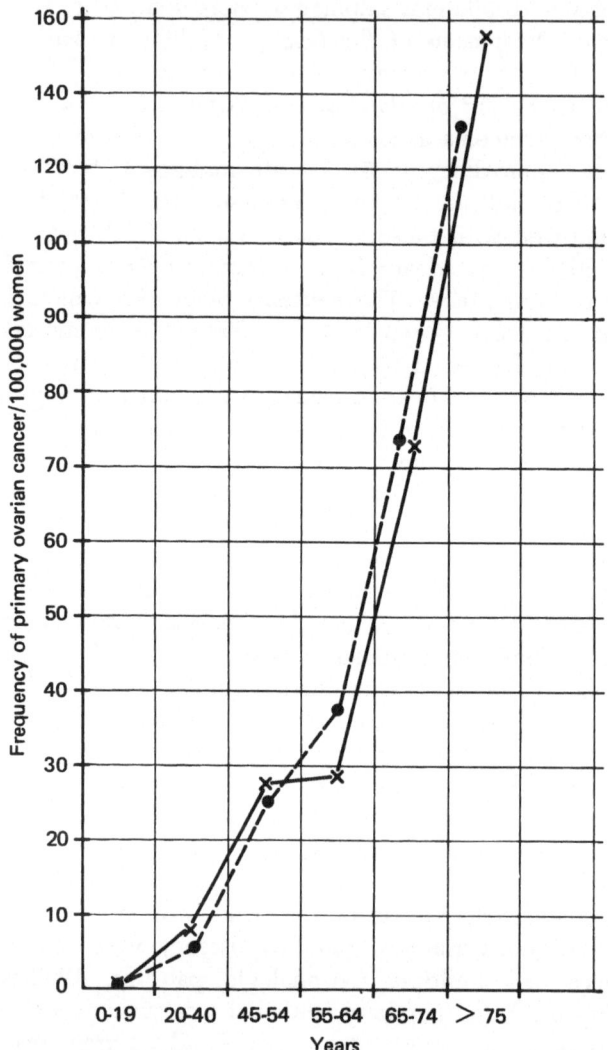

Fig. 181. Frequency of primary ovarian carcinoma in relation to age, based on 100,000 women. ●———● Oakland Kaiser Hospital (Bennington et al., *Obstet. Gynec.* 32:627, 1969), ×———× New York 1960 (Gerber et al.: Cancer in New York State exclusive of New York City. Bureau of Cancer Control, N.Y. Dept. of Health, 1962).

carcinoma of the ovary may be derived from embryonic celomic epithelium or from malignant transformation of an endometrioma of the ovary (page 269).

Diagnosis of a primary ovarian endometrioid carcinoma requires the exclusion of a primary carcinoma of the endometrium. This differential

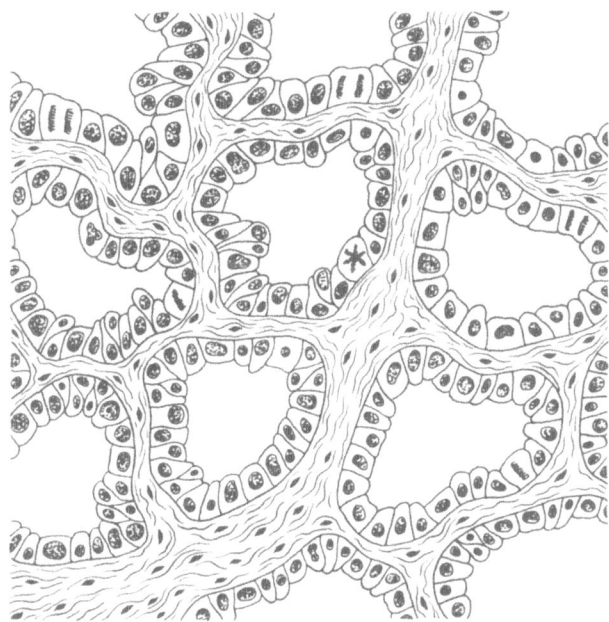

Fig. 182. Solid adenocarcinoma of the ovary. The malignant cells are arranged in glandular formation. The connective tissue is abundant in some places and sparse in others. The glandular epithelium is of variable thickness with varying degrees of cellular atypia.

diagnosis is sometimes difficult. Ovarian cancer tends to spread early to the endometrium and, conversely, the ovaries are involved early by metastases from endometrical cancer (page 322).

A classification of epithelial ovarian tumors has been suggested by FIGO in 1964 based on clinical aspects, prognosis, and potentially malignant course of serous, mucinous, and endometrioid ovarian tumors. The undifferentiated carcinomas have been grouped separately in Table 27.

About one-third of ovarian carcinomas are bilateral. The size varies considerably. The tumor grows rapidly and infiltrates the pelvis, predominantly the posterior cul-de-sac. Early spread to neighboring organs and the other ovary is common. The tumor then spreads to the surfaces of the fallopian tube, uterus, bladder, vagina, and rectum. At the same time, malignant cells may reach the uterus via the oviducts. Large fixed tumors result from extensive adhesions and infiltrating growth. Early spread to the peritoneum is common, with implantation on the omentum and development of ascites. Spread occurs via the lumbar lymph nodes and the lymphatics of the diaphragm, whence the tumor involves the pleural cavity. Liver, lungs, and bones may be involved by hematogenous spread. The multiple pathways of metastasis and rapid proliferation result in generalized carcinomatosis.

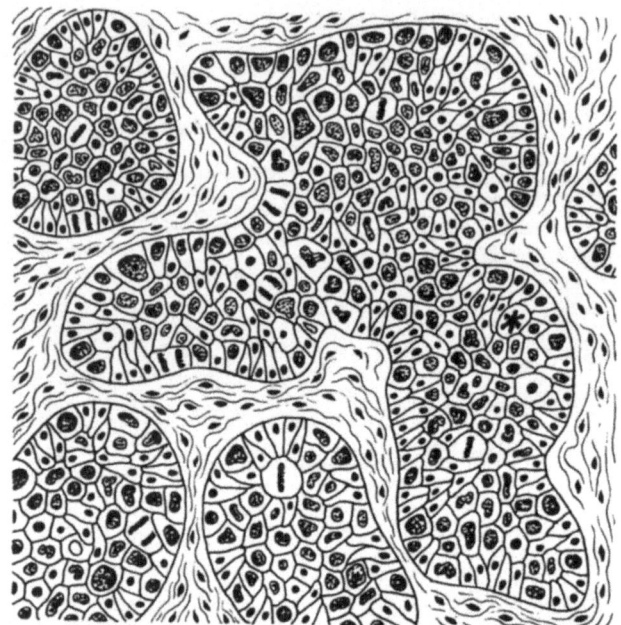

Fig. 183. Solid ovarian carcinoma of the alveolar type. The malignant cells form broad bands and alveoli. Abundant connective tissue is seen. Cells are polygonal and densely packed. They meet all of the criteria of malignancy.

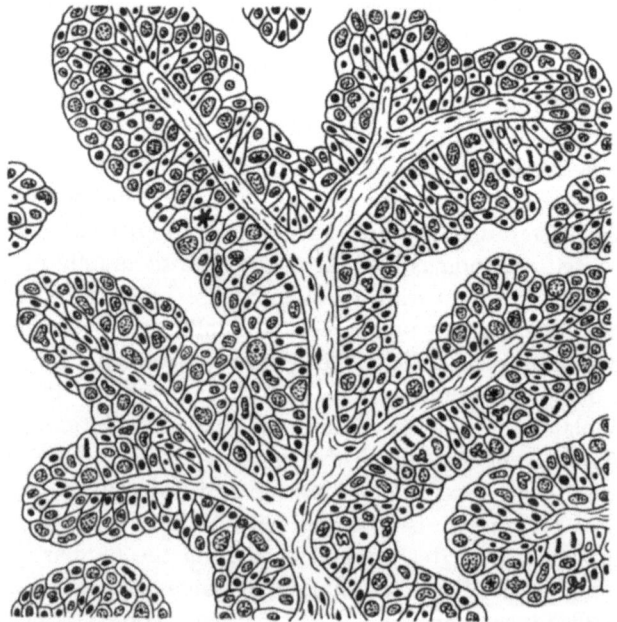

Fig. 184. Papillary serous cystadenoma. In contrast to the benign papillary serous tumor (Fig. 177) the epithelium is more extensively folded and multi-layered. The epithelial cells show all the criteria of malignancy; there are numerous atypical mitoses; the connective tissue is sparse.

Table 27

I. Serous cystomas
  A. benign serous cystadenoma
  B. proliferating papillary serous cystadenoma with nuclear abnormalities but no infiltrating growth (possibly malignant)
  C. serous cystadenocarcinoma
II. Mucinous cystoma
  A. benign mucinous cystadenoma
  B. proliferating mucinous cystadenoma with nuclear abnormalities but no infiltrating growth (possibly malignant)
  C. mucinous cystadenocarcinoma
III. Endometrioid ovarian tumors
  A. benign endometrioid cyst
  B. proliferating endometrioid adenoma and cystadenoma with nuclear abnormalities but no infiltrating growth (possibly malignant)
  C. endometrioid adenocarcinoma
IV. Mesonephroid ovarian tumors
  A. benign mesonephroid cyst
  B. possibly malignant mesonephroid tumor
  C. mesonephroid adenocarcinoma
V. Undifferentiated carcinoma that cannot be classified in I, II, III, or IV

Malignant ovarian teratoma is a rare tumor (0.015% of all ovarian tumors). Seventy-five percent of the patients are below the age of 25. These tumors appear as small, firm, solid masses that penetrate the capsule early and spread widely in the pelvis.

Malignant teratomas are thought to arise from primordial germ cells and contain derivatives of all three germ layers. Since the mesodermal elements often predominate, teratomas histologically and clinically resemble sarcomas such as rhabdomyosarcoma or chondrosarcoma. In contrast to the components of well-differentiated benign teratomas (dermoid cysts), the cells and tissues of the malignant variants are poorly differentiated and embryonal. Rapid proliferation and infiltration lead to early lymphatic and hematogenous spread. Treatment is the same as that of carcinoma of the ovary. Some malignant teratomas are moderately sensitive to radiation.

Primary choriocarcinoma of the ovary is extremely rare. The trophoblastic tissue is also derived from germ cells. This tumor in contrast to gestational uterine choriocarcinoma does not respond well to methotrexate.

Dysgerminoma is an ovarian tumor of variable malignancy. It occurs in children and young women (75% of all patients are under the age of 26), usually begins unilaterally (83%), and rapidly grows to attain a diameter of 20 cm or more. The tumor is found occasionally with intersexes (20%). The dysgerminoma also is thought to arise from germ cells (Fig. 185) and is histologically identical with the seminoma in men. The tumor usually has a firm capsule and is grayish on section.

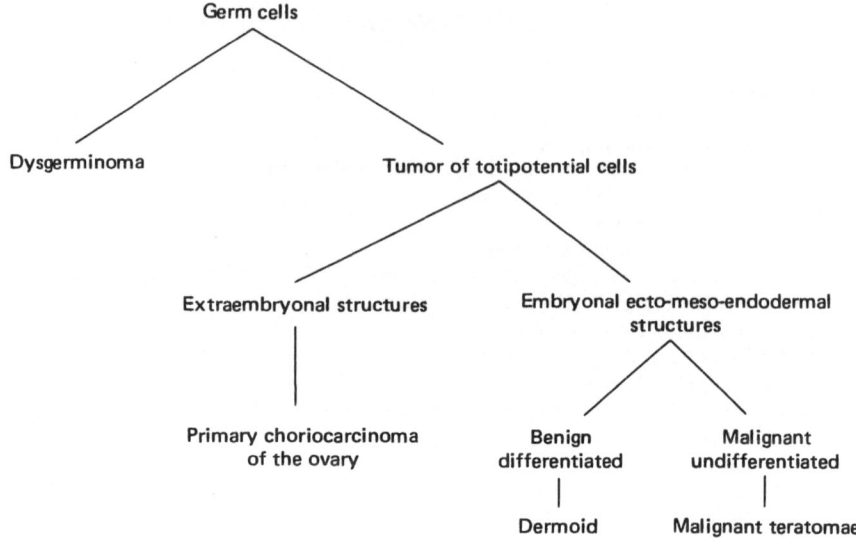

Fig. 185. Histogenesis of embryonal tumors of the ovary (according to G. Teilum, 1971).

Histologically, typical large ovoid or polygonal cells are seen arranged in nests separated by thin connective tissue strands often with interspersed lymphocytes. Bilaterality and penetration of the capsule render the prognosis poor. Treatment is that of ovarian carcinoma in general (page 352) except that the tumor is sensitive to radiation. Distant metastasis is less frequent than in malignant teratomas.

Metastatic Ovarian Carcinomas. Of all malignant ovarian tumors 25 to 30% are metastatic. Since the ovarian metastasis is often larger than the primary tumor, it may first call attention to cancer in other organs. Spread to the ovary is usually by hematogenous or lymphatic routes and less frequently by direct spread. Metastatic ovarian tumors are seen predominantly before the menopause, 60% occurring under the age of 50. In general, they are bilateral. Shape, size, and consistency vary considerably. They may be microscopic or large enough to fill the entire abdomen. The tumor consists of solid structures with cystic areas. Pedunculated metastatic ovarian tumors are occasionally seen.

Breast carcinoma commonly metastasizes to the ovary, as indicated by examination of tissue removed at the time of prophylactic bilateral oophorectomy. Malignant tumors of the stomach, bowel, and occasionally gall bladder are the next most frequent extragenital sites of primary cancers that metastasize to the ovary.

Ovarian metastases may proliferate rapidly and reach large sizes in a very short time. The term Krukenberg tumor was originally used in reference to an ovarian metastasis from the gastrointestinal tract with "signet ring" cells. Tumors of the same histological appearance may be seen occasionally as primary lesions.

Among genital cancers, endometrial carcinoma and the rare carcinomas of the fallopian tubes spread to the ovaries, as does cancer from one ovary to the other. Surgical management is hysterectomy and bilateral salpingo-oophorectomy, together with the appropriate treatment of the primary tumor.

Clinical Aspects of Ovarian Tumors

Symptomatology: Since only very large ovarian tumors cause pain and other symptoms, cancer of the ovary is often found accidentally, and frequently in an advanced stage. Table 28 lists the percentage frequency of complaints in ovarian carcinoma Stages Ia and Ib. It is obvious that the signs and symptoms of localized ovarian carcinoma are nonspecific.

Among these complaints only postmenopausal vaginal bleeding is of diagnostic significance. In 7% of cases it is caused by benign tumors and in 3 to 4% of cases by malignant ovarian tumors. Regardless of the patient's age, irregular bleeding occurs in about one-third of all women with ovarian tumors. Uterine bleeding is a major sign of hormonally active tumors.

Diffuse lower abdominal discomfort may occasionally be caused by pressure on neighboring organs. Malignant tumors often spread to the peritoneal cavity, the serosa of the intestine, and the omentum before a change in the patient's general condition becomes apparent. Symptoms of unilateral pressure occur early when the tumor develops within the broad ligament. Extremely large tumors may affect the patient's general condition, whether they are benign or malignant, and may cause a feeling

Table 28. Symptomatology of ovarian carcinoma Stages Ia and Ib. (According to Spechter.)

Symptoms	Incidence (%)
Nonspecific lower abdominal complaints	34
Increase in abdominal girth	27
Bleeding abnormalities and dysmenorrhea	21.5
Vaginal discharge	3
Weight loss	1.5
Uncommon complaints such as elevated temperature, nausea, vomiting, fullness, urinary tract complaints, or no complaint at all	

of generalized weakness. Progressive anorexia, cachexia, increase in abdominal girth with or without ascites, a feeling of fullness, and constipation are typical of a malignant tumor. These signs and symptoms indicate an advanced stage of the disease.

The lack of early symptoms explains why in 50 to 80% of all cases ovarian tumors at the time of laparotomy have already spread beyond the ovaries. Every ovarian tumor must therefore be considered potentially malignant until proved otherwise by histologic examination.

Acute signs and symptoms may result from the following complications of an ovarian tumor:

(1) torsion of the pedicle
(2) rupture
(3) hemorrhage
(4) incarceration
(5) infection

Torsion of the pedicle leads to sudden onset of severe abdominal pain, rebound tenderness, nausea, vomiting, ileus, and sometimes shock The temperature is slightly elevated; in general leukocytosis is moderate. Elevation of the erythrocyte sedimentation rate may occur.

Complete torsion with obstruction of the arterial blood supply may lead to gangrene of the tumor. Extensive infarction and necrosis increase the risk of rupture, which may occur spontaneously or during examination or laparotomy. When torsion is incomplete, the signs and symptoms are less striking. Obstruction of venous return may result in a decreased blood flow, with edema and extravasation of blood into the cyst. The tumor may then appear to grow rapidly in size. At the time of laparotomy, it is usually edematous and cyanotic.

Spontaneous rupture of a benign or malignant ovarian tumor is a rare event. It may occur, however, during pelvic examination under anesthesia or when the tumor is delivered from the abdomen at laparotomy. Thin-walled cysts are more likely to rupture. The signs of acute rupture are similar to those accompanying torsion of the pedicle. Rupture may also cause subacute progressive signs if the contents of the cysts produce a chemical peritonitis. Rupture of a mucinous cystoma may result in pseudomyxoma peritonei (page 332). How often spillage of malignant cells, after rupture of an ovarian carcinoma, results in implantation of cancer is not yet clear.

Hemorrhage into the tumor (whether benign or malignant) occurs rather frequently even without previous torsion or trauma. Rarely is the bleeding extensive enough to cause acute signs or symptoms. Incarceration of a mobile ovarian tumor in the posterior cul-de-sac is a rare complication. The signs it characteristically produces are acute urinary retention, caused by impingement on the bladder, and pressure on the urethra by the anteriorly displaced cervix. The rectum also may be compressed.

Infection of an ovarian tumor is also an unusual complication. A cystoma may become infected by spread from an acute salpingitis. Spread may also occur from local inflammation of neighboring organs such as appendicitis or diverticulitis. Symptoms are related to the primary disease and the accompanying peritonitis.

Ovarian Tumors in Pregnancy. Ovarian tumors, with the exception of corpus luteum cysts, are rather uncommon in pregnancy. In the absence of acute symptoms, laparotomy may ordinarily be delayed until the second trimester, since pregnancy with coexisting carcinoma of the ovary is extremely rare (about one in 100,000 deliveries). An exploratory laparotomy with removal of the tumor is performed after vaginal delivery. If the tumor interferes with normal labor and vaginal delivery, cesarean section with oophorectomy is the procedure of choice. If there are reasons to suggest malignancy, immediate surgical exploration is required.

Diagnosis of an ovarian tumor: Functional cysts rarely exceed 6 cm in diameter and are elastic, freely mobile, and smooth. A tumor with a doughy consistency located anterior to the uterus is likely to be a dermoid cyst. Ovarian tumors, cystic or solid, may be pedunculated. The external hand on pelvic examination must be placed high on the abdomen to avoid missing an ovarian tumor. Large tumors filling the abdomen sometimes may not be outlined either vaginally or rectally.

The site of origin may be difficult to ascertain because of the size and mobility of the tumor. Solid tumors seem firm on palpation but occasionally may feel slightly doughy because of central necrosis. If the tumor fills the posterior cul-de-sac, the uterus is pushed anteriorly and the cervix is displaced forward toward the symphysis. Early ovarian carcinoma is not detected on palpation. A slightly enlarged firm ovary may arouse suspicion but a small lesion within the ovary will be missed on manual examination. In most instances such small lesions are discovered accidentally after oophorectomy.

Vaginal cytology is not a valuable diagnostic method for ovarian carcinoma. Ovarian cancer cells can be detected in the vagina only in exceptional cases, when metastasis has already occurred. In postmenopausal women hormonal cytology may suggest a hormonally active tumor.

The only presently available diagnostic test for the detection of preclinical ovarian carcinoma is transvaginal injection of physiologic saline into the posterior cul-de-sac, with subsequent aspiration and search for tumor cells in the sediment of the aspirated fluid. This technique is, unfortunately, not a good screening measure.

Irregular firm or cystic masses filling the cul-de-sac are suggestive

Table 29. Incidence and clinical characteristics of ovarian tumors

Type of tumor	Proportion of all ovarian tumors	Frequency of bilaterality	Frequency of malignant change	Special remarks
Serous cystadenoma	20–25%	20% in benign forms 50% in malignant forms	50%	Occurs most frequently in reproductive years. Highest frequency of malignancy after the menopause. Bilaterality suggests malignancy. Penetration of the cyst wall and ascites may also be seen in benign forms.
Mucinous cystadenoma	10–18%	5% in benign forms 23% in malignant forms	12–15%	Most frequently seen between ages of 30 and 60. May grow to huge size. Rupture of the cyst wall or its penetration by tumor tissue may lead to pseudomyxoma peritonei even in apparently benign forms.
Benign cystic teratoma (dermoid cyst)	15%	25%	1–3%	Occurs mainly between 20 and 30 years of age. Well-differentiated tissues of all three embryonic layers, predominantly ecto-dermal structures. Cyst wall formed of skin, hair, and sebaceous and sweat glands.
Struma ovarii	1–3% of benign dermoids		5–10%	Seen during reproductive years. Occasionally cause symptoms of hyperthyroidism.
Brenner tumor	1.7%	Very rare	Rare; only 23 malignant cases in world literature	Mainly in patients above age of 40. Solid epithelial tumor, Walthard cell nests; 10% estrogen-producing with hyperplasia of endometrium and endometrial carcinoma.
Fibroma	1–5%	10%	Less than 1% (fibrosarcoma)	Occurs mainly after the menopause. Solid tumor consisting of fibrous tissue. Slow-growing. In 30% Meigs' syndrome develops.
Granulosa cell tumor	1–3%	5%	10–30%	Occurs most often after menopause, but may be seen at all ages, including adoles-

Tumor	Incidence		Malignancy	Description
				cence. Accounts for about 10% of all solid ovarian tumors. May be very small. Estrogen-producing. May be related to precocious puberty, cystic glandular hyperplasia, and endometrial carcinoma.
Theca cell tumor (thecoma)	1-2%		1%	Occurs predominantly after the menopause. Never occurs before puberty; 3-5% of all solid ovarian tumors. Estrogen-producing; sometimes androgen-producing. Small tumors called granulosa-theca cell tumors. Mixed forms called granulosa-theca cell tumors. Increased incidence of carcinoma of the endometrium reported by some investigators.
Arrhenoblastoma	Rare; only a few hundred cases described	5%	20-25%	Seen mainly during reproductive years but also in children or after menopause. Androgen-producing: defeminization and virilization. Very small to moderate-sized. Differentiated form is Pick's adenoma.
Hilus cell tumor	Uncommon		1%	Occurs after menopause. Androgen-producing.
Dysgerminoma	Fairly uncommon	17%	Always at least potentially malignant	Mainly in patients under 20. Occasionally in intersexes. Derived from germ cells.
Malignant teratoma	0.015%		Malignant	Predominantly in children and young women. Derived from germ cells. Rapid course; bad prognosis.
Primary ovarian carcinoma	About 15%	33%	Malignant	Seen usually after menopause; 15-20% of all gynecological cancers. Usually grows rapidly. Early metastases. Ascites. Of all malignant ovarian tumors 25-30% are metastatic.

of a malignant ovarian tumor. The suspicion is strengthened by the presence of ascites and nodular masses in the abdomen.

Doubtful findings on palpation require examination under anesthesia and laparoscopy. Carcinoma of the ovary is ruled out by laparotomy (page 352). Histological diagnosis by frozen section, obtained during laparotomy, provides the basis for treatment.

In the presence of ascites or hydrothorax, an attempt should be made to demonstrate malignant cells in the fluid preoperatively.

Differential diagnosis: Differential diagnosis includes enlargement or displacement of normal organs. Large ovarian tumors extending beyond the true pelvis may be mistaken for tumors originating from abdominal organs.

To avoid a common error in diagnosis, the bladder should be emptied before any examination. Reexaminations may be required after an enema.

Uterus. Occasionally a retroverted uterus may be mistaken for a solid ovarian tumor. Enlargement of the uterus, caused by pregnancy or myomas, may be difficult to differentiate from an ovarian tumor. The myomatous uterus is usually irregular, and as a rule, myomas are firmer than ovarian tumors. An ovarian tumor may, however, be as firm as a myoma. Motion of the cervix is frequently transmitted to the pelvic mass, and vice versa, if the mass arises from the uterus. When a myoma arises from the lateral aspect of the uterus, the ovary may be palpated separately if the uterus is elevated. A degenerating soft myoma or a myoma developing in the broad ligament in most instances cannot be differentiated by palpation from an ovarian tumor. Diagnostic problems are also presented by pedunculated subserous and soft myomas. Twisted myomas are indistinguishable by symptoms or palpation from a twisted ovarian cyst.

During the reproductive years, an intrauterine pregnancy must always be considered. Rapidly developing hydramnios in the second trimester of pregnancy also may occasionally be mistaken for an ovarian cyst.

Signs and symptoms of pregnancy, a positive pregnancy test, sonar diagnosis, and auscultation of fetal heart sounds may provide the diagnosis.

Tubes. Compared with an ovarian tumor, the unilateral mass of a tubal pregnancy is softer, irregular, and less clearly delineated. Pain on motion of the cervix is characteristic only in tubal pregnancy. Irregularities of vaginal bleeding, symptoms of pregnancy, and increasing anemia are important diagnostic signs. Since the pregnancy test is negative in about 50% of tubal abortions, diagnostic measures such as culdocentesis, colpotomy, or laparoscopy are often employed to confirm the diagnosis.

It may be difficult to differentiate a twisted or ruptured ovarian cyst from a ruptured tubal pregnancy. In a woman in the reproductive years with an adnexal mass and an acute abdomen, a ruptured tubal pregnancy should be considered first.

Rapidly increasing anemia, a fall in blood pressure, and an increase in pulse rate support the diagnosis of ruptured ectopic pregnancy. The absence of an adnexal mass is of no significance in the differential diagnosis since an ovarian tumor may also apparently disappear after its rupture.

It is not ordinarily possible to differentiate a carcinoma of the fallopian tube from an ovarian tumor by palpation.

Inflammatory diseases of the adnexa, acute or subacute, usually present no problem in differentiation from an ovarian tumor. The history and symptoms differ considerably. Inflammatory masses are softer, barely mobile, and usually much more tender. Chronic inflammatory adnexal masses bound down in the posterior cul-de-sac, particularly when they are nodular, may occasionally be mistaken for an ovarian carcinoma. Endometrial implants in the cul-de-sac and uterosacral ligaments, and on the posterior aspect of the cervix (page 269) also may feel like ovarian carcinoma. The generally good physical condition of the patient, the characteristic secondary dysmenorrhea, and the tender nodularities favor the diagnosis of endometriosis. In questionable cases, laparoscopy may be indicated.

Intestinal Tract. Firm fixed masses in the pelvis in any woman over 50 years may be caused by an ovarian carcinoma or a malignant tumor of the lower intestinal tract. History, signs, and symptoms such as constipation, diarrhea, colicky pain, and rectal bleeding suggest intestinal disease. Carcinoma of the rectum is palpable on rectal examination as a firm tumor extending over the surface of the mucosa or as an ulcer with indurated edges. The examining finger may be stained with blood. Proctoscopy and barium enema confirm the diagnosis. It is difficult to differentiate an ovarian carcinoma fixed to the pelvic side wall from a carcinoma of the sigmoid. Diverticulosis of the sigmoid must also be considered. A posterior ill-defined mass of doughy consistency located high in the pelvis suggests diverticulosis.

Ileocecal tumors are located high in the abdomen and can hardly be reached by rectal or vaginal examination. Gurgling bowel sounds may be heard after displacing the mass with the palpating hand.

Acute appendicitis and appendiceal abscesses are seldom confused with ovarian tumors. Ruptured or twisted ovarian cysts may occasionally be mistaken for a perforated appendix, however, and vice versa.

Retroperitoneal tumors may extend into the pelvis. An immobile, firm resistant mass near the posterior wall of the pelvis is easily differentiated from an ovarian tumor as long as the genital organs can be palpated separately. Questionable findings at palpation may be clarified by laparoscopy. A pelvic kidney appears as a rather doughy, ovoid-shaped mass. An intravenous pyelogram confirms this diagnosis.

Differential Diagnosis of Extremely Large Ovarian Tumors. Large ovarian neoplasms may not be palpable through the vagina or the rectum. The uterus is elevated because of the mobile cystoma and is located high in the pelvis. Percussion is useful to delineate the tumor. Pancreatic cysts, mesenteric cysts, abscess of the gall bladder, and hepatomegaly must be ruled out.

Ascites. Differentiation between ovarian cysts and ascites is made on the basis of the findings on abdominal percussion before and after change of position. Since ovarian cysts push the intestines laterally, dullness on percussion is found over the medially located tumor, and tympany laterally in the flanks. A change in position is not followed by a change in the findings on percussion. With ascites, the intestinal loops float medially on top of the abdominal fluid and create tympany around the umbilicus and dullness in the flanks. When the patient is positioned on one side, the intestinal loops float upward, resulting in tympany in the flanks. If ascites and ovarian cysts coexist, the findings on percussion are confusing.

Suggestions for the Treatment of Ovarian Tumors. In general, all ovarian tumors should be removed surgically. Because of the malignant potential of many benign neoplasms of the ovary, surgical treatment is also effective prophylaxis (Fig. 186). The extent of the operation depends on whether the tumor is benign or malignant, the age of the patient and her reproductive history, and the malignant potential of the tumor. Only functional cysts may be observed for 4 to 8 weeks. If they do not regress or if they increase in size during the period of observation, laparotomy is indicated.

For a benign neoplasm, unilateral oophorectomy is sufficient for women in the reproductive age. Small areas of normal functioning ovarian tissue may be preserved after removal of cysts. Inspection and bisection of the apparently normal ovary on the other side is preferable in most tumors, especially in benign cystic teratomas and hormone-producing tumors (page 336). Whenever histological examination of the specimen raises doubt about the potential malignancy of the lesion, a secondary operation with removal of the remaining ovary and the uterus is indicated.

Bilateral tumors require extirpation of both adnexa and uterus. In menopausal women, bilateral oophorectomy with hysterectomy is performed even if the neoplasm involves only one ovary.

Treatment and prognosis of carcinoma of the ovary depend mainly on the extent of spread at the initiation of treatment. To compare different methods of treatment and their respective cure rates, the interna-

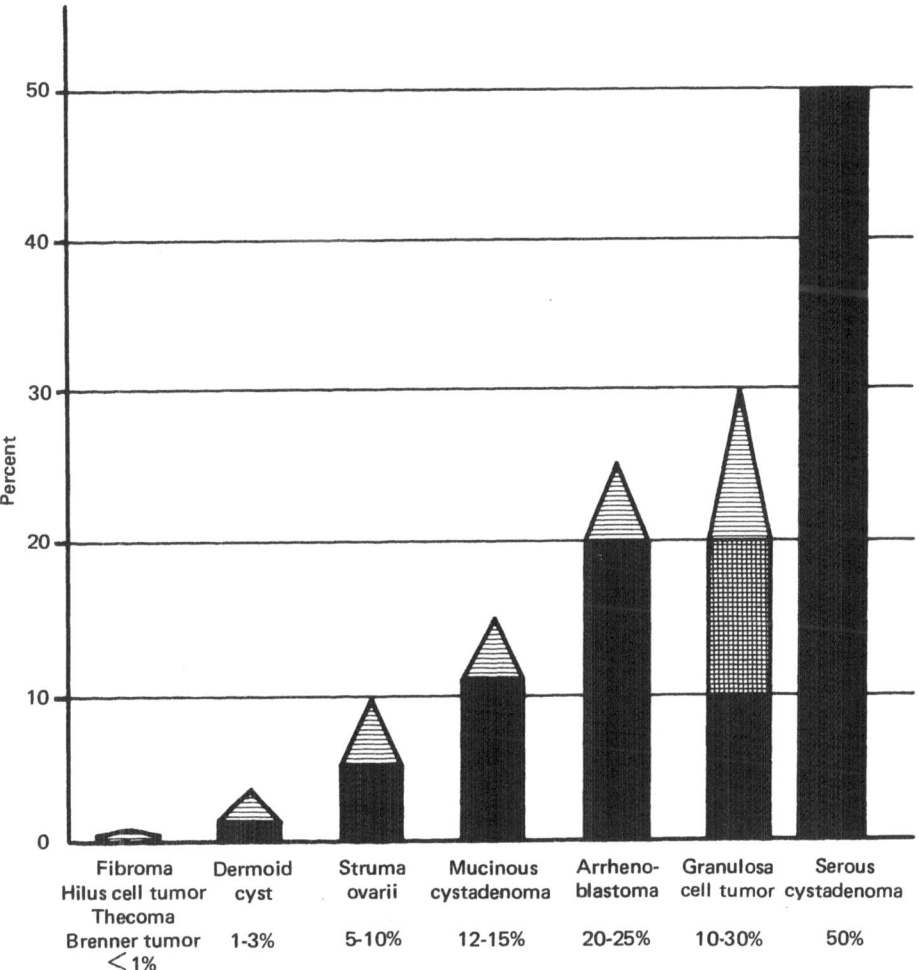

Fig. 186. Schematic representation of frequency of malignant transformation of various ovarian tumors.

tionally accepted classification should be used for the staging of primary ovarian carcinomas.

In 1964, FIGO suggested a classification based on the findings at clinical examination and laparotomy:

Stage I: tumor limited to the ovaries
Stage Ia: only one ovary involved; no ascites
Stage Ib: both ovaries involved; no ascites
Stage Ic: both ovaries involved; ascites with tumor cells
Stage II: tumor in one or both ovaries with spread elsewhere in pelvis, with or without ascites

Stage IIa: direct spread or metastasis only to the uterus, tubes, or both
Stage IIb: spread to other pelvic organs
Stage III: tumor in one or both ovaries with extensive intraperitoneal metastasis (omentum, intestine, mesentery) with or without ascites
Stage IV: tumor in one or both ovaries with distant metastasis outside the peritoneal cavity, with or without ascites

In 1968 a new classification based on the TNM System was suggested by the U.I.C.C. (Union International Contre le Cancer), as follows:

TP: extension of the primary tumor
TP I: tumor mobile, limited to one ovary
TP II: both ovaries involved, tumor mobile
TP III: uterus, tubes, or both affected
TP IV: spread to other neighboring anatomical structures. Ascites is not considered.
N: regional lymph nodes. (This symbol was introduced for carcinoma of the ovary as *NX*, which means that in ovarian carcinoma the pelvic nodes usually cannot be evaluated. Positive or negative lymph nodes found at laparotomy are listed as *NX+* or *NX—*.)
M: distant metastasis
MO: no distant metastasis
MI: implantation or metastasis present
a: metastasis in the pelvis
b: metastasis in the peritoneal cavity, involving peritoneum, omentum, intestine, mesentery, liver, or other organs
c: metastasis outside the peritoneal cavity

The advantage of the TNM system is that the extent of the tumor can be defined more precisely. Consequently, comparisons of treatment results are easier.

The methods of treatment of ovarian carcinoma are:

(1) surgical
(2) radiotherapeutic
(3) chemotherapeutic

There is general agreement about the prime role of surgical treatment. The value of radiation and chemotherapy is somewhat less clearly defined.

Cure of ovarian carcinoma is possible only after total removal of the tumor. If the operation is incomplete, that is, if the tumor is left behind, the prognosis becomes much worse. Removal of the uterus is preferable to removal of the adnexa only.

The primary objective of surgical exploration is to ascertain the extent of the tumor. The decision as to whether the tumor can be removed completely can be made only during laparotomy. In most instances, it is not known definitely whether the tumor is benign or malignant before laparotomy. Frozen sections, cytology, and inspection and biopsy of the other ovary (page 347) provide the basis for deciding the best procedure. The opposite ovary was found to be microscopically involved in 12% of carcinomas. Unilateral oophorectomy without hysterectomy is justifiable only in young women when the tumor is unilateral, well-encapsulated, and benign.

Malignant tumors always require hysterectomy and bilateral salpingo-oophorectomy. Spillage of contents of the cyst into the abdominal cavity may not be so dangerous as previously thought, since the capacity of these cells to implant is questionable. If it is not possible to deliver a large cystic tumor through the abdomen, it may be necessary to reduce its size by aspiration.

Prophylactic partial or total removal of the omentum is advised in ovarian cancer since microscopic metastases are frequently found there.

If total removal of the tumor is not possible because of adhesions or extensive infiltration of neighboring organs, only a biopsy is performed. Complete extirpation may be successful in a so-called second-look operation after a trial of chemotherapy or radiation. The objective of the adjunctive methods of treatment is to transform an inoperable into an operable tumor by isolating the tumor from the surrounding tissues.

The simultaneous use of several chemotherapeutic agents that interfere with different steps in cell metabolism is recommended by many investigators. The effect is somewhat independent of the histological type and degree of differentiation of the tumor. Possible side-effects of chemotherapy include leukopenia, thrombocytopenia, anemia, nausea, and vomiting. Patients receiving chemotherapeutic agents require constant supervision. Certain effects such as depression of bone marrow may be greatly increased when chemotherapy and radiation are employed simultaneously.

Chemotherapy often prolongs life and provides symptomatic relief although it is not curative. With few exceptions most ovarian tumors are fairly resistant to radiation. Because of the extent of the tumor, an effective dose usually cannot be administered without seriously damaging abdominal organs.

In children, the necessity for radiation or chemotherapy must be evaluated very carefully since it entails the risk of damage to growing

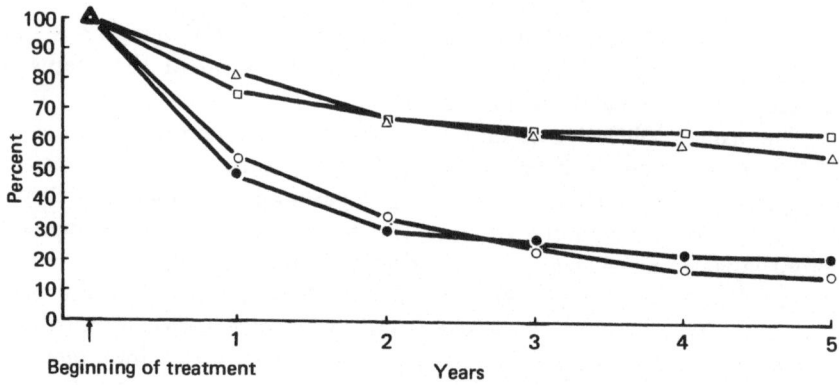

Fig. 187. Prognosis of ovarian carcinoma in relation to histologic type (after Kottmeier). △ endometrioid carcinoma; □ mucinous carcinoma; ○ serous carcinoma; ● unclassified carcinoma.

bones and germ cells and the induction of tumors that may manifest themselves decades later.

Prognosis of Ovarian Carcinoma. Although significant progress in treatment has been achieved, the prognosis of ovarian carcinoma is still poor.

The 5-year cure rates for all malignant neoplasms range between 20 and 30%. This low survival rate is explained mainly by the fact that in 50 to 80% of women, the tumor has already spread beyond the ovary at the onset of therapy (Fig. 187).

Prognosis is closely related to the stage of the disease at the beginning of therapy. Survival drops in Stage I (when only one ovary is involved) from 62.2 to 38.8% when tumors are found in both ovaries. In addition to extent of tumor spread, the histological type also is important for the prognosis. The less differentiated the carcinoma, the worse is the prognosis. The 5-year cure rates of the mucinous and endometrioid carcinomas are definitely better (51.3%) than those of the serous cystadenocarcinomas (20.6%).

Follow-up of Patients. Examinations at regular intervals are required because of the high incidence of recurrences. Conservative operations in young women are acceptable only if close supervision is guaranteed.

During long-term chemotherapy, platelet and leukocyte counts are important methods of monitoring the patient. If therapy is unsuccessful, as indicated by further spread of the tumor, palliative measures are employed: narcotics for pain, repeated peritoneal taps to reduce accumulations of ascitic fluid, and antibiotics for prevention or treatment of urinary tract infections. The management of these incurable patients requires considerable skill and effort of the doctor and nursing staff.

References

1. Gusberg, S. B., Frick, H. C. II. Corscaden's Gynecologic Cancer, 4th Ed., Baltimore, Williams and Wilkins, 1970.
2. Jeffcoate, T. N. A. Principles of Gynaecology, 3rd Ed., New York, Appleton-Century-Crofts, 1967.
3. Novak, E. R., Woodruff, J. D. Novak's Gynecologic and Obstetric Pathology, 6th Ed., Philadelphia, W. B. Saunders, 1967.
4. Te Linde, R. W., Mattingly, R. F. Operative Gynecology, 4th Ed., Philadelphia, J. B. Lippincott, 1970.

CHAPTER 25

Acute Abdomen: Differential Diagnosis

The acute onset of intraabdominal pain with the signs and symptoms of the so-called acute abdomen requires consideration of gynecological lesions as well as those outside the genital tract.

An acute abdomen requires speedy decisions, for the patient's life depends on the correct diagnosis.

The diagnosis of an acute abdomen is made on the basis of the following signs and symptoms:

(1) Severe, acute pain that is either continuous or colicky

(2) Acute distress (The patient is restless or shocky and the face is pale or flushed.)

(3) Rebound tenderness, localized or diffuse over the entire abdomen

Generalized peritonitis produces the following signs and symptoms:

(1) Cold sweat

(2) Hypotension and other signs of shock

(3) Fever

(4) Vomiting

(5) Dry tongue and wrinkled skin (indicating dehydration)

(6) Shallow breathing

(7) Diffuse abdominal tenderness

(8) Lack of peristalsis

An acute abdomen is the result of three major disorders:

(1) Perforation or rupture of an organ or tumor

(2) Acute occlusion of blood flow, involving an organ or a tumor, or the acute obstruction of a hollow organ

(3) Acute inflammation in the abdomen

Gynecological disease is suspected if the pain is most severe in the lower abdomen. The following lesions must be considered:

(1) Intraabdominal bleeding caused by ectopic pregnancy (especially if ruptured) or rupture of a follicle cyst.

(2) Torsion of an ovarian tumor or a subserous myoma.

(3) Rupture of a cystic ovarian tumor or a pelvic abscess.

(4) Acute degeneration of a myoma.

(5) Diffuse peritonitis stemming from acute salpingitis.

The following signs and symptoms are characteristic of some cases of *ectopic pregnancy,* although none is invariably found.

(1) The history: a missed menstrual period with subsequent spotting and acute onset of pain.

(2) Pregnancy test may be positive or negative.

(3) Tachycardia and hypotension.

(4) Hematocrit may be low (but the drop may be delayed for more than 12 hours). Sedimentation rate and white count are normal or a low grade leukocytosis (up to 12,000 mm^3) may be found.

(5) Palpation of a unilateral tumor.

(6) Diffuse tenderness over the lower abdomen.

(7) Culdocentesis reveals dark, unclotted blood.

Differential diagnosis includes ruptured spleen, liver, or cyst—all of which require laparotomy.

The diagnosis of so-called tubal abortion is more difficult. The symptoms develop slowly and are usually not associated with an acute abdomen.

Acute torsion of an ovarian tumor or subserous myoma is associated with the following signs and symptoms: Acute onset of pain in the left or right lower quadrant. Abnormality of the menses is not to be expected. Temperature, sedimentation rate, and white count are usually normal. The abdomen is tender, with generalized rebound tenderness. The lower pole of the tumor can often be felt on pelvic examination.

After rupture of a tumor, the mass can no longer be felt. A *leaking pelvic abscess* leads to progressive peritonitis, which is not improved by bedrest and antibiotics.

Acute degeneration of a myoma is uncommon, occurring usually in association with pregnancy or torsion of a pedunculated mass.

Extragenital lesions producing signs and symptoms in the lower abdomen. If the rebound tenderness is localized to the right lower quadrant, the physician must think immediately of appendicitis. Leukocytosis, tenderness over McBurney's point, or right-sided pain after releasing pressure in the left lower quadrant are characteristic. Diagnosis of a perforated appendix is more difficult, for it can be mistaken for pelvic inflammatory disease. Progressive peritonitis may complicate both diseases. The cul-de-sac may be tense and painful in both cases.

When the pain is maximal in the left lower quadrant, diverticulitis must be considered. Pelvic examination may sometimes reveal a mass high on the left, unrelated to the adnexa or the uterus.

Ureteral calculi may occasionally cause signs and symptoms suggestive of an acute abdomen. The characteristic pain is colicky, originating in the left or right flank and radiating to the back or following the anatomical course of the ureter. Rebound tenderness is rare. Fever is usually absent and the white count and sedimentation rate are usually normal. Hematuria may be found and an intravenous pyelogram may be diagnostic.

Peritonitis, regardless of its cause, may produce an adynamic ileus characterized by absence of bowel sounds and flatus and increasing leukocytosis and sedimentation rate. An upright abdominal roentgenogram often shows generalized air and fluid within the bowel. Paralytic ileus may occasionally be produced by cholelithiasis, ureteral calculus, or acute pyelonephritis, and rarely, by acute occlusion of a mesenteric artery.

Mechanical obstruction of the bowel is characterized by high-pitched bowel sounds and abdominal distension. On x-ray examination, the bowel is distended, with definite fluid levels. Pain becomes colicky and the hyperactivity of the bowel may sometimes be felt abdominally. A painful mass located in the inguinal canal may be indicative of an obstructed inguinal hernia, which may be associated with a bloody mucous diarrhea. Untreated mechanical obstruction rapidly progresses to irreversible ileus, with its accompanying high mortality.

CHAPTER 26

Physiologic Disturbances and Treatment of Shock

Definition

The modern definition of shock implies inadequate tissue perfusion. Oxygenated blood in the microcirculation is reduced. The emphasis is, therefore, on blood flow rather than blood pressure. Shock can result from a variety of individual mechanisms or a combination of etiological factors (Table 30).

The end results of inadequate tissue perfusion are identical regardless of the etiological factors. The most frequent varieties of shock in gynecology are hemorrhagic and endotoxic.

Hemorrhagic shock: A deficit of 20% of the circulating blood volume can reduce cardiac output by 25 to 45%, depending on the baseline value. One of the initial mechanisms to compensate for hypovolemia is the release of catecholamines with resultant peripheral vasoconstriction and shunting of the remaining blood volume to vital organs such as the brain, liver, kidney, and heart. The responses of the microcirculation and the cardiovascular system modify the subsequent course of the disorder. Temporarily, an increase in cardiac output and a decrease in resistance may compensate for blood loss, but after an initial period of compensation, a vicious cycle is generated.

The sequence of functional derangements is not always clear, but it is generally accepted that hypoperfusion induces sludging of blood, metabolic acidosis, and finally ischemic necrosis of organs. Amines such

Table 30. Mechanisms of shock

Pathogenesis	Clinical examples
(1) Hypovolemia	Hemorrhagic shock, dehydration caused by loss of plasma, trauma, burns
(2) Interference with cardiac systole or diastole	Cardiac shock
(3) Toxic metabolic products	Endotoxic shock, anaphylactic shock
(4) Neurogenic vascular collapse	Loss of sympathetic control, as in spinal anesthesia and orthostatic collapse
(5) Obstruction of major vessels	Pulmonary embolism, amniotic fluid embolism, supine hypotensive syndrome

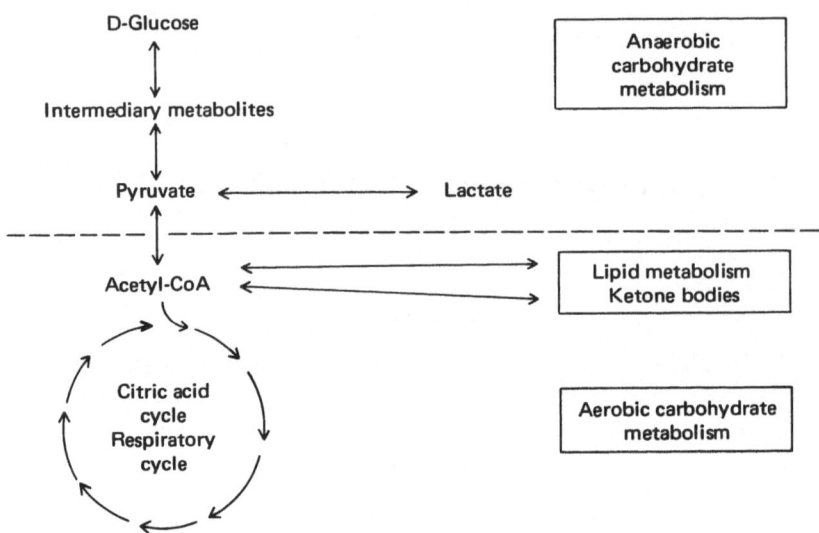

Fig. 188. Intermediary metabolism of lactic acid.

as serotonin and the kinin system, as well as vasoactive polypeptides, are activated in the initial phase of shock. The release of catecholamines, originally a defense mechanism, is also responsible for constriction or contraction of precapillary and postcapillary sphincters in the microcirculation. The result is trapping of blood with dilatation of capillaries. The decrease in blood flow results in sludging of leukocytes and erythrocytes and an increase in viscosity. The resulting metabolic acidosis is associated with a shift from aerobic to anaerobic metabolism. The capillary hydrostatic pressure increases and plasma is shifted to the extravascular compartment. This is the explanation for increasing hypovolemia without blood loss (Figs. 188 and 189).

The increasing ischemia of involved organs is characterized by a rise in lactate or lactic-pyruvate ratio, indicating progressive tissue damage. Since the pyruvate is shifted to lactate, rather than into the Krebs cycle, metabolic energy is reduced and the organism is starved.

In the later stages of shock, the blood coagulation system is activated, resulting in a consumption coagulopathy and fibrin deposition in the microcirculation (disseminated intravascular coagulation). This event ultimately leads to necrosis of organs and presumably irreversible shock.

Hypovolemia is the principal pathogenic factor. The vicious cycle just described develops over a longer period of time, but, if the final phase of hemorrhagic shock (decompensation) is reached, it is not much different from that of any other type of shock.

Endotoxic shock: The incidence of septic shock complicating abortion varies from state to state. In areas with restrictive abortion policies, it

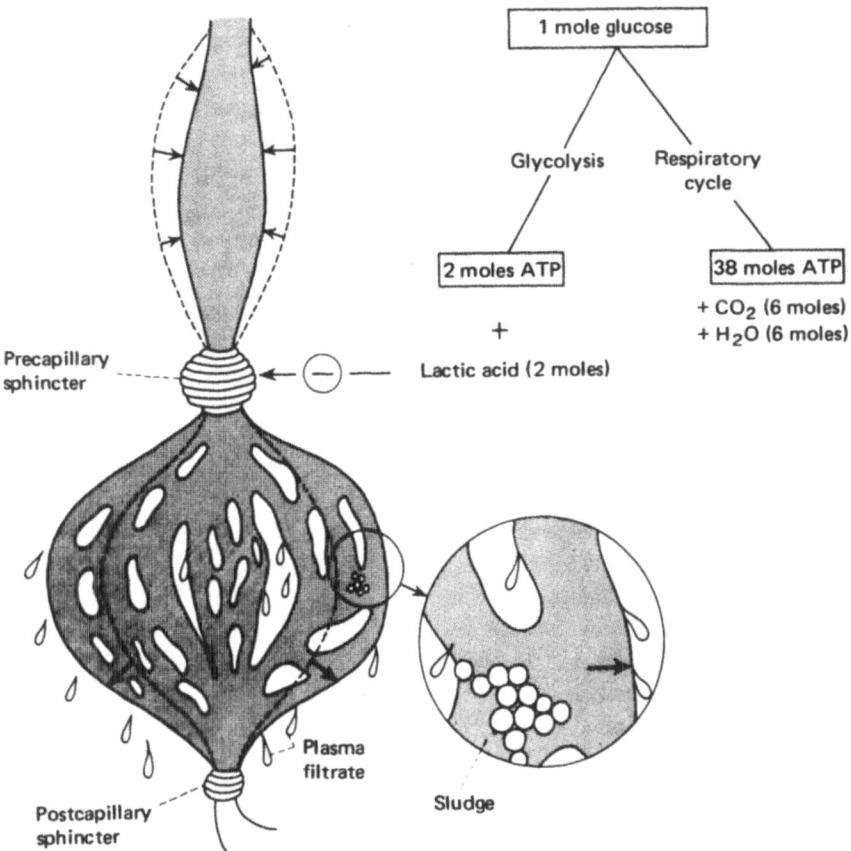

Fig. 189. Disturbances of the microcirculation.

is still approximately 3%. Endotoxic shock is characterized by progressive hemodynamic and hemostatic failure. The mortality was as high as 80% 10 years ago but it has been reduced to as low as 10% in some series. Improvement depends largely on the early elimination of the source of infection (endotoxin) by surgical means.

Endotoxins are lipopolysaccharides from cell walls of gram-negative bacteria. They cause a wide variety of toxic reactions, including the potentiation of catecholamines and therefore of vascular hyperreactivity and the triggering of the coagulation system. These two effects result in early hemodynamic alterations in major organ systems.

Potentiation of catecholamines may explain the hyperreaction of the microcirculatory bed. What was originally a defense mechanism of the organism thus becomes pathological, particularly when acting in conjunction with deposition of fibrin in the microcirculation. Hypercapnea, bronchospasm, and increase in pulmonary arterial pressure result from pulmonary involvement. Fibrin deposition in the renal glomeruli is another early development. Hypoperfusion in the hepatic sinusoids decreases clearance and hence the capacity for detoxification.

Arteriovenous shunting reduces tissue perfusion. The organism tries to compensate initially by an increase in cardiac output. Later the characteristic decrease in cardiac output and increase in peripheral resistance develop. The result is irreversible shock with tissue necrosis.

Symptoms and diagnosis: The peripheral blood pressure is only a poor indication of the degree of hypoperfusion. The status of peripheral circulation is more important.

(1) Skin temperature and color, especially of the extremities (cold, clammy extremities indicate peripheral vasoconstriction).

(2) Collapse of the peripheral veins

(3) Color of the nail beds

(4) Blood pressure and pulse rate. The "100/100 rule" may be of some help in speedy orientation. A pulse rate above 100 combined with a systolic pressure below 100 mg Hg suggests shock.

(5) Hourly urinary output by means of catheter. Volumes below 30 ml/hr indicate renal shutdown and osmolality measurements may indicate tubular necrosis.

(6) Comparisons between body (rectal) and peripheral (skin) temperatures

(7) Central venous pressure. This is not a measurement of blood volume per se, but a CVP below 4 cm water is a strong indication of hypovolemia.

(8) Hematologic measurements: platelet counts below 150,000 mm^3 with normal fibrinogen values suggest gram-negative sepsis (direct effect of endotoxins on platelets). Low platelets together with low fibrinogen indicate consumption coagulopathy. An increase in plasma hemoglobin by 100% suggests fibrin deposition in the microcirculation. The hematocrit is of little value because a decrease can be caused by hemolysis, hypovolemia, or in most instances, both.

(9) Biochemical assays in serum. (A significant increase in lactate may indicate irreversible shock.)

(10) Blood gas analysis.

Treatment: Any therapeutic regimen is dependent on the stage and the degree of shock. Pressor substances are generally contraindicated since they treat only one symptom: the peripheral blood pressure. There is no increase in cardiac output and the principal physiologic disturbances are not affected. An increase in blood pressure may, therefore, temporarily mask the deleterious effect of shock.

The results of treatment of shock largely parallel the knowledge and experience of the physician in charge. Only the principles will be outlined in this context.

(1) Volume replacement is imperative in both hemorrhagic and septic forms of shock. The order of preference is blood, plasma, and human albumin. The amount is determined in hemorrhagic shock by

(a) the pulse rate
(b) the CVP (at least 4 cm water)
(c) the urinary output
(d) the peripheral blood pressure.

An early rise in systolic blood pressure without volume replacement is a good prognostic sign. It indicates reactivity of the vascular system.

After every four or five units of old blood, a unit of blood not older than 6 hours should be transfused in order to provide new platelets and prevent a secondary hemorrhagic diathesis. Each unit of fresh blood raises the platelet count by 10 to 15,000 mm³.

Volume replacement is more complicated in septic shock, because renal shutdown may be caused by glomerular fibrin deposition, partial renal cortical necrosis, or both. No improvement can be expected in this condition by mere replacement of blood volume.

(2) Respiratory support by oxygen or, if necessary, endotracheal intubation.

(3) Control of infection. High doses of a combination of penicillin and a broad-spectrum antibiotic, effective against gram-negative organisms, are popular (chloramphenicol, kanamycin, cephalothin, or colistimethate). A "pseudo-Herxheimer "reaction" has occasionally been observed, whereby patients develop shock in response to massive destruction of gram-negative organisms in the circulation. This reaction can usually be prevented by high doses of corticosteriods.

(4) Removal of infection: early curettage is imperative. If no improvement is noted, hysterectomy may be required 6 to 12 hours later. Immediate surgical drainage of abscesses is mandatory. It is striking to note how well patients in endotoxic shock often tolerate surgical procedures.

(5) Alpha-adrenergic blockers such as phenoxybenzamine have been shown in some investigations to exert a favorable effect on the microcirculation.

(6) The effect of pharmacological doses of corticosteroids is poorly understood, but many experts in the field of shock favor this form of treatment. If the vital signs do not improve after steroid therapy, hysterectomy may be considered a final life-saving resort.

(7) Beta-adrenergic stimulators such as isoproterenol are of value when used in a patient with a normal pulse rate. The drug can produce ventricular arrhythmias, tachycardia, and fibrillation.

(8) Digitalization is recommended by some experts, even in the absence of significant cardiac involvement.

In addition, a variety of controversial measures have been recommended. Osmotic diuretics such as mannitol have been used to maintain urinary flow. They should be used, however, only after proper volume replacement has been achieved. Buffers, such as sodium bicarbonate and Tris (tromethamine), may correct blood pH. The high lactic acid, however, reflects tissue damage that is not reversed by buffers. Low-molecular-weight dextran has been used because of its alleged capacity to counter sludging.

Heparin has been used to inhibit disseminated intravascular coagulation (DIC). The results have been disappointing in gynecologic practice. After DIC has developed, heparin may do little except to superimpose another coagulation defect upon an already existent consumption coagulopathy. Heparin, however, has been reported to be of some prophylactic value. A patient with a febrile abortion (temperature above 101°F, or 38°C) and subnormal platelet count (below 150,000 mm ^{2}) may be a candidate for this treatment. A low-dose continuous infusion of 10 to 20,000 units over 24 hours is used. This treatment should raise the clotting time from 10 to 20 minutes. Such minimal anticoagulation does not interfere with any surgical procedure and even hysterectomy can be performed without increasing blood loss.

Suggested Readings

Beller, F. K.: Pathophysiologic Aspects of Circulating Endotoxins in Septic Abortion. *Int. J. Gynec. Obster.* 8:617 (1970).

Hershey, S. G., Del Guerico, L. R. M., and McCann, R.: *Septic Shock in Man.* Boston: Little Brown Company (1971).

Hardway, R. M.: *Clinical Management of Shock,* Springfield, Ill.: Thomas (1968).

Mills, L. C., and Moyer, J. H.: *Shock and Hypotension.* New York: Grune & Stratton (1965).

Principles of operative management.

Kaeser, O., and Ikle, F. A.: *Gynecologic Operations.* Stuttgart: Thieme (1973).

CHAPTER 27

Principles of Surgical Treatment in Gynecology

Gynecologic surgery can be conveniently divided into diagnostic and therapeutic procedures.

Diagnostic Procedures

The most frequently performed diagnostic procedure is dilatation and curettage (D & C). Anesthesia is required. The size of the uterine cavity is measured by a uterine sound and abnormalities such as submucous myomas and malformations are outlined. After dilatation of the cervical canal, usually with Hegar dilators, the endometrium is scraped with a curette, which is available in various sizes. The material obtained is submitted for histologic diagnosis.

Most frequent indications are: pathological uterine bleeding and diagnosis of intrauterine lesions (page 320).

Suspicion of adenocarcinoma of the endometrium requires a so-called fractional curettage, in which cervical canal and endometrial cavity are scraped separately in order to identify the extent of the lesion (whether it is in the corpus only, the cervix only, or both). Material from both scrapings are submitted separately to the pathologist. In some instances, the diagnostic procedure is also therapeutic, as in the case of an endometrial polyp (page 315).

Cone biopsy is a most important procedure to identify the character and extent of lesions (cervical cancer and precancerous lesions) around the squamocolumnar junction (page 292).

Laparoscopy is the inspection of intraabdominal organs through an endoscope inserted into the peritoneal cavity. Techniques have recently been developed to perform tubal interruptions throught the laparoscope.

Culdoscopy is another method of inspection through the cul-de-sac. A culdoscope is similar to but shorter than the laparoscope. It is inserted through the cul-de-sac. Frequent indications for both laparoscopy and culdoscopy are uterine or ovarian malformations, sterility, unidentified tumors of the adnexa, endometriosis, pelvic tuberculosis, and pelvic inflammatory disease.

Insertion of a needle through the cul-de-sac is called *culdocentesis*. The aspiration of dark unclotted blood indicates intraabdominal bleeding, as in ruptured ectopic pregnancy. If frank pus is obtained, the diagnosis of an abscess is confirmed. A modified form of culdocentesis is cul-de-sac lavage for obtaining tumor cells.

The surgical incision of the posterior fornix of the vagina and sharp entrance into the peritoneum is called *colpotomy*. It also allows inspection of the tubes and ovaries. Occasionally an ectopic pregnancy can be removed by this procedure. More recently it has provided surgical access to the tubes for ligation or resection.

Therapeutic Surgical Procedures

The therapeutic procedures are performed to achieve two goals:

(1) removal of diseased organs or parts of organs
(2) restoration of function (plastic procedures)

Modern anesthesia and improvement in postoperative care have reduced surgical mortality and morbidity to a degree that patients at extreme risk are becoming increasingly uncommon. Elderly patients in particular are now able to tolerate extensive procedures well, especially through the vaginal route.

Despite the reduction in surgical risk, every abdominal operation requires a clear indication. Such a procedure is indicated only if the benefit to the patient exceeds that of conservative measures.

The risk of a given procedure is determined by a statistical evaluation of postoperative mortality and morbidity. Statistics, however, will be only a guide. The decisions regarding indications and types of procedures, rather than mere technical skill, determine the success of the surgeon.

Gynecological surgery can be performed by two routes, vaginal and abdominal. In general, the *vaginal route* has a lower morbidity and mortality in the hands of a experienced surgeon. The disadvantage is related to the greater difficulty in identification of anatomical structures and the smaller operative field. The *abdominal approach* allows greater access and better anatomical dissection, although operative trauma may be increased. Furthermore, a visible scar usually results and the anesthesia has to be deeper.

Certain procedures are performed preferably or exclusively by the abdominal route:

(1) removal of large myomas exceeding the size of 8 weeks' gestation or myomectomy; plastic correction of most uterine anomalies
(2) ovarian tumors
(3) conservative operations for endometriosis or tubal restoration
(4) operations for pelvic inflammatory disease
(5) operations after previous abdominal surgery
(6) most operations for cancer, especially radical procedures
(7) unclear findings on pelvic examination

The vaginal approach is normally used for the following procedures:

(1) simple removal of a normal-sized uterus, with or without anterior and posterior colporrhaphy

(2) trachelectomy (removal of a cervical stump after previous supracervical hysterectomy)

(3) repair of many vesicovaginal and rectovaginal fistulas

Radical surgical procedures are used to treat certain operable carcinomas of the female genital tract. The abdominal radical hysterectomy is employed for certain early stages of invasive carcinoma of the cervix. The original Wertheim operation was later extended by Meigs, who added pelvic lymphadenectomy. The full radical hysterectomy removes the entire uterus, the adnexa, and the upper third of the vagina. The parametria are removed after mobilization of the ureters and a pelvic lymphadenectomy is performed, dissecting en bloc bilaterally the iliac, hypogastric, obturator, and periaortic lymph nodes.

In some European clinics a radical vaginal hysterectomy (Schauta procedure) is performed. This operation removes the uterus, a wide vaginal cuff, and the parametria. Occasionally an extraperitoneal abdominal lymph node dissection is added. The operation is rarely performed in the United States.

Radical vulvectomy is employed mainly for treatment of invasive carcinoma of the vulva. It involves wide removal of all structures of the vulva together with adjacent skin and subcutaneous fat down to the deep fascia and muscles. It is usually accompanied by regional lymph node dissection through single or separate incisions.

Every surgical procedure, minor or major, requires the written consent of the patient. A frank discussion is of utmost importance in regard to the purpose of the procedure, the expected result, morbidity, loss of function, removal of organs, and expected recovery time. The written permission should not be requested after the patient has already had sedatives. The surgeon is also well advised to discuss fees and cost with the patient before the operation. Following this advice literally may help to reduce the number of law suits.

Selected Readings

Simmons, St., Luck, R. J.: *General Surgery in Gynecological Practice*. Oxford: Blackwell (1971).

TeLinde, R. W., and Mattingly, R. F.: *Operative Gynecology*. 4th ed., Philadelphia: Lippincott, (1970).

Kaeser, O., and Ikle, F. A.: *Gynecologic Operations*. Stuttgart: Thieme (1973).

Index

Excretion
  gonadotropin, 70
  hormone, 62
  pregnanediol, 69, 70
Exstrophy, bladder, 259
External genitalia, 67
  differentiation, 119

Factor X, 18
Fallopian tube, 14, 36, 327
  carcinoma, 351
  ciliated cell, 62
  glycogen, 62
  neoplasm, 327
  occlusion, 187
Family planning, 189
Feminine hygiene sprays, 75
Fetus, adrenocortical hyperplasia, 131
Fern pattern, 63, 185
Fertility, 68
Fertilization, 40, 41
Fibrin, clots, 63
Fibrinogen, 366
Fibrinogenolysis, 63
Fibromas, 336
Fistulas, urogenital, 261
Fixation, alcohol/ether, 101
Fluid, follicular, 59
Fluorescence, 4
Foams, spermicidal, 94
Follicle, 59
  atretic, 38
  cell, 8
  cysts, 71, 329
  fluid, 59
  graafian, 38, 40
  primary, 14, 68
  primordial, 39
  tertiary, 39, 41
Follicle stimulating hormone, 55
Forbes-Albright syndrome, 148
Fornices, lateral, 33
Fourchette, 26
Four-quadrant biopsy, 302
Fractional curettage, 117
Frei test, 242
Frenulum, 26
Frigidity, 83, 88, 94
Fungi, 115

Galactorrhea, 146
Gartner's duct, 16
Gender orientation, 131
Gene, sex specific, 1ᴐ

Genital
  development, 1
  ridges, lateral, 17
  tract, malformations, 133
  tubercle, 17
  tuberculosis, 115
Genitalia, 41, 42
  anomalies, 119
  external, 67
  injuries, 267
  neoplasma, 279
Germ cell, primordial, 9
Germinal epithelium, 38, 42
Gland
  adrenal, 52
  Bartholin's, 26, 77
  hyperplasia, 159
  lesser vestibular, 27
  racemose, 33
  sawtooth, 37
Glans, 26
Glucose
  metabolism, 64
  6-phosphate-dehydrogenase deficiency,
    121
Glycogen, fallopian tube, 62
Glycoprotein, 54
Gonad, 8, 67, 119
  adenomas, 172
  agenesis, 126
  anlage, 9, 11
  blastema, 8
  development, 125
  dysgenesis, 112, 126, 164
    pure, 127
  dysgerminomas, 172
  seminomas, 172
  streak, 127, 166
  testosterone, 58
  undifferentiated, 19
Gonadotrope, 57
Gonadotropin, 55, 72, 179
  excretion, 70
  human chorionic (HCG), 163
  test, 72, 151
Gonorrhea, 239
Graafian follicle, 38, 40
Gram-negative sepsis, 366
Granuloma inguinale, 243
Granulosa cell, 39, 59
  layer, progesterone, 59
  lutein cells, 40, 52, 59
  tumors, 336
Graves speculum, 99
Gravindex, 73

carcinoma, 371
condyloma, 216, 279
cysts, 279
dysplastic changes, 281
furunculosis, 216
herpes zoster, 219
inflammation, 215
mesenchymal tumor, 279
neurodermatitis, 216
papillomas, 279
psoriasis, 219
seborrheic dermatitis, 219
staging, cancer, 285
tumor, 279
Vulvectomy, radical, 371
Vulvitis, 215
  Candida albicans, 219
  secondary, 219
  simple, 215

Walthard rest, 336
Water retention, 66
Wedge resection, 161
Wertheim operation, 371
Wet smear, 111
Wolffian duct, 14
World population, 190

X chromosome, 1
X factor, 18
Xg blood factor, 121
X inactivation, 8

Y chromosome, 1, 13
Yolk sac, 8, 9

Zona functionalis endometrium, 62
Zona recticularis, 52
Zone, transformation, 105